Energy and Engineering Science Series

MODELLING OF FURNACES AND COMBUSTORS

ENERGY AND ENGINEERING SCIENCE SERIES

Editors:

Titles Already Published or in Preparation

Flame Combustion Processes in Industry
A. Stambuleanu

Swirl Flows
A. K. Gupta, D. G. Lilley, and N. Syred

Flowfield Modeling and Diagnostics
A. K. Gupta and D. G. Lilley

Fluidic Flow Control
J. R. Tippetts

Gas Turbine Fuels and their Influence on Combustion
J. Odgers and D. Kretschmer

Spatially Precise Laser Diagnostics for Combustion Temperature and Species
A. C. Eckbreth

Wind Turbines
R. E. Peacock

Modelling of Furnaces and Combustors

E. E. Khalil

Department of Mechanical Engineering
Faculty of Engineering
Cairo University
Cairo

ABACUS PRESS

First Published in 1982 by Abacus Press,
Abacus House, Speldhurst Road, Tunbridge Wells, Kent TN4 0HU

British Library Cataloguing in Publication Data

Khalil, Essam Eldin
Modelling of furnaces and combustors.
(Energy and engineering science series)
1. Furnaces – Combustion – Mathematical models
I. Title II. Series
621.402'318 TJ320

ISBN 0-85626-303-6

Typeset and printed in Great Britain by
Pintail Studios Ltd, Ringwood, Hampshire

TO THE MEMORY OF MY PARENTS

TO MY WIFE HALA WITH LOVE

CONTENTS

PREFACE

In turbulent combustion, the interaction of fluid mechanics, chemical kinetics and heat transfer continues to be both a challenging and a complicated area of research and application. Once its central importance in modern energy and pollution problems has been recognized, it has experienced a dramatic growth over the past decade in both breadth and depth of practical and fundamental studies. This development is attributed to the advent of computers and the modelling approaches, coupled with the new diagnostic techniques used to measure accurately the flow properties and heat transfer characteristics. These advances lay the foundation for a better understanding of turbulent combustion.

There is an increasing need to acquire the ability to predict the behaviour of flames in furnaces and combustors and their performance as a function of input and boundary conditions. Although significant progress has been achieved in predictive modelling techniques, the shortcomings observed in certain modelling aspects highlight the lack of detailed information on the turbulent-combustion characteristics of the flow. Detailed knowledge of the flow pattern, turbulent nature of the flow, flame characteristics and heat flux distributions is required in order to optimize furnace performance for high combustion effeiciency and low pollutant emission.

The material contained in this book is intended to describe the prediction methods used in furnace and combustion modelling approaches. The main aspects of flow, reaction, heat transfer, emission and the corresponding governing equations and modelling assumptions are given. Comparisons of predicted flame characteristics with the experimental data in combustors support the application of the predictive methods to furnaces and combustors design. These lead to obtaining the necessary information at a much lower cost than performing full scale experiments.

Many people have provided invaluable guidance and assistance, others have provided copies of their published work during the preparation of this volume, and to them I owe a debt of gratitude. I am grateful to Prof. J. H. Whitelaw for his continuing support and encouragement, and to my wife for performing the arduous task of revising the text. Thanks are also due to Mr. K. Emam for typing the draft of this book.

Essam Eldin Khalil
August 1981

NOMENCLATURE

Symbol	Meaning	Equation of first mention
A	area	
A_0	pre-exponential coefficient	(4.2.5)
A_x, A_r, A_z	expansion coefficients in Taylor series expansion	(6.2.37)
A_p^Φ	coefficient of convection and diffusion terms at nodal point P	(7.2.5)
$A_N^\Phi A_S^\Phi$, A_E^Φ, A_W^Φ, A_L^Φ, A_H^Φ	coefficients of convection and diffusion terms at the grid nodes N, S, E, W, L and H	(7.2.5)
B_x, B_r, B_z	expansion coefficients in Taylor series expansion	(6.2.37)
C_D	drag coefficient	(5.2.18)
C_{g_1}, C_{g_2}	constants of combustion model	(4.2.25)
C_n	convective term at grid node 'n'	(7.2.7)
C_p	constant pressure specific heat	(3.3.24)
C_R	constant of combustion model	(4.2.34)
C_s	constant of R–s turbulence model	(3.3.11)
C_μ	constant in effective viscosity expression	(3.3.2)
C_1, C_2	constants of the $K - \varepsilon$ model	(3.3.7)
C_3, C_4	constants of the probability equation	(4.2.59)
C_{ϕ_1}, C_{ϕ_2}	constants of the R–s turbulence model	(3.3.14)
C_{ε_1}, C_{ε_2}, C_ε	constants of the R–s turbulence model	(3.3.16)
C_{vol}	concentration of volatiles	(5.2.38)
D, d	diameter	
D_n^+	diffusion term at grid node 'n'	(7.2.7)
D_l	diffusion coefficient	(4.2.36)
D_t	turbulent diffusion coefficient	(4.2.36)
D_{32}	droplet Sauter mean diameter	(5.2.7)
E	constant of the log law of the wall	(3.3.18)
E	activation energy	(4.2.5)
E	emissive power	(6.2.9)
f	mixture fraction defined as $f = (\xi - \xi_A)/(\xi_F - \xi_A)$	(4.2.9)

Symbol	Meaning	Equation of first mention
F	function of turbulence–chemistry interaction	(4.2.8)
G	turbulence production term	(3.3.7)
g	concentration fluctuations	(4.2.20)
g_{fu}	square of fuel mass fraction fluctuations	(4.2.39)
H	stagnation enthalpy	(3.2.4)
H_{fu}	calorific value of fuel	(4.2.14)
H_a	heat of reaction of species a	(3.2.4)
h	fluctuating component of enthalpy	(3.2.8)
h	heat transfer coefficient	(5.2.15)
i	stoichiometric oxygen to fuel mass ratio	(4.2.9)
i, j, k, l	indices	
$\mathbf{i}, \mathbf{j}, \mathbf{k}$	unit vectors in the x, r, z coordinate directions	(6.2.37)
k	kinetic energy of turbulence	(3.3.1)
K, K_a	absorption coefficient	(6.2.1)
K_b	rate coefficient in reverse direction	(4.2.43)
K_f	rate coefficient in forward direction	(4.2.4)
K_c	equilibrium constant	(4.2.43)
K_s	scattering coefficient	(6.2.1)
L	latent heat of vapourization	(5.2.14)
L	optical path length	
l	length	
M_a	mass fraction of species a	(3.2.3)
m_{a}	fluctuating component of species mass fraction a	(3.2.7)
$\dot{m}_{\mathrm{evap}}$	evaporation rate	(5.2.16)
N_D	Damkohler number	(4.2.53)
Nu	Nusselt number	(5.2.20)
P	pressure	(3.2.2)
$P(M_a)$	probability distribution of species mass fraction	(4.2.1)
$P(\Phi)$	probability distribution of Φ	(4.2.16)
Pe	Péclet number	
q	heat flux density	
Q	total heat flux	
R	Universal gas constant	(4.2.5)
r	radius, radial coordinate	
$\mathbf{r}$	position vector	
R_{CO}	rate of CO formation	(6.3.11)
Re	Reynolds number	

Symbol	Meaning	Equation of first mention
R_{fu}	rate of fuel consumption	(4.2.33)
R_{NO}	rate of NO formation	(6.3.8)
R_x, R_r, R_z	total radiation fluxes in the x, r and z directions	(6.2.31)
$S_\Phi, S(\Phi)$	source term of the entity Φ	(3.2.9)
$S_{L\Phi}$	liquid gas interaction source term	(5.2.28)
S	swirl number	(A1.1)
S_U, S_p	components of linearized source term	(7.2.2)
T	temperature	(3.3.24)
t, t_0	time	
U	axial velocity component	(3.3.25)
u	fluctuating component of an axial velocity	
U_j	velocity component in the x_j coordinate direction	(3.2.1)
u_j	fluctuating component of velocity	(3.2.5)
V	radial velocity component	(3.3.25)
v	fluctuating component of radial velocity	
W	tangential velocity component	(3.3.26)
w	fluctuating component of tangential velocity	
$\dot{W}_a$	molecular weight of species a	(7.4.4)
x	axial coordinate direction	
x_j	coordinate directions, $j = 1, 2, 3$	(3.2.1)
y	coordinate direction	
y_p	distance from the wall	(3.3.18)
z	coordinate direction	

GREEK SYMBOLS

α, β	constants in droplet size distribution	(5.2.26)
β	under relaxation factor	(7.4.1)
$\gamma(\Phi)$	variance of Φ	(4.2.19)
Γ	exchange coefficient	
Γ_Φ	turbulent exchange coefficient	(3.2.13)
Γ_a	diffusion coefficient of specie a	(3.2.3)
Γ_k	turbulent exchange coefficient of k	
Γ_ε	turbulent exchange coefficient of ε	
Γ_f, Γ_{fu}	turbulent exchange coefficients of f and M_{fu}	(4.2.10)
Γ_H	turbulent exchange coefficient of H	
δt	increment of t	(4.2.16)
$\delta\Phi$	increment of Φ	(4.2.16)

Symbol	Meaning	Equation of first mention
δ_{ij}	Kronecker delta	(3.2.2)
μ_0, μ_i	directional cosine	(6.2.2)
η_0, η_i	directional cosine	(6.2.2)
θ	polar angle	(6.2.2)
θ	coordinate direction	(6.2.2)
λ	thickness of parcel	(5.2.9)
ε	dissipation rate	(3.3.12)
$\varepsilon(T, L)$	emissivity of gas	(6.2.3)
ζ	azimuthal angle	(6.2.2)
μ	viscosity	(3.2.2)
ξ_0, ξ_i	directional cosine	(6.2.2)
ν	kinematic viscosity	(3.3.7)
ρ	density	(3.2.1)
σ	Stefan–Boltzmann constant	(6.2.3)
σ_f	effective Schmidt number of f	(4.2.10)
σ_{fu}	effective Schmidt number of fuel	(4.2.10)
σ_h	effective Prandtl number	
σ_k	effective Schmidt number of turbulence energy	(3.3.6)
σ_ε	effective Schmidt number of dissipation rate	(3.3.7)
σ_Φ	effective Schmidt/Prandtl number of Φ	
Φ	general dependent variable	(3.2.9)
ϕ	fluctuating component of Φ	(3.2.9)
τ	time scale	
τ_w	wall shear stress	(3.3.18)
κ	constant of the log law of the wall	(3.3.18)
ψ_0	pressure gradient coefficient	(3.3.20)
Ω	solid angle	(6.2.1)

SUBSCRIPTS

A	air stream
D	droplet
eff	effective (including effects of turbulence)
evap	evaporation
eq	equivalent
fu	fuel
F	fuel stream
h	heating
i	initial

Symbol	Meaning
l	laminar
m	mixing
P, N, S, E, W	grid node locations
ox	oxygen
Pr	product
St	stoichiometric
t	turbulent
W	wall
α, β, γ	species

SUPERSCRIPTS

$\bar{}$	time averaged
$\tilde{}$	Favre averaged
'	fluctuations
*	guessed values

CHAPTER 1

INTRODUCTION

1.1 THE TOPIC

The steady state flow of gases and liquids plays a vital role in many kinds of engineering application and device. Much of the world's electric power was, and still is, generated as a result of the flow of steam in boilers, turbines and condensers which form the components of a power station. The steam is heated by the steady flow of flue gases through the combustion chamber, which is contained in the boiler. In fire tube boiler furnaces, fuel and air are mixed to burn; the resultant heat release from the flame is transferred to the jacketed walls which contain water, and hence steam is generated. Most of the gas turbine power plants are, likewise, dependent for their success on the proper aerodynamic and thermal characteristics of the products of combustion in gas turbine combustion chambers. In rotary cement kilns and cyclone combustors, the aerodynamic flow pattern is of prime importance in the performance of such applications.

In this book, the topic dwelt on is fluid flow, chemical reaction and heat transfer phenomena in furnaces and combustion chamber configurations. Due to the present energy crisis, efforts have been made to optimize the amount of fuel burnt per unit of useful power, through improved combustion efficiency. Such a formidable task requires an accurate description, which can lead to better understanding, of the flow pattern in combustors. Zones of forward flow and recirculated flow at the wall and along the centre line influence the stabilization of the flame; hence detailed information about the flow pattern is of prime importance. An objective of the present book, however, is to provide and describe schemes that predict the flow pattern in combustors, with the aid of the governing conservation equations.

In reacting flows, the flame properties, the geometry of the reaction zone, rates of fuel consumption and heat release are quite important. A description of how the reaction proceeds and the rates at which it proceeds are among the aspects of combustion modelling. The different preparation techniques of the air and fuel result in diffusion, premixed and arbitrarily fired flames. Each of these flames differs from the others in the ratio of time scales of turbulence and chemistry, and in the role of the aerodynamic properties and turbulence characteristics.

Heat transfer to the chamber walls, in many situations, is the main requirement from the combustion process. This transfer is utilized to heat up a certain medium, such as steam in boilers, or crude oil in petrochemical heater furnaces. An adequate representation of radiative and convective heat tansfer modes naturally leads to the proper design figures. Models that predict heat transfer differ in their assumptions and vary in complexity. Proper specification of products of combustion yields reliable computation which is used in furnace design procedure.

1.2 OUTLINE OF THE BOOK

In Chapter 2 a description of furnace and combustor flows, their types and applicability in boilers, gas turbine combustors, cement kilns and cyclone combustors is given. Available approaches that describe the flow pattern, chemical reaction and heat transfer in these configurations are described, namely experimental techniques and calculation procedures. A brief review of the various measurement techniques and numerical computations of furnace and combustor flows is given in the same chapter with examples of flows considered.

In Chapter 3, the conservation equations appropriate to steady, two- and three-dimensional, confined, turbulent, recirculating and reacting flows are discussed. The turbulence models embodied in the computational scheme are described together with the appropriate modifications for the present flow situations. Samples of validation tests of the various turbulence model closures are given.

Chapter 4 provides a description of the various modelling approaches to gas-fired flames, and these are discussed with respect to their physical significance, formulation and limitations. Assessment of the various models appropriate to diffusion, premixed and arbitrarily fired flames is provided with appropriate comments on the usefulness of each model. In Chapter 5, mathematical models for two-phase combustion are described with emphasis on liquid fuel combustion models in furnaces and gas turbine combustors. A brief description of the available models for coal-fired appliances is also provided. Examples of calculated flow characteristics in oil- and coal-fired furnaces are given at the end of Chapter 5.

Chapter 6 deals with the last of the three features of combustor modelling that are tackled in this book, which is heat transfer, by radiation and convection. The various modelling methods for thermal radiation are described and their applicability to furnace and combustor design is discussed. This chapter also tackles the problem of nitric oxides formation; the method employed to evaluate NO_x emission in combustor flows is described.

Chapter 7 sets out numerical procedure and describes the numerical algorithm, grid arrangement, stability and convergence. Due to the elliptic nature of the flows, the specification of inlet profiles and boundary conditions is a must. Recommendations for these conditions and the effect of specifying improper ones are discussed.

Several case studies of furnace and combustion chamber flow design are provided in appendices; the calculations are compared, wherever possible, with available experimental data.

CHAPTER 2

FURNACE AND COMBUSTOR FLOWS

2.1 THE PROBLEM

2.1.1 General

The present book is concerned with the turbulent flow in furnaces and combustion chambers. The characteristics of various flame types with and without swirl are investigated. A typical example of the type of the flows considered is the flow downstream in the confined axisymmetric jet arrangement illustrated in Figure 2.1, in which the regions of forward and reverse velocities are indicated. In the initial region, where the mixing of the coaxial jets occurs, the flow pattern is similar, in terms of the decay of the centreline velocity and spreading rate, to that of a coaxial jet issuing into stagnant surroundings. The sudden enlargement creates reverse flow regions, whose size is governed by the expansion ratio, at the corners of the confining cylinder; these zones influence the entrainment to the developing coaxial jet. The introduction of swirl in the annular stream shortens the developing region, and may create at the same time a central recirculation zone, whose size depends on the ratio of the tangential to axial momentum at the jet exit. The influence of combustion on the flow characteristics is appreciable, and the attempt to quantify it forms a major part of the contribution of the present work.

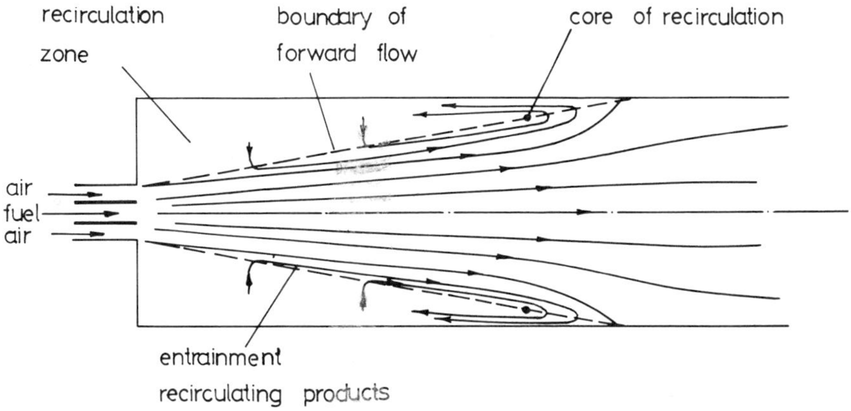

Figure 2.1 Flow field in a confined axisymmetric jet arrangement

In this configuration, if the central jet flow is fuel and the annulus flow is combustion air, the mixture will ignite when the temperature is high enough for the reaction to occur. The unpremixed flame can be stabilized on the wall between the two concentric jets. The introduction of swirl and the consequent creation of a region of recirculation along the centreline causes hot gases to recirculate, thus raising the temperature of the reactants entering this stabilization region.

In general, a well designed furnace has a stabilized flame with maximum combustion efficiency, and minimum pollutant emission. The mean flow and turbulence fields have a major influence on the flame structure, and the formation of pollutants is strongly temperature dependent.

In furnaces, and particularly in fire-tube arrangements, the burner consists of coaxial jets, and the heat release from the resulting flame is transferred to the containing walls by convection and radiation. The stabilization of the flame is governed by many factors such as, the dimensions of the furnace and burner, the stream velocities of the air and fuel, and the turbulence levels. In order to stabilize the flame, a disc (baffle) may be used to create a central recirculation zone, or swirl is imparted to the air stream to create the same effect. The first method of flame stabilization is more suitable for premixed flames, where the premixed reactants flow through the annular space to the combustion chamber as shown in Figure 2.2. In this arrangement, the chemical reaction rate may have a strong influence on the reaction and on the properties of the resulting flame. In gas turbine combustors and after burners, flames are commonly stabilized on baffles and V-gutters as shown in Figure 2.3.

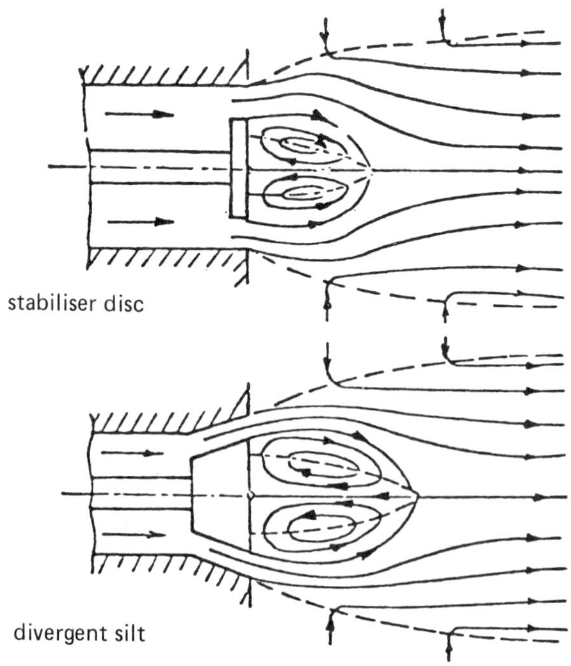

Figure 2.2 Flame stabilization behind bluff bodies

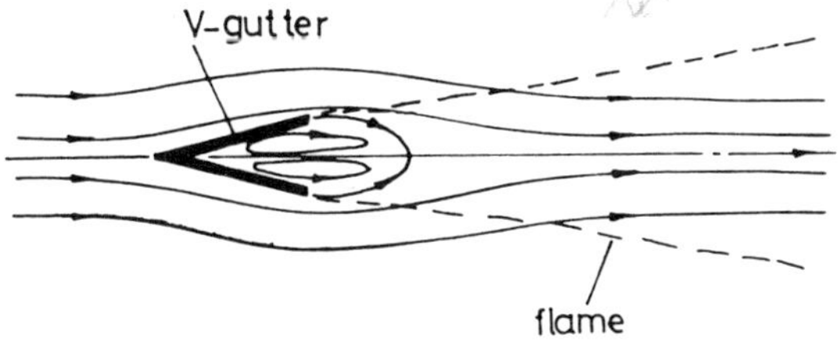

Figure 2.3 Flame stabilization in premixed flames

In diffusion flames, separate streams of air and fuel enter the furnace and the flame is controlled mainly by the rate of mixing of the reactants. The swirl, generated in the air stream, causes the hot gases to recirculate and preheat the fresh reactants, hence preventing blow off. A typical flow pattern in the vicinity of the central recirculation zone is shown in Figure 2.4.

It is true that the investigations of this book cover mainly the flow in furnaces and combustion chamber; nevertheless these investigations can also be relevant when discussing the flow in rotary cement kilns, where the cement ore is poured in at one end of a slowly rotating axisymmetric enclosure, and the burner located at the other end. The heat released from the flame is transferred to the molten cement which is discharged from the flame end as shown in Figure 2.5. Further applications of cyclone combustors are also relevant to this work.

Various types of fossil fuels are commonly used in combustors and vary in their

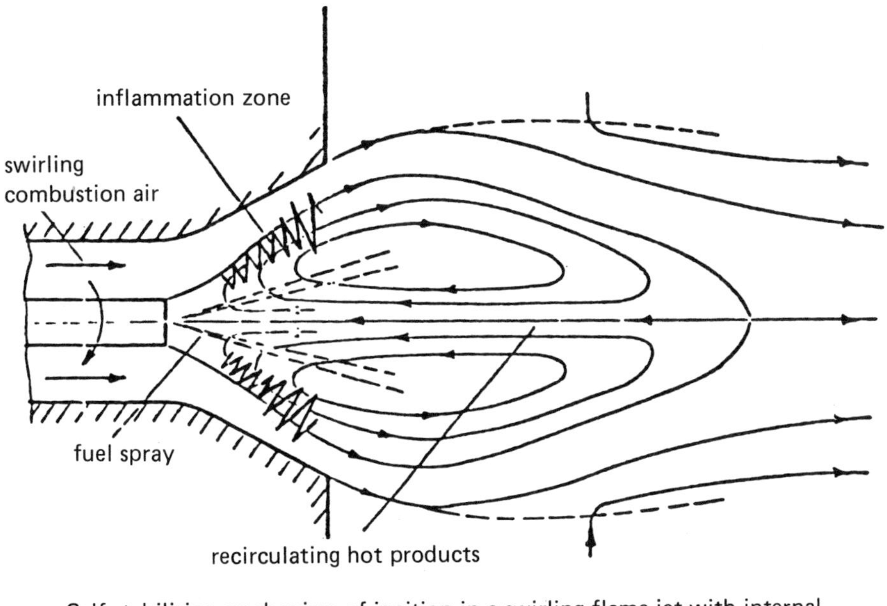

Self-stabilising mechanism of ignition in a swirling flame jet with internal recirculation (Leuckel, 1969)

Figure 2.4 Flow pattern near central recirculation zone

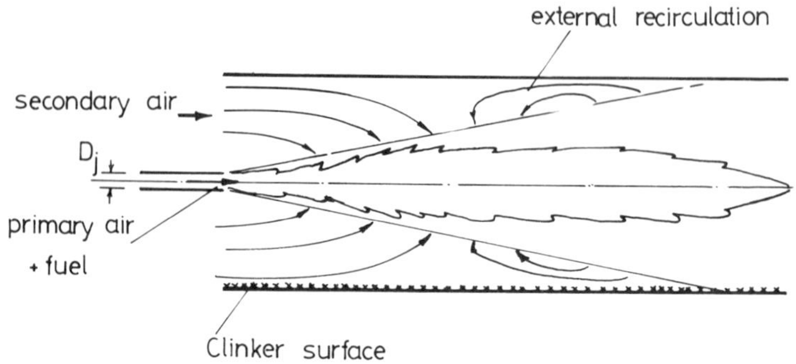

Figure 2.5 Flow pattern in a cement kiln

method of preparation; their mixing with air and consequent combustion burners, therefore, vary from the simple coaxial pipe arrangements, commonly used in gas-fired applications, to burner arrangements with liquid fuel atomizers to replace the central gas pipe. A third arrangement is one that would be appropriate to the solid fuel combustion.

2.1.2 How to Tackle Design Problems

The various aspects of flow, reaction and heat transfer in practical combustors can be investigated and the flame and heat release characteristics can be better understood through one of two approaches.

2.1.2.1 *The Experimental Approach.* Experimental information on the flow pattern, reaction and heat transfer characteristics help to quantify the influence of the various geometrical factors, swirl intensity and fuel-air mixing, on the aerodynamic and heat transfer properties. Such experimental information comprises mainly:

The mean flow velocity components and the corresponding turbulence levels.

The mean flow velocity components and corresponding turbulence levels for two-phase combustion.

The mean gas temperatures and mean species concentrations and the corresponding turbulence intensities.

Heat flux measurements in the flame and along the combustion chamber walls.

Measurements in configurations of direct relevance to the present work are listed in Table 2.1, where the measured properties, measuring techniques and flow configurations are displayed. Chedaille *et al.*[1] provided a review of earlier measurements of mean flow velocities and concentrations in coaxial jets and furnace geometries; these investigations are not reconsidered in this review. The investigations summarized in Table 2.1 encompass a comparatively narrow range of flows and these data are useful in yielding an overall picture of the furnace flow characteristics.

Velocity measurements were commonly carried out with water-cooled Pitot probes which can interfere with the flow, particularly in recirculation zones, producing

TABLE 2.1

Ref.	Velocities	Concentration	Temperature	Wall heat flux	NO_x	Flow type
2	Pitot	–	suction pyrom.	–	–	jet flame
3	Pitot	gas sampling	suction pyrom. thermocouples	radiation flux meters	chemi-luministic	jet flame
4	Pitot	gas sampling	thermocouples	radiation flux meters	–	jet flame
5	Pitot	gas sampling	thermocouples	radiation flux meters	–	flame
6	–	gas sampling	thermocouples	–	–	flame
7	LDA	–	–	–	–	jet flame
8	Pitot	gas sampling	thermocouples	–	–	jet flame
9	LDA	–	–	–	–	jet flame
10	LDA	–	suction pyrom.	calorimeters	–	flame
11	LDA	–	suction pyrom.	calorimeters	–	flame
12	LDA	–	suction pyrom.	calorimeters	–	flame
13	LDA	–	suction pyrom.	calorimeters	–	flame
14	Pitot	gas sampling	suction pyrom.	–	chemi-luministic	flame
15	Pitot	gas sampling	double sonic orifice	–	chemi-luministic	flame
16	Pitot	gas sampling	suction pyrom.	–	chemi-luministic	flame
17	Pitot	gas sampling	suction pyrom.	–	–	flame
18	Pitot	gas sampling	thermocouples	–	–	flame
19	Pitot	gas sampling	suction pyrom.	heat flow meters	–	flame
20	Pitot	gas sampling	suction pyrom. thermocouples	–	–	flame

erroneous velocity information. Moreover, Pitot probes and hot wire anemometers are unable to resolve the flow directions and their use near recirculation zone boundaries can lead to large errors.

The technique of Laser Doppler Anemometry (LDA), in contrast, does not interfere with the flow and allows the direction of the flow to be resolved; as a result, precise measurements can be obtained in recirculation zones. This non-disturbing velocity measuring technique was applied to furnace flows, as indicated in Table 2.1 (references 7, 9–13). The method also makes possible the determination of turbulence levels in the combustion chamber. Critical reviews of the velocity measuring techniques in combustors were reported by Gupta *et al.*[21] and Self *et al.*[22]

Measurements of local gas temperature were obtained with thermocouples and suction pyrometers with a likely precision of $\pm 5\%$. The related uncertainties depend on the probe size, coating material, number of shields, etc ... In most of the references cited in Table 2.1, the gas temperature was reported. The measurements of local gas concentrations have been obtained with iso-kinetic sampling techniques, which are discussed in detail in the original papers of Table 2.1; the assessment of such techniques was investigated by Gupta *et al.*[21] Measured radiative heat fluxes have been determined by ellipsoidal radiometers which were inserted in the combustor, and an overall balance on the combustor cooling sections was carried out to obtain the convective portion of the total load.

It can be seen that experiments provide an overall and detailed description of the flow field, flame structure and heat transfer characteristics. Such a facility is appropriate to small-scale experimental furnaces, but would require modifications when used in large-scale furnaces. This would necessitate a relatively large laser anemometer and problems of optical alignment and beam attenuation emerge. One must not also put aside the usually high cost of setting up an experimental rig.

2.1.2.2 *Calculation Approach.* The vast amount of money involved in the combustion of fossil fuels in industrial furnaces and combustors has for a long time stimulated efforts to optimize furnace performance.

As the achievement of this goal via large-scale experiments on individual furnaces is expensive and therefore undesirable, the developing of mathematical prediction procedures for combustor heat transfer has been an alternative for the last fifty years. The large number of models (23–28) and their individual range of applicability often make it difficult to decide which model suits a particular purpose best. To some extent, these difficulties can be overcome when the models are classified into various groups characterized by well defined specific limitations and application ranges.

The different model types vary both in their degree of complexity and the information obtained upon their application. Three different types exist:

Type I (0-Dimensional Models)

These are models to predict an average combustor heat transfer, furnace efficiency and other overall averaged properties.

Type II (1-Dimensional Models)

These models allow the prediction of one dimensional heat flow distributions and other properties. A representative dimension is commonly chosen as the furnace (flame) axis.

Type III (2- and 3-Dimensional Models)

These models allow the determination of the spatial distribution of fluid and heat flow and other properties.

Figure 2.6 shows an input–output representation of the various types (Michelfelder[29]).

Models of Type I are commonly used for design and are an over-simplification of the real physical situation. This type determines the total heat transfer as a function of the fuel and air inputs.

A zero dimensional model does not take account of the spatial and temporal variations of the flow and flame properties within the furnace. The distributions of all properties in the three coordinate directions are assumed uniform, i.e. zero gradients of velocities, temperature, enthalpies etc . . . Assumptions made in this model represent the major processes of an ideal well stirred reactor and the obtained information is in terms of averaged representative values of:

1. Flow pattern U_G
2. Gas temperature T_G
3. Wall and reactants temperatures T_W, T_i
4. Gas concentrations M_{fu}, M_{ox}, M_{pr}
5. Radiative and convective heat fluxes Q_R, Q_C

Design of furnaces, petrochemical heaters, radiant sections of boilers, has been carried out for a large number of years with the aid of such a model; such investigations were reported by Hottel *et al.*[30] and Jeschar *et al.*[31] Correct trends of the effect of operational

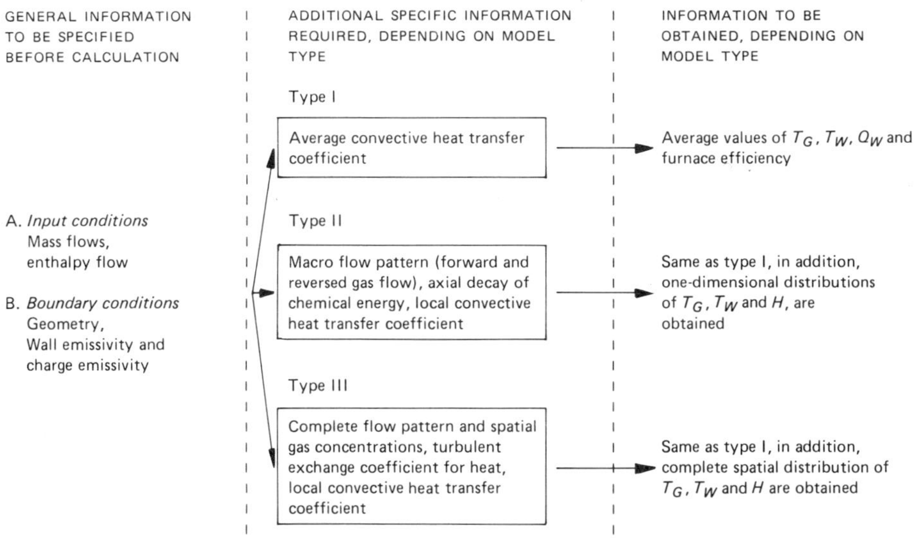

Figure 2.6 Input–output representation of furnace flow modelling

variables on the overall combustor performance were calculated, in many cases, and the method's accuracy deteriorates as the furnace flow situation increasingly differs from the zero dimensionality assumptions.

Models of Type II introduce the effect of dimensions like the time or length. One-dimensional models consider the variation of flow and flame properties in one dimension. In large furnaces, heat flow and temperature distributions along the furnace are commonly required. That implies the presence of a property gradient along the flame axis, while the gradients in the two other coordinate directions are neglected. This assumption is valid for long narrow furnaces and combustors; the furnace is subdivided into sections with uniform cross-stream properties. This type of model is restricted to flows with no recirculation of gases and assumes completed combustion at burner exit. These models can be modified to account for recirculation by considering a combination of well stirred reactors and plug flows. Models of this type have been reported in the literature by, among others, Michelfelder *et al.*,[32] Lucas *et al.*,[33] Thring *et al.*[34] and Craya *et al.*[35]

Applications of Models of Type III in furnace configurations were reported recently; they are relatively advanced due to the advent of digital computers. These models allow the solution of simultaneous partial differential equations which represent the conservation of mass, momentum, energy and species. The solution of unsteady, three-dimensional equations appropriate to a turbulent flame cannot be obtained due to the limitations of storage and computing time. As a consequence, equations based on the time averaged values of velocity, density, temperature and chemical species concentrations have been solved. These equations are supplemented by model approximations to represent the turbulent nature of the flow, reaction characteristics and heat transfer.

Previous computational investigations of turbulent furnace flows have been carried out with the aid of procedures that solve equations for the conservation of mass, momentum and scalar entities as shown in references 10, 11, 12, 13, 23–28, 36 and 37. In these investigations, calculated and measured results were compared to illustrate the agreement which can be obtained from procedures of this type. Various mathematical models to represent the turbulent nature, flame characteristics and heat transfer properties were proposed and their validity assessed.

The present work describes a computational approach of Type III discussed above. One object of the book is to review and bring together available information on numerical procedures applied to furnace and combustor flow predictions. The reader is made acquainted with the governing conservation equations of mass, momentum, energy and species in their general form. Naturally, the reader is bound to ask such questions as: how well is the model expected to fulfil its task and what are the implications of the mathematical models proposed and utilized to yield the conservation equations solvable?

To deal with the last question first, the book presents the state of the art of modelling of the following areas:

1. Turbulence nature of the flow.
2. Chemical reaction mechanisms and heat release.
3. Radiative and convective heat transfer.
4. Pollutant formation.

The validity of the models to various furnace and combustion chamber configurations is assessed with the aid of comparisons with the corresponding experiments. Limitations and implications of these models are discussed together with appropriate modifications and suggestions to improve their physical reality.

As to how well the numerical method does its job, one must distinguish between the computational sources of error and those which result from inadequate knowledge of the basic physical processes. For turbulent reacting flows, the available knowledge is incomplete as will be discussed later. For some time to come, therefore, a stream of proposed models and assumptions is expected. Comparisons with experiments to assess the validity of the models are highly appreciated.

Computational accuracy can be improved with mathematical errors reduced to less than a few percent. Methods to improve computational stability and accuracy are discussed, and surely at least some of the modifications will, in the course of development, prove to be compatible with the requirements of generality, fast convergence and ease of use. Various case studies of furnace and combustion chamber flows are given at the end of the book in an attempt to illustrate the extent to which the present procedure can readily predict local flow pattern, flame characteristics and heat transfer properties.

References to Chapter 2

1. Chedaille, J., Leuckel, W. and Chesters A. K. (1966) Aerodynamic studies carried out on turbulent jets by the International Flame Research Foundation, *J. Inst. Fuel*, **39**, p. 311.
2. Afrasimova, N. V. (1967) Study of the areodynamics of a furnace space. *Thermal Engineering*, **14**, p. 10.
3. International Flame Research Foundation reports IFRF-Doc. K20/a/65, F35/a/4 and F36/a/4 (1974).
4. Wu, H. L., and Fricker, N. (1971) An investigation of the behaviour of swirling jet flames in a narrow cylindrical furnace. *2nd Members Conference of the I.F.R.F.* See also: *J. Inst. Fuel*, **49**, p. 144, 1976.
5. Steward, F. R., Osuwan, S. and Picot, J. J. (1972) Heat transfer measurements in cylindrical test furnaces. *Proc. 14th Symp. (Int.) on Combustion*, p. 651.
6. Gunther, R. and Lenze, B. (1972) Exchange coefficients and mathematical models of jet diffusion flames, *Proc. 14th Symp. (Int.) on Combustion*, p. 687.
7. Baker, R., Hutchinson, P., Khalil, E. E. and Whitelaw, J. H. (1974) Measurements of three velocity components in a model furnace with and without combustion. *Proc. 15th Symp. (Int.) on Combustion*, p. 553.
8. Beltagui, S. and Maccallum, N. R. L. (1976) Aerodynamics of vane swirled flames in furnaces. *J. Inst. Fuel*, **49**, p. 183.
9. Owen, F. K. (1975) Laser velocimeter measurements in free and confined coaxial jets with recirculation. AIAA Paper 75–120.
10. Hutchinson, P., Khalil, E. E., Whitelaw, J. H. and Wigley, G. (1976) The calculation of furnace flow properties and their experimental verification. *J. Heat Transfer*, **98**, p. 276.
11. Hutchinson, P., Khalil, E. E. and Whitelaw, J. H. (1976) The calculation of wall heat transfer rate and pollutant formation in axisymmetric furnaces. *Proc. 4th Members Conference of the IFRF.*
12. Hutchinson, P., Khalil, E. E. and Whitelaw, J. H. (1977) Measurement and calculation of furnace flow properties. *J. Energy*, **1**, p. 212.
13. Khalil, E. E. (1977) Flow and combustion in axisymmetric furnaces, Ph.D. Thesis, London University.
14. Shoffstal, D. R. (1976) Burner design criteria for control of NO_x from natural gas combustion, EPA-600/2-76-098.
15. Bowman, C. T. and Cohen, L. S. (1975) Influence of aerodynamic phenomena of pollutant formation in combustion, Vol. 1 EPA-650/2-75-061-2.

16. Paauw, T. N. T. A. (1974) Some measurements and numerical calculations on turbulent diffusion flames. AGARD Meeting, CP 164.
17. Syred, N., Dahmen, K. R., Styles, A. C. and Najm, S. A. (1977) A review of combustion problems associated with low calorific value gases. *J. Inst. Fuel*, **50**, p. 195.
18. Bilger, R., and Kent, J. H. (1972) Concentration fluctuation in turbulent jet diffusion flame. Sydney University, Dept. of Mechanical Engineering, Report F. 36.
19. Arscott, J. A., and Godridge, A. M. (1978) Flame data from a large residual fuel oil-fired burner, *Proc. 17th Symp. (Int.) on Combustion*, p. 737.
20. Agenew, W. G. (1978) Fifty years of Combustion research at General Motors. *Prog. in Energy and Combustion Science*, **4**, p. 115.
21. Gupta, A. K. Swithenbank, J. and Beer, J. M. (1977) Modern diagnostics for the measurement of fluctuating quantities in flames. *J. Inst. Fuel*, **50**, p. 163.
22. Self, S. A. and Whitclaw, J. H. (1976) Laser anemometry for combustion research. *Combustion Science and Technology*, **13**, p. 171.
23. Pun, W. M. and Spalding, D. B., (1967) A procedure for predicting the velocity and temperature distribution in confined, steady, turbulent, gaseous, diffusion flame. *Proc. 18th International Astronautical Congress, Belgrade.*
24. Pai, B. R. and Lowes, T. M. (1972) The predictions of flow, mixing and heat transfer in the Ijmuiden furnace. IFRF Doc. G 02/2/22.
25. Anasoulis, R. F., McDonald, H. and Bugglen, R. C. (1973) Development of a combustor flow and analysis, Part 1: Theoretical studies AF-APL-TR-73-98.
26. El Ghobashi, S. E. and Pun, W. M., (1974) A theoretical and experimental study of turbulent diffusion flames in cylindrical furnaces. *Proc. 15th Symp. (Int.) on Combustion*, p. 1353.
27. Abou Ellail, M. M., Gosman, A. D., Lockwood, F. C. and Megahed, I. (1977) Description and validation of a three-dimensional procedure for combustion chamber flows. AIAA paper 77–138 (15th Aerospace Sciences Meeting).
28. Khalil, E. E., Hutchinson, P. and Whitelaw, J. H. (1980) The calculation of the flow and heat transfer characteristics of gas fired furnaces. *Proc. 18th Symp. (Int.) on Combustion*, p. 1927.
29. Michelfelder, S. (1976) Survey on prediction procedures for the calculation of furnace performance. *Proc. 4th Members Conference of the IFRF.*
30. Hottel, H. C. and Sarofim, A. F. (1967) *Radiative Transfer* (McGraw-Hill).
31. Jeschar, R. and Schupe, W. (1974) Simplified mathematical models for convective and radiative heat transfer with special reference to heat recovery. *Proc. 3rd Members Conference of the IFRF.*
32. Michelfelder, S. and Lowes, T. M. (1973) Strahlungsaustausch in Glasschmelzofen. *Glastechn. Ber.*, p. 99.
33. Lucas, D. M. and Lockett, A. A. (1972) Mathematical modelling of heat flux and temperature distribution in shell boilers. *Proc. 4th Symp. on Flames and Industry.*
34. Thring, M. W. and Newby, M. P. (1953) Combustion length of enclosed turbulent jet flames. *Proc. 4th Symp. (Int.) on Combustion*, p. 789.
35. Craya, A. and Curtet, R. (1953) Sur l'évaluation d'un jet en espace confine *C.R.A.S.*, **24**, p. 621.
36. Lilley, D. G. (1979) Flowfield modelling in practical combustors: a review. *J. Energy*, **3**, No. 4, p. 193.
37. Lilley, D. G. (1980) Prospects for computer modelling in ramjet combustors. *Proc. AIAA/SAE/ASME 16th Joint Propulsion Conference*. AIAA Paper AIAA-80-1189.

CHAPTER 3

GOVERNING EQUATIONS AND TURBULENCE MODELS

3.1 INTRODUCTION

3.1.1 Purpose of the Chapter

The problems of flow and heat transfer in boiler furnaces and combustors are of a multidisciplinary nature, and hence, attention must be focused on the general principles which govern the behaviour of the flow in such complex configurations. The well established conservation equations of mass, momentum, species and energy can be expressed through partial differential forms; it is these equations, therefore, that form the foundation upon which all prediction procedures are based, either directly or indirectly.

The first purpose of this chapter is to provide clear statements of the governing differential equations in a form which is amenable to solution by the numerical procedures of Chapter 7. Before proceeding any further, it is necessary to explain the different terms contained in these equations, especially the terms that are involved with the modelling. The second purpose of this Chapter is to provide such an explanation.

The governing equations do not of themselves provide the complete specification of the mathematical problem; additional information is necessarily required.

This information covers the turbulent nature of the flow necessary to solve the set of momentum equations in turbulent flows; generally it forms *turbulence models*. The second type of information is that associated with the reactive flows; i.e. the specification of the rate of fuel consumption and probability of reaction; this information is classed under *combustion models*. The radiation contribution to the energy balance is affected by the radiation model, which is needed to complete the specification of the energy balance equation.

Other information such as the specification of the thermodynamic properties is needed, as discussed later in Chapter 7. To yield the equations soluble, boundary and inlet conditions should be specified; this is postponed to Chapter 7, where the calculation procedure is discussed.

3.1.2 Outline of the Chapter

The remainder of the chapter is divided into two main parts; in section 3.2, the differential equations are introduced. The necessary information required to yield these equations amenable to solution on digital computers are classified into turbulence, combustion and radiation models. In section 3.3, the first of these models, i.e., the turbulence model, is discussed; models derivation, formation and assessment are given in this section and a few samples of the predicted flow patterns are displayed to demonstrate the models' applicability to furnace flow geometries. This section aims at giving the necessary information about the most promising turbulence models which are used in furnace flow computations. The reader's attention is drawn especially to the time-averaged conservation equations, to be used in the calculation procedure.

3.2 DIFFERENTIAL EQUATIONS

The starting step is, therefore, to express the partial differential equations, governing the motion of the fluid, in the tensor form[1] as follows:

3.2.1 Mass Conservation (Continuity)

$$\frac{\partial}{\partial x_j} \rho U_j = -\frac{\partial \rho}{\partial t} \qquad (3.2.1)$$

This equation states merely that the net rate of inflow of mass into unit volume equals the rate of increase of density. The continuity equation can also be expressed in general orthogonal coordinates as:

$$\frac{1}{l_1 l_2 l_3}\left[\frac{\partial}{\partial \xi_1}(l_2 l_3\, \rho U_1) + \frac{\partial}{\partial \xi_2}(l_1 l_3\, \rho U_2) + \frac{\partial}{\partial \xi_3}(l_1 l_2 \rho U_3)\right] = -\frac{\partial \rho}{\partial t}$$

where ξ_1, ξ_2, ξ_3 are the coordinates, l_1, l_2 and l_3 are the control volume dimensions while U_1, U_2, U_3 are the velocity components in the three orthogonal coordinate directions.

3.2.2 Momentum Conservation

The momentum equations are expressed in the tensor form as:

$$\frac{\partial}{\partial t} \rho U_i + \frac{\partial}{\partial x_j} \rho U_j U_i = -\frac{\partial \sigma_{ij}}{\partial x_j} + S_{Ui} \qquad (3.2.2)$$

where

$$\sigma_{ij} = P\delta_{ij} - \mu\left(\frac{\partial U_i}{\partial x_j} + \frac{\partial U_j}{\partial x_i}\right) + \frac{2}{3}\mu \frac{\partial U_i}{\partial x_j}\delta_{ij}$$

and

$$\delta_{ij} \text{ is the Kronecker-Delta function} = \begin{cases} 0, & i \neq j \\ 1, & i = j \end{cases}$$

The term S_{U_i} includes body force and resistance components in the i-direction. Again here the equations state that the rate of increase of the i-direction momentum per unit volume equals the net rate of inflow i-direction momentum into the unit volume plus the net force on the unit volume.

3.2.3 Chemical Species Conservation

This equation can be expressed in tensor form as follows:

$$\frac{\partial}{\partial t}(\rho M_a) + \frac{\partial}{\partial x_j}(\rho U_j M_a) = \frac{\partial}{\partial x_j}\Gamma_a \frac{\partial M_a}{\partial x_j} + R_a \tag{3.2.3}$$

For the chemical species, the above equations state that the rate of increase in mass of chemical species 'a' equals the net inflow of species 'a' into unit volume plus the rate of reaction in unit volume by reaction.

3.2.4 Energy Conservation

The energy conservation equation is expressed in the tensor form as:

$$\frac{\partial}{\partial t}(\rho \tilde{H} - P) + \frac{\partial}{\partial x_j}(\rho U_j \tilde{H}) = \frac{\partial}{\partial x_j}\left[\Gamma_n c \frac{\partial T}{\partial x_j} + \Sigma_a \Gamma_a H_a \frac{\partial M_a}{\partial x_j}\right]$$

$$-\mathrm{div}\, \bar{W}_s + S_{\mathrm{rad}} + S_{\mathrm{elect}} \tag{3.2.4}$$

where

$$H = I + \frac{P}{\rho}, \quad \tilde{H} = H + \vec{U}.\vec{U}/2$$

Expressing the same equation in words, then the rate of increase in the internal energy plus the kinetic energy equals the net rate of inflow of stagnation enthalpy by both convection and diffusion plus the net rate of inflow of heat and shear work plus the radiation, electrical and other sources of energy, all per unit volume.

The above Equations 3.2.1 to 3.2.4 are time dependent and describe the transport of mass, three components of momentum, chemical species and energy in a general form. The analytical simultaneous solution of such equations, since they stand for complicated furnace and combustor geometries, is not feasible, due to their complexity. With the advent of digital computers, however, it has been possible to solve them numerically in a reasonable number of relatively simple flow configurations[2, 3, 4, 5, 6, 7, 8].

3.2.5 Time Averaging the Differential Equations

In turbulent flows, rapid fluctuations are exhibited and the flow properties have mean and fluctuating components. Usually, in engineering applications, interest is mainly in the mean flow properties. The influence of turbulent fluctuations on the mean flow characteristics, however, has been proved of utmost importance, especially in determining the pollutant formation rates.

The instantaneous velocities, densities, mass fractions and enthalpies in Equations 3.2.1 to 3.2.4 can be decomposed into their respective mean and fluctuating components as:

$$U_j = \bar{U}_j + U_j + u_i \text{ and } \bar{U}_j = \frac{1}{t_0}\int_0^{t_0} U_j dt$$

where t_0 is a large time, as shown in Figure 3.1

$$\rho = \bar{\rho} + \rho'$$

and

$$\bar{\rho} = \frac{1}{t_0}\int_0^{t_0} \rho dt$$

The instantaneous mass fraction M_a is decomposed as:

$$M_a = \bar{M}_a + m_a \quad \text{and} \quad \bar{M}_a = \frac{1}{t_0}\int_0^{t_0} M_a dt$$

for the energy H, where

$$H = \bar{H} + h \quad \text{and} \quad \bar{H} = \frac{1}{t_0}\int_0^{t_0} H\,dt$$

The introduction of these definitions into the relevant conservation equations produces a more desirable form of these equations, i.e. the time-averaged form.

3.2.5.1 *Time-Averaged Continuity*

$$\frac{\partial\bar{\rho}}{\partial t} + \frac{\partial}{\partial x_j}(\bar{\rho}\bar{U}_j + \overline{\rho' u_j}) = 0 \tag{3.2.5}$$

Equation 3.2.5 is very similar to Equation 3.2.1 except for the additional term which is the density-velocity correlation representing a source/sink of mass. This term is commonly neglected when modelling furnace flow. Nevertheless, it will be shown later that it

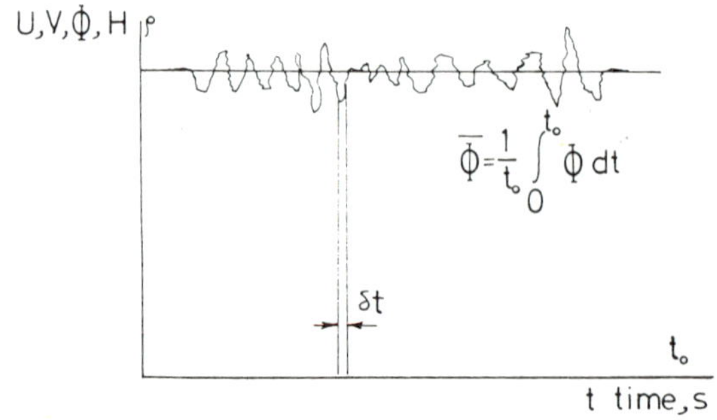

Figure 3.1 Temporal variations of mean flow properties

has a significant role in the vicinity of reaction zones, where rate of change in density is at its highest.

3.2.5.2 *Time-Averaged Momentum*

$$\frac{\partial}{\partial t}(\bar{\rho}\bar{U}_i + \overline{\rho' u_i}) + \frac{\partial}{\partial x_j}(\bar{\rho}\bar{U}_i\bar{U}_j + \bar{\rho}\overline{u_i u_j} + \bar{U}_i\overline{\rho' u_j} + \bar{U}_j\overline{\rho' u_i} + \overline{\rho' u_i u_j}) = -\frac{\partial\bar{\sigma}_{ij}}{\partial x_j} + \bar{S}_{U_i} \quad (3.2.6)$$

where

$$\bar{\sigma}_{ij} = \bar{P}\delta_{ij} - \bar{\mu}\left(\frac{\partial\bar{U}_i}{\partial x_j} + \frac{\partial\bar{U}_j}{\partial x_i}\right) + \frac{2}{3}\bar{\mu}\frac{\partial\bar{U}_i}{\partial x_j}\delta_{ij} + \frac{2}{3}\overline{\mu'\frac{\partial u_i}{\partial x_j}}\delta_{ij}$$

$$-\overline{\mu'\left(\frac{\partial u_i}{\partial x_j} + \frac{\partial u_j}{\partial x_i}\right)}$$

Neglecting the fluctuations in the laminar viscosity, the expression for $\bar{\sigma}_{ij}$ takes the same form as σ_{ij} in Equation 3.2.2 but with all the variables time-averaged.

3.2.5.3 *Time-Averaged Chemical Species Conservation Equation.* The chemical species concentration equation, no different from the two former equations, is expressed as:

$$\frac{\partial}{\partial t}(\bar{\rho}\bar{M}_a + \overline{\rho' m_a}) + \frac{\partial}{\partial x_j}(\bar{\rho}\bar{U}_j\bar{M}_a + \bar{\rho}\overline{u_j m_a} + \bar{U}_j\overline{\rho' m_a} + \bar{M}_a\overline{\rho' u_j} + \overline{\rho' u_j m_a})$$

$$= \frac{\partial}{\partial x_j}\Gamma_a\frac{\partial\bar{M}_a}{\partial x_j} + \bar{R}_a \quad (3.2.7)$$

3.2.5.4 *Time-Averaged Energy Equation.* Following the same line as for the continuity and momentum, the energy equation takes the form:

$$\frac{\partial}{\partial t}(\bar{\rho}\bar{H} - \bar{P} + \overline{\rho' h}) + \frac{\partial}{\partial x_j}(\bar{\rho}\bar{U}_j\bar{H} + \bar{\rho}\overline{u_j h} + \bar{U}_j\overline{\rho' h} + \bar{H}\overline{\rho' u_j} + \overline{\rho' u_j h})$$

$$= \frac{\partial}{\partial x_j}\left(\Gamma_h c\frac{\partial\bar{T}}{\partial x_j} + \sum_a \Gamma_a H_a\frac{\partial\bar{M}_a}{\partial x_j}\right) + \bar{S}_H \quad (3.2.8)$$

where $\bar{S}_H$ is the time-averaged source term.

The equations describing the transport of any scalar fluid property $\bar{\Phi}$ can, in a similar manner, be expressed as

$$\frac{\partial}{\partial t}(\bar{\rho}\bar{\Phi} + \overline{\rho'\phi}) + \frac{\partial}{\partial x_j}(\bar{\rho}\bar{U}_j\bar{\Phi} + \bar{\rho}\overline{u_j\phi} + U_j\overline{\bar{\rho}\phi} + \bar{\Phi}\,\overline{\rho' u_j} + \overline{\rho' u_j\phi})$$

$$= \frac{\partial}{\partial x_j}\Gamma_{\phi,l}\frac{\partial\bar{\Phi}}{\partial x_j} + \bar{S}_\phi \quad (3.2.9)$$

here $\bar{S}_\phi$ is the time-averaged source/sink term of the entity $\bar{\Phi}$ and ϕ is the fluctuating component of the entity Φ.

The number of time-averaged conservation equations is less than the number of the unknown terms contained in these equations. Thus the correlations $\overline{\rho' u_j}$, $\overline{u_i u_j}$, $\overline{u_j\phi}$ and $\overline{\rho'\phi}$ must be at the best either predetermined or modelled; otherwise, and that would be taking the shortest way out, they are to be neglected. For non-reacting flows, the equations of interest are those governing the mass and momentum; hence, modelling of the term $\overline{u_i u_j}$ is needed in order to solve this set of differential equations. In reacting flows, the specification of the density fluctuation correlations is also required. To estimate these correlation terms, which appear in Equations 3.2.5 to 3.2.9, different approaches are available. They can be summarized as follows:

1. In this first approach, all the terms involving the density fluctuations are ignored and hence the only correlation to be specified is $\overline{u_i u_j}$. This approach is more appropriate to non-reacting and isothermal flows where densities are mostly uniform or at the worst vary by a small amount.
2. Secondly, for the situations where combustion and density gradients are involved, equations for all correlations involving density fluctuations are solved. This implies the solution of transport equations for each correlation, thus involving further modelling.
3. The third approach is to express the Equations 3.2.5 to 3.2.9 in a form such that density fluctuation correlations do not appear. Hence the equations have a form similar to those of non-reacting flows. This involves the well known Favre averaging[9, 10] which follows this definition:

$$U_j = \bar{U}_j + \overline{\rho' u_j}/\bar{\rho} \quad (3.2.10)$$

3.2.5.5 *The Favre-Averaging Approach.* It is worthwhile at this point to go into some detail about the Favre-averaging approach. Despite its merits, it will be shown that its disadvantages are numerous, discouraging researchers from using it frequently. To arrive at the Favre-averaged forms of the different equations, definition 3.2.10 is substituted into the conservation equations:

1. Continuity Equation

$$\frac{\partial\bar{\rho}}{\partial t} + \frac{\partial}{\partial x_j}\bar{\rho}\tilde{U}_j = 0 \quad (3.2.11)$$

2. *Momentum Conservation Equations*

$$\frac{\partial}{\partial t}\bar{\rho}\tilde{U}_i + \frac{\partial}{\partial x_j}(\bar{\rho}\tilde{U}_j\tilde{U}_i + \bar{\rho}\widetilde{\overline{u_iu_j}}) = -\frac{\partial\tilde{\sigma}_{ij}}{\partial x_j} + \tilde{S}_{U_i}$$

where

$$\tilde{\sigma}_{ij} = \tilde{P}\delta_{ij} - \tilde{\mu}\left(\frac{\partial\tilde{U}_i}{\partial x_j} + \frac{\partial\tilde{U}_i}{\partial x_i}\right) + \frac{2}{3}\tilde{\mu}\frac{\partial\tilde{U}_i}{\partial x_j}\delta_{ij} \qquad (3.2.12)$$

3. *Scalar Property Conservation Equation*

$$\frac{\partial}{\partial t}\bar{\rho}\tilde{\Phi} + \frac{\partial}{\partial x_j}(\bar{\rho}\tilde{\Phi}\tilde{U}_j + \bar{\rho}\widetilde{\overline{u_j\phi}}) = \frac{\partial}{\partial x_j}\Gamma_{\phi,l}\frac{\partial\tilde{\phi}}{\partial x_j} + \tilde{S}_\Phi \qquad (3.2.13)$$

The nature of the Favre averaging which is to produce mass-averaged properties constitutes its major problem. These produced results cannot be compared to any experimental data as most measurements are time-averaged properties rather than mass-averaged. Several investigations[10–14] were concerned with reporting Favre-averaged entities and argued about the usefulness of such an approach in reacting flows. To be able to compare the Favre-averaged entities with the experimentally measured time-averaged ones, it is essential to know the values of the various correlations; this constitutes an added burden to this method.

It is an undoubted fact, therefore, that the time- or ensamble-averaged variables are different from the Favre-averaged ones; for reacting flows the magnitude of this difference is yet undiscovered. Attempts to quantify this difference were made by Khalil[12], Kolbe *et al.*[15] and Hutchinson *et al.*[16], through solving the equations for the various correlation terms. Unfortunately, various terms in the correlation equations have to be modelled, and, as will be discussed later, the absolute accuracy of the result is difficult to quantify.

3.2.6 Turbulence-Reaction and Heat Transfer Characteristics of the Flow

Having fulfilled the first purpose of this chapter, i.e. exhibiting the conservation equations, we will now turn towards grouping the unknown terms which constitute each of the three above-mentioned characteristics.

3.2.6.1 *Turbulent Characteristics of the Flow*

1. Reynolds and shear stresses

$$\bar{\rho}\,\overline{u_iu_j}$$

2. Density–velocity correlation terms

$$\overline{\rho' u_i} \text{ and } \overline{\rho' u_j}$$

3. Triple correlation

$$\overline{\rho' u_iu_j}$$

The Reynolds and shear stresses which constitute the closure of the aerodynamic equations are obtained by different approaches, through turbulence models. In general, these models relate the Reynolds and shear stresses to the mean flow entities; they also contain physical information about the statistical behaviour of the turbulent flows. The more information the model contains, the more complex it becomes, as more and more modelling assumptions are made. Details of turbulence models are given later in this chapter.

The density–velocity terms are, more often than not, neglected in turbulent reacting flows. To account for these correlations, transport equations for the entities $\overline{\rho' u_i}$, expressed in form similar to Equation 3.2.9 with $\rho' u_i$ replacing Φ, are solved. There exists a simpler solution than solving these transport equations, and that is to model them algebraically. Triple-order correlation terms are nearly always neglected.

3.2.6.2 *Turbulent Reacting Characteristics of the Flow.* Naturally, like all the other entities, the conservation equation governing the transport of fuel mass fraction is expressed in the form 3.2.9. The rate of chemical reaction $\bar{R}_{fu}$ is the source/sink term $\bar{S}_\phi$ when the fuel mass fraction M_{fu} replaces Φ. In homogeneous reactions, the rate of chemical reaction is defined as the rate of disappearance of one of the reactants, in order to form the products, or else it is the rate of formation of these products. If the reaction is an oxidation, then the fuel concentration decreases and the rate of formation of the products of combustion is positive. Such a rate is a function of the temperature, pressure and composition. In modelling the reacting flows, a sufficient amount of attention should be given to $\bar{R}_{fu}$, i.e. the rate of fuel consumption, which is strongly influenced by the turbulence and mixing characteristics of the flow[13, 16].

Other terms that appear in the fuel mass fraction equation such as, $\overline{u_i m_{fu}}$, $\overline{\rho' m_{fu}}$ and $\overline{\rho' u_j m_{fu}}$, are more difficult to model. The correlation term $\overline{u_i m_{fu}}$ is generally replaced by a turbulent exchange coefficient and the mean gradient of $\bar{M}_{fu}$. The term $\overline{\rho' m_{fu}}$, is an unknown quantity, and there are even strong doubts about whether ρ and m_{fu} are correlated or not. Triple-order correlations are nearly always neglected.

The effect of turbulence on combustion and reaction rate is contained in the mass fraction fluctuations $\overline{m_{fu}^2}$ terms, the fuel-oxidant fluctuation correlation $\overline{m_{fu} m_{ox}}$ terms. the temperature-species correlation terms and the probability density function of fuel mass fraction. These parameters and their effect on the reaction will be discussed in depth in Chapters 4 and 5.

3.2.6.3 *Heat Transfer Characteristics and Pollutant Formation.* Heat transfer by radiation in flows, such as those discussed here, is of high importance. Consequently source terms which account for it should be included in Equation 3.2.8; $\bar{S}_H$ denotes the radiative source term in reacting flames. Different radiation models have been proposed,[17–22] and are applied to furnace flows in order to calculate the radiative heat fluxes and the radiative source term. The details of the radiation models and convection heat transfer calculations are described in Chapter 6.

Calculations of pollutant formation rates in turbulent reacting flames are given in Chapter 6.

3.3 TURBULENCE MODELLING

3.3.1 Preview

As this chapter is mainly devoted to the subject of turbulence modelling, it is worth while to start with a brief review of the various models. To solve the set of the momentum equations, the Reynolds stresses must be related to the mean hydrodynamic properties of the flow field. There are many ways of relating these stresses to the mean quantities, either by algebraic expressions or by the more complicated partial differential equations.

The eddy or turbulent viscosity concept was first introduced by Boussinsq,[23] who suggested that it is possible to replace the effective turbulent shear stress, $-\bar{\rho}\overline{uv}$, by μ_t $\partial\bar{U}/\partial y$. Since then it has been the task of many workers to determine the turbulent viscosity μ_t. For example, a simple algebraic expression for μ_t has been proposed by Prandtl.[24] His proposal was simply that μ_t is proportional to l^2 and $\partial\bar{U}_i/\partial x_j$, where l is defined as the mixing length appropriate to each flow of the boundary layer type. Other expressions for μ_t have also been proposed, for example by Launder *et al.*[25, 26]

Progress in the field of evaluating μ_t advanced from merely formulating algebraic expressions to the more difficult task of solving the partial differential equation, for the kinetic energy of turbulence k as reported;[24–30] all assumed that the local state of the fluid depends on one or more turbulence quantities which can be determined from the solution of their corresponding transport equations. This concept allows the turbulent flows to be characterized by two turbulence quantities, for example the kinetic energy of turbulence k and a characteristic length scale. The solution of the differential equations for these two properties was proposed by Harlow *et al.*[30] and Jones *et al.*;[31] and it is now commonly known as the two-equation turbulence model.

TABLE 3.1

Main Turbulence Models

Reference	Author(s)	Model
23	Boussinisq	eddy viscosity concept
24	Prandtl	mixing length
27	Bradshaw	one-equation turbulence model
30	Harlow *et al.*	two-equation turbulence model
31	Jones *et al.*	two-equation turbulence model – low Reynolds number
33	Launder *et al*	Reynolds stress model
34	Rebiero	probability approach
35	Pope *et al.*	Reynolds stress

Reviews of turbulence models are given in references 3, 25 and 34.

An even more complicated approach, to the evaluation of the Reynolds stresses $\overline{u_i u_j}$, is to solve the transport equations for the stresses themselves as proposed, for example, by Daly and Harlow,[32] Launder, Reece and Rodi,[33] Rebiero[34] and Pope and Whitelaw.[35] These equations represent the turbulent transport, generation, dissipation and redistribution of Reynolds stresses.

More complicated turbulence models were also proposed and are reviewed by Launder *et al.*[3]

Table 3.1 displays the main turbulence models available in the current literature.

3.3.2 Mean Flow Closure

The effective viscosity hypothesis connects the Reynolds stresses to the mean velocity gradients through the relation:

$$\bar{\rho}\,\overline{u_i u_j} = \frac{2}{3}\bar{\rho}k\,\delta_{ij} - \mu_{\text{eff}}\left(\frac{\partial \bar{U}_i}{\partial x_j} + \frac{\partial \bar{U}_j}{\partial x_i}\right) \tag{3.3.1}$$

The velocity gradient term appearing in this equation offers the only possible linear combination of velocity gradients, which provides the same tensor properties as the Reynolds stresses. Both sides of the equation are symmetric and contract to zero. It must be noted that this expression is analogous to the stress/rate-of-strain relation which applies to a laminar flow. The major difference, however, is that the effective viscosity is a function of the flow rather than of the fluid.

To satisfy dimensional requirements at least two scaling parameters are needed to relate the Reynolds stresses to the rate of strain. There are several choices for such scales, e.g. a velocity scale v and a time scale τ. The time scale τ_m is deduced from the velocity gradients as:

$$\tau_m = \left(\frac{\partial \bar{U}_i}{\partial x_j} \times \frac{\partial \bar{U}_i}{\partial x_j}\right)^{-\frac{1}{2}}$$

However, as the macro-time scale of turbulence, $\tau_t = (k/\varepsilon)$, is independent of τ_m in simple shear flows, as was proved by Lumley[36], the two scaling parameters must be independent of the mean velocity field. Launder and Spalding[3] reviewed a great number of other authors' choices of the two scaling parameters. A feasible choice is the kinetic energy of turbulence k and its dissipation rate ε. Thus:

$$\mu_{\text{eff}} = C_\mu \bar{\rho} k^2/\varepsilon \tag{3.3.2}$$

where C_μ is a constant, and the kinetic energy k is defined as:

$$k = \frac{1}{2}\overline{u_i u_i} \tag{3.3.3}$$

and where ε is defined as:

$$\varepsilon = \frac{\mu}{\bar{\rho}}\left(\overline{\frac{\partial u_i}{\partial x_j}\frac{\partial u_i}{\partial x_j}}\right) \tag{3.3.4}$$

To calculate $\overline{u_i u_j}$, the values of k and ε must be known; those values are obtained from the solution of their respective transport equations. The transport equation for k is obtained by subtracting Equation 3.2.6 from 3.2.2, and then multiplying the result by u_k. A second equation is formed by commuting the suffixes k and j; adding the two equations, averaging the result and neglecting the density fluctuations results in the final form of the k equation, which is:

$$\underset{\text{(A)}}{\frac{\partial k}{\partial t}\bar{\rho} + \frac{\partial}{\partial x_j}\bar{\rho}\bar{U}_j k} = \underset{\text{(B)}}{-\bar{\rho}\overline{u_i u_j}\frac{\partial \bar{U}_i}{\partial x_j}} \underset{\text{(C)}}{- \frac{\partial}{\partial x_j}\left(\frac{1}{2}\bar{\rho}\overline{u_i u_i u_j} - \mu\frac{\partial k}{\partial x_j} + \overline{P' u_j}\right)} + \underset{\text{(D)}}{\mu\overline{\left(\frac{\partial u_i}{\partial x_j}\right)^2}} \tag{3.3.5}$$

In this equation, term (A) denotes the change along an averaged stream line, while term (B) is the 'production of turbulent kinetic energy'; its definition implies that it is a positive term. This fact is strengthened by the measurements that show that the 'production' is invariably positive, although there are regions in some flows where it is negative as in asymmetric channels and annuli. Term (C) represents the transport of k due to turbulence and viscosity actions. The last term (D) is the dissipation of k due to viscous action. By analogy to the laminar transport of k, its turbulent transport is modelled as

$$\frac{1}{2}\bar{\rho}\,\overline{u_i u_i u_j} - \mu\frac{\partial k}{\partial x_j} + \overline{P' u_j} = -\frac{\mu_{\text{eff}}}{\sigma_k}\frac{\partial_k}{\partial x_j}$$

Thus, finally the modelled kinetic energy equation for steady state stands as

$$\frac{\partial}{\partial x_j}\bar{\rho}\bar{U}_j k = \frac{\partial}{\partial x_j}\left(\frac{\mu_{\text{eff}}}{\sigma_k}\frac{\partial k}{\partial x_j}\right) - \bar{\rho}\,\overline{u_i u_j}\frac{\partial \bar{U}_i}{\partial x_j} - \bar{\rho}\varepsilon \tag{3.3.6}$$

The kinetic energy equation is complemented by an equation for the transport of k. To arrive at the equation of ε, first the momentum equations are differentiated with respect to x_l, then multiplied by 2 $\nu\,\partial U_i/\partial x_j$ and then averaged. After mathematical manipulation, the equation can be expressed as follows:

$$\frac{D\varepsilon}{Dt} = -2\nu\left[\overline{\frac{\partial u_i}{\partial x_l}\frac{\partial u_j}{\partial x_l}} + \overline{\frac{\partial u_l}{\partial x_i}\frac{\partial u_l}{\partial x_j}}\right]\frac{\partial \bar{U}_i}{\partial x_j} - 2\nu\overline{\frac{\partial u_i}{\partial x_l}\frac{\partial u_j}{\partial x_l}\frac{\partial u_i}{\partial x_j}}$$
$$-2\nu\,\overline{u_j\frac{\partial u_i}{\partial x_l}}\frac{\partial^2 \bar{U}_i}{\partial x_j \partial x_l} - 2\overline{\left(\nu\frac{\partial^2 u_i}{\partial x_j \partial x_l}\right)^2} - \frac{\partial}{\partial x_j}\left[\overline{u_j\varepsilon'} + \frac{\nu}{\bar{\rho}}\overline{\frac{\partial P}{\partial x_l}\frac{\partial u_j}{\partial x_l}} - \nu\frac{\partial \varepsilon}{\partial x_j}\right]$$

The net rate of change of ε for a fluid element is equal to the net imbalance due to the following interactions:

1. Generation due to vortex stretching by mean flow
2. Generation due to vortex stretching by turbulence
3. Secondary generation by mean flow
4. Destruction of ε due to tendency of viscosity to smear out velocity variations

5. Diffusive transport arising from velocity and pressure fluctuations and from molecular diffusion.

The final modelled form of the equation, according to Jones *et al.*[31], is expressed for steady state as:

$$\frac{\partial}{\partial x_j}\bar{\rho}\varepsilon\bar{U}_j = \frac{\partial}{\partial x_j}\left(\frac{\mu_{\text{eff}}}{\sigma_\varepsilon}\frac{\partial\varepsilon}{\partial x_j}\right) - \bar{\rho}\frac{\varepsilon}{k}(C_1G + C_2\varepsilon) \tag{3.3.7}$$

In the above equation, $G = \overline{u_iu_j}(\partial\bar{U}_i/\partial x_j)$ and C_μ, C_1 and C_2 are the constants of the turbulence model. C_μ was obtained from equilibrium flows, C_1 from turbulence decay behind grids, and C_2 from computer optimization. Launder *et al.*,[3] after extensive examination of free turbulent flows, recommended that the constants of Equations 3.3.2, 3.3.6 and 3.3.7 should take the values of Table 3.2.

It is now evident that the present closure comprises the transport equations for k and ε, i.e. Equations 3.3.6 and 3.3.7, together with the expressions for the Reynolds stresses and the effective viscosity, i.e. Equations 3.3.1 and 3.3.2.

The validity of the above model in simple non-reacting flows was assessed by way of comparison with experimental data for a wide range of flow configurations by several researchers, among them, Launder *et al.*,[3] Spalding,[37] Khalil *et al.*,[5] Gosman *et al.*,[6] Pai *et al.*,[38] Rodi,[39] Hanjalic *et al.*[40] and Khalil.[12]

Although the two-equation model is widely applied to many engineering applications as it stands, it must be modified to account for variations in density and non-similarity of the flows. Hence the exact solution for the transport equation of k in a turbulent reacting flow is different from that for the non-reacting flow due to the heat release phenomenon and the presence of the density fluctuation correlations. However, and for the sake of simplicity, previous investigators in turbulent combustion modelling, e.g. El Ghobashi,[41] used the form of Equation 3.3.6, which is appropriate to a non-reacting incompressible flow.[39] More recently, the exact equation in turbulent reacting flows was described by Bray[14,42] for boundary layer flows, and then was extended to elliptic flows by Hutchinson *et al.*[11] and Khalil.[12]

Additional terms emerged in the conservation equation of the kinetic energy of turbulence, Equation 3.3.6, due to the effect of density changes, and are included as additional source terms of k as follows:

$$\frac{\partial}{\partial x_j}\bar{\rho}\bar{U}_jk = \frac{\partial}{\partial x_j}\left(\frac{\mu_{\text{eff}}}{\sigma_k}\frac{\partial k}{\partial x_j}\right) - \bar{\rho}\overline{u_iu_j}\frac{\partial\bar{U}_i}{\partial x_j} - \bar{\rho}\varepsilon + \frac{\mu_{\text{eff}}}{\sigma_\rho}\left(\frac{\partial k}{\partial x_j}\frac{\partial\bar{\rho}}{\partial x_j}\right) \tag{3.3.8}$$

where σ_ρ equals 0.9. In a typical flame, where the density of the hot gases changes, Equation 3.3.1 does not hold and should be modified to:

$$\bar{\rho}\overline{u_iu_j} = \frac{2}{3}\left(\bar{\rho}k - \frac{\mu_{\text{eff}}}{\bar{\rho}}\bar{U}_l\frac{\partial\bar{\rho}}{\partial x_l}\right)\delta_{ij} - \mu_{\text{eff}}\left(\frac{\partial\bar{U}_i}{\partial x_j} + \frac{\partial\bar{U}_j}{\partial x_i}\right) \tag{3.3.9}$$

This modification is done to satisfy the definition of k and the continuity equation. To give a clear idea about the influence of the density variation in reacting flows, Equations 3.3.6 and 3.3.8 have been applied separately. In a typical flame, where the density of the hot products is nearly 20% of that of the fresh air/fuel mixture, the additional terms in

TABLE 3.2

Two-equation Turbulence Model Constants

C_μ	C_1	C_2	σ_k	σ_ϵ
0.09	1.44	1.92	1.0	1.22

These constants are obtained from reference 25.

Equation 3.3.8, resulted in not more than 20% difference in the local values of k, and only 3% difference in the values of the mean velocities, concentrations and gas temperatures.[12,43] These figures stem from the calculations performed for the flow in a confined coaxial flame configuration and co-flowing flame situation.

This model was assessed by means of comparing the results it produced with experimental data. However the presentation of such an assessment is postponed to the end of the chapter, i.e. after discussing the Reynolds stresses model and one or two complementary topics.

3.3.3 The Reynolds Stresses Closure

This more complicated model needs a book in its own right so that every detail can be given. However, in this work, it is discussed briefly in order that the reader is presented with a clear idea of the model. This model solves the equations for the Reynolds stresses themselves. These equations are derived by subtracting Equation 3.2.6 from 3.2.2. Ignoring the density fluctuations, multiplying by u_k, forming a second equation by commuting the suffixes k and i and finally adding the two equations and averaging the result, yield the Reynolds stresses equation:

$$\underbrace{\bar{\rho}\frac{\partial}{\partial t}\overline{u_i u_j} + \bar{\rho}\bar{U}_j\frac{\partial}{\partial x_j}\overline{u_i u_k}}_{(A_0)} = \underbrace{-\bar{\rho}\,\overline{u_j u_k}\frac{\partial \bar{U}_i}{\partial x_j} - \bar{\rho}\,\overline{u_i u_j}\frac{\partial \bar{U}_k}{\partial x_j}}_{(B_0)}$$

$$\underbrace{-\frac{\partial}{\partial x_j}\left(\bar{\rho}\,\overline{u_i u_j u_k} - \mu\frac{\partial}{\partial x_j}\overline{u_i u_k} + \overline{P' u_i}\,\delta_{jk} + \overline{P' u_k}\delta_{ij}\right)}_{(C_0)}$$

$$\underbrace{+\overline{P'\frac{\partial u_i}{\partial x_k}} + \overline{P'\frac{\partial u_k}{\partial x_i}}}_{(D_0)}\underbrace{-2\mu\overline{\frac{\partial u_i}{\partial x_j}\frac{\partial u_k}{\partial x_j}}}_{(E_0)} \qquad (3.3.10)$$

The term (A_0) denotes the change along an averaged streamline, while the term (B_0) denotes the 'production' due to Reynolds stress and mean velocity gradient interaction. The term (C_0) represents the transport due to turbulence and viscosity. The term (D_0) denotes redistribution or pressure scrambling while the last term (E_0) denotes the dissipation due to viscous action. All these terms have to be modelled in terms of known flow quantities.

Launder *et al.*[33] presented a form of the equation for the Reynolds stresses which provided a modelled substitute of every term. The term (C_0) which represents the transport, as mentioned earlier, was substituted as follows:

$$-\frac{\partial}{\partial x_j}\bar{\rho}\,\overline{u_i u_j u_k}=\frac{\partial}{\partial x_j}\left(C_s\,\bar{\rho}\frac{k}{\varepsilon}\,\overline{u_j u_l}\frac{\partial}{\partial x_l}\,\overline{u_i u_k}\right) \tag{3.3.11}$$

In this form the pressure velocity correlation term is ignored, and the magnitude of the diffusive coefficient depends on the direction. The tensor properties of each side of Equation 3.3.11 are the same; however, if the term inside the derivative is considered to represent the triple correlation, then the tensor properties of the exact term and the modelled term differ, as discussed by Launder *et al.*[33]

The dissipation term (E_0) was modelled assuming isotropic small scale turbulence at high Reynolds number. The modelled expression for this term is:

$$-2\mu\,\overline{\frac{\partial u_i}{\partial x_j}\frac{\partial u_i}{\partial x_j}}=-\frac{2}{3}\bar{\rho}\varepsilon\,\delta_{ik} \tag{3.3.12}$$

The redistribution (pressure scrambling) term (D_0) is very important since it governs the level of isotropy of the Reynolds stresses; this term was modelled in two parts:

$$\overline{P'\frac{\partial u_j}{\partial x_i}}=\phi_{ij,1}+\phi_{ij,2}$$

The first terms contained in $\phi_{ij,1}$ were modelled after Rotta[44], and represent a trend towards isotropy at the rate of the turbulent time scale:

$$\phi_{ij,1}+\phi_{ji,1}=-C_{\phi 2}\,\bar{\rho}\frac{\varepsilon}{k}\left[\overline{u_i u_j}-\frac{2}{3}k\,\delta_{ij}\right] \tag{3.3.13}$$

The second terms contained in $\phi_{ij,2}$ were modelled by Launder *et al.*[33] as

$$\phi_{ij,2}+\phi_{ji,2}=-C_{\phi_2}\left(P_{ij}-\frac{2}{3}P\,\delta_{ij}\right) \tag{3.3.14}$$

where

$$P_{ij}=-\left(\bar{\rho}\overline{u_j u_k}\frac{\partial \bar{U}_i}{\partial x_k}+\bar{\rho}\,\overline{u_i u_k}\frac{\partial \bar{U}_j}{\partial x_k}\right)$$

and

$$P=\frac{1}{2}P_{ii}\,.$$

TABLE 3.3

C_{ϕ_1}	C_{ϕ_2}	C_{ϵ_1}	C_{ϵ_2}	C_s	C_ϵ
2.5	0.4	1.45	1.9	0.25	0.15

The above expression is the counterpart of Rotta's proposal,[44] for $(\phi_{ij,2} + \phi_{ji,2})$ tends to isotropize the turbulence production tensor. To make C_{ϕ_2} able to act as a useful approximation for general flows, the best value for it should be near 0.3.

Now that the various terms of the transport equation of the Reynolds stresses have been modelled, the final result is an equation that reads as follows:

$$\bar{\rho}\frac{\partial}{\partial t}\overline{u_iu_j} = -P_{ij} + C_s\frac{\partial}{\partial x_l}\left[\bar{\rho}\frac{k}{\varepsilon}\overline{u_l u_k}\frac{\partial \overline{u_iu_j}}{\partial x_k}\right]$$

$$-\left[C_{\phi_1}\bar{\rho}\frac{\varepsilon}{k}\left(\overline{u_iu_j} - \frac{2}{3}k\delta_{ij}\right) + C_{\phi_2}\left(P_{ij} - \frac{2}{3}P\delta_{ij}\right)\right] - \frac{2}{3}\bar{\rho}\varepsilon\delta_{ij} \qquad (3.3.15)$$

The dissipation equation used with the Reynolds stresses model is the same as that of mean closure model, with the exception that the transport term is modelled in terms of the stresses as

$$\bar{\rho}\frac{\partial}{\partial t}\varepsilon + \frac{\partial}{\partial x_j}\bar{\rho}\bar{U}_j\varepsilon = \frac{\partial}{\partial x_j}\left(C_\varepsilon\bar{\rho}\frac{k}{\varepsilon}\overline{u_iu_j}\frac{\partial\varepsilon}{\partial x_j}\right) + \frac{\varepsilon}{k}(C_{\varepsilon_1}P - C_{\varepsilon_2}\bar{\rho}\varepsilon) \qquad (3.3.16)$$

Six constants appear in the model, C_{ϕ_1}, C_{ϕ_2}, C_{ε_1}, C_{ε_2}, C_s and C_ε; their values, according to Launder *et al.*,[33] are given in Table 3.3.

3.3.4 Wall Functions

The above turbulence models are intended for flow regions with high turbulence Reynolds number of the order of

$$\mathrm{Re}_w = k^{\frac{1}{2}}\, l_0/\nu > 5000 \qquad (3.3.17)$$

where

$$l_0 = k^{3/2}/\varepsilon$$

In the regions of the flow which are near a wall the velocities tend to zero and hence these are zones where the local values of Re_w are so small that the viscous effect becomes dominant. If the exact momentum equations are to be solved in this region, then an excessively large number of grid nodes is needed; i.e. up goes the cost of the solution. One can resort, of course, to the alternative solution of 'laminarizing' the model in these regions.

However, there is a third path, and perhaps a more accessible one; the flow in the region near the wall is represented by the approximated Couette flow solutions based on

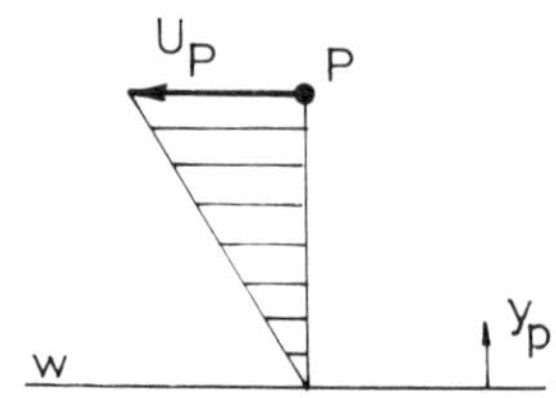

Figure 3.2a Near wall grid node

the logarithmic law of the wall. Near a wall, at point P shown in Figure 3.2a, the flow is predominantly parallel to the wall and the shear stress is assumed constant, as described by Launder *et al.*[3] Thus the shear stress is related to normal distances from the wall by:

$$\tau_{w_0} = \mu \left(\frac{\partial \bar{U}}{\partial y} \right)_w = \frac{\bar{U}_p}{y_p} \frac{y_+}{u_p^+} \tag{3.3.18}$$

where $u^+ = (1/\kappa) \ln E\, y_+; y_+ = \bar{\rho}(k_p C_\mu^{1/2})^{1/2} y_p/\mu$; k_p is the value of the kinetic energy of turbulence at P and y_p is the distance normal to the wall.

The above expression is modified for flows with pressure gradients according to McDonald[45] as

$$\mu \frac{\partial \bar{U}}{\partial y} = \frac{\tau}{1 + \kappa y^2}$$

where

$$y = y_p \left(\frac{\bar{\rho}}{\mu} \frac{\partial \bar{U}}{\partial y} \right)^{1/2}$$

The total shear stress is nearly constant across the layer as

$$\frac{\partial}{\partial y} \tau = \frac{\partial P}{\partial x} + \bar{\rho} U_\tau \frac{\partial U_\tau}{\partial x} u^{+2} \tag{3.3.19}$$

where

$$U_\tau = (\tau_w/\bar{\rho})^{1/2}$$

Integrating the above equation yields

$$\tau_w = \tau_{w_0} + \frac{\partial P}{\partial x} y_p$$

Another form of expressing the pressure gradient is through utilizing the pressure coefficient ψ_0, i.e.

$$\psi_0 = \frac{dP}{dx} (C_\mu^{1/2} y_p/\bar{\rho} k).$$

The effects of ψ_0 on the functional relationship were experimentally and theoretically determined by several investigators among them, McDonald,[45] Mellor[46] and Borgnakke[47]. This effect can be introduced in the log law of the wall as

$$u_p^+ = \frac{1}{\kappa} \ln E y_+ (1 + \psi_0) \tag{3.3.20}$$

Equation 3.3.20 is appropriate to boundary layer flows with non-zero pressure gradient, and has been used for separated boundary layer flows by Cebeci *et al.*[48]. The velocity parallel to the wall and at a distance y_p from the wall is then expressed as

$$\frac{\bar{U}_p}{\tau_w} \bar{\rho} C_\mu^{1/4} k^{1/2} = \frac{1}{\kappa} \ln [E y_+ (1 + \psi_0)] \tag{3.3.21}$$

The constants of the logarithmic law of the wall, κ and E, depend on the wall roughness and are 0.42 and 8.8 respectively. This wall function represents the dependence of the flux of momentum to the wall on the turbulence characteristics at a point P remote from it.

The value of k_p is calculated from the transport equation for k with the diffusion of turbulent kinetic energy to the solid wall set equal to zero. The value of ε used to obtain k_p was obtained from the expression[3]:

$$\varepsilon_p = (C_\mu^{1/2} k_p)^{3/2} / \kappa y_p \tag{3.3.22}$$

and the integral of ε from the wall to y_p, which is required for the kinetic energy and Reynolds stress equations, is given by

$$\int_0^{y_p} \varepsilon \, dy = C_\mu^{3/4} k_p^{3/2} u_p^+ \tag{3.3.23}$$

The heat flux to the wall is represented, in a similar manner; with $\bar{T}_w$ and q''_w denoting the wall temperature and heat flux per unit area respectively as:[3]

$$\frac{C_p(\bar{T}_p - \bar{T}_w)}{\left(\dfrac{q''_w}{\bar{\rho}}\right)} C_\mu^{\frac{1}{4}} k_p^{\frac{1}{2}} = \frac{\sigma_{h,t}}{\kappa} [\ln E y_+] + P_J \tag{3.3.24}$$

where P_J is defined as:

$$P_J = \sigma_{h,t} \frac{\pi/4}{\sin \dfrac{\pi}{4}} \left(\frac{A}{\kappa}\right)^{\frac{1}{2}} \left[\frac{\sigma_h}{\sigma_{h,t}} - 1\right] \left[\frac{\sigma_{h,t}}{\sigma_h}\right]^{\frac{1}{4}}$$

A = van Driest's constant = 26 for smooth walls, $\sigma_{h,t}$ = effective Prandtl number for fully turbulent flows, and σ_h = laminar Prandtl number.

Inclined Walls

Not all the flow situations one encounters in practice are bounded by parallel walls or in cylindrical ducts; therefore provision for inclined walls must be included. The treatment

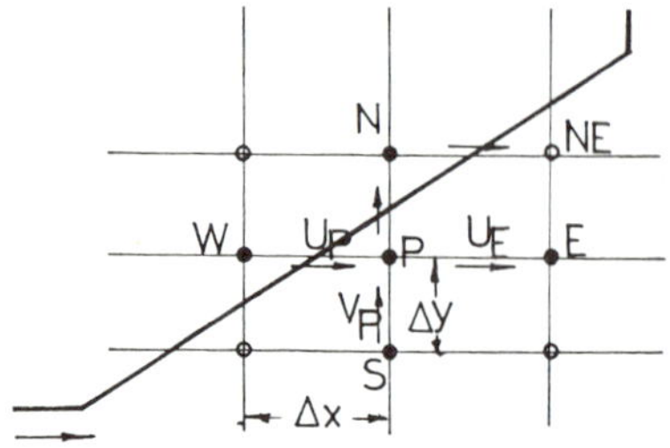

Figure 3.2b Grid arrangement near inclined wall

of irregular boundaries and obstacles of various shapes was carried out with the appropriate wall functions. The various grid nodes near an inclined wall are shown in Figure 3.2b and the resultant velocity $\bar{U}_R$ is expressed[12] as

$$U_R = \left(\left(\frac{1}{2}(\bar{U}_p + \bar{U}_E)\right)^2 + \left(\frac{1}{2}(\bar{V}_p + \bar{V}_N)\right)^2\right)^{1/2} \qquad (3.3.25)$$

The local velocity gradients near the wall were calculated taking into account the wall angle $\theta = \tan^{-1}(\partial Y/\partial x)$ and the flow angle $\theta_0 = \tan^{-1}[(\bar{V}_p + \bar{V}_N)/(\bar{U}_p + \bar{U}_E)]$. The resultant velocity parallel to the wall is $\bar{U}_R \cos(\theta_0 - \theta)$ and is used in the wall function to obtain the resultant shear stress. In the presence of swirl, the resultant velocity $\bar{U}_{R\theta}$ is

$$\bar{U}_{R\theta} = [(U_R \cos(\theta_0 - \theta))^2 + \bar{W}_p^2]^{1/2} \qquad (3.3.26)$$

This method of representing the inclined and irregular walls is a simplified means of attacking the complicated problem of specifying the boundary conditions; nevertheless it is acceptable for engineering design purposes. Modifications to the above method are needed for fine representation of near wall flows. Among the possible modifications is the use of analytical and numerical transformation methods.[49] Both approaches solve transformed conservation partial differential equations and boundary conditions in a more convenient rectangularized computational domain, where the boundaries coincide with coordinate lines. Finite element techniques offer an advantage over the widely used finite difference techniques, by more accurate representation of the irregular boundaries, but their use in turbulent flows is still in the early stages of development, and has been, as yet, confined to simple non-recirculating flows.

3.3.5 Assessment of the Models

No model can be claimed either useful or useless, until it has been rigorously tested and tried. In this section an assessment is given of the success and failure of the previously discussed models. From the previous discussion it is clear that these models are of general nature, and produce, as will be proved, satisfactory information; their economical advantage will be evident later in the book. As these models are intended to be used in furnace and combustor flow predictions, it is essential to test them in a wide range of flow configurations.

The credit of the two-equation turbulence model has been established by its usage in numerous free shear flows. Many investigators reported calculated results that were in good agreement with corresponding measurements, for example Launder *et al.*,[3, 25] Mathews *et al.*,[50] and El Ghobashi *et al.*[51] The tested flows comprised plane jet flow, wall jets, co-flowing jets and pipe flows. Table 3.4 shows a number of the non-reacting flows which were tested and exhibits the reasonable agreement that was reached. These same flows are shown in Figures 3.3, 3.4, 3.5, 3.6 and 3.7; the figures show how well the original $k-\varepsilon$ model of Launder *et al.*,[3] i.e Equations 3.3.1, 3.3.2, 3.3.6 and 3.3.7, performed for such flows.

The level of agreement obtained for such comparatively simple flows suggested the use of the $k-\varepsilon$ model for more complicated flows. Before discussing such complicated flows, we note that the constants of the $k-\varepsilon$ model shown in Table 3.2 were reported to be not perfectly appropriate for strongly swirling jets, by Morse.[52] He found that the constant C_1 should be modified to be able to predict correctly the rate of spread of the coaxial swirling jet which he investigated. He eventually modified the value of C_1 to the

TABLE 3.4

Reference	Flow
3	plane jet in a moving stream
3	flow in a pipe
38	wall jets
3, 6	co-flowing jets
3	flow along a twisted tape
51	coaxial jets with concentration fluctuation

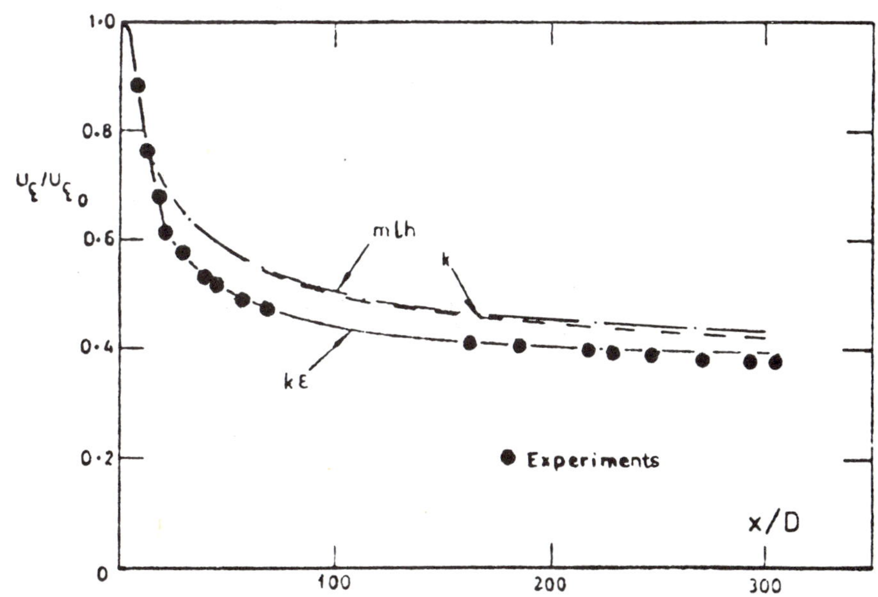

Figure 3.3 Decay of centreline velocity in jet in moving stream[3]

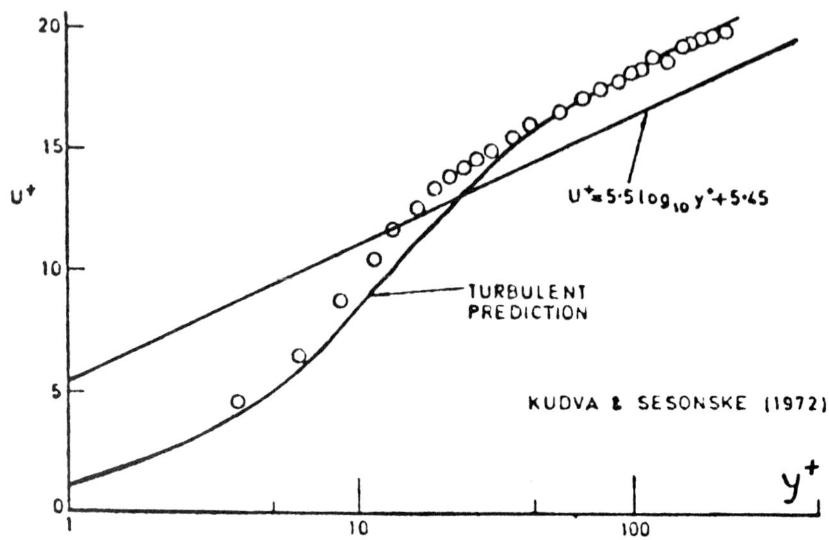

Figure 3.4 Measured and calculated pipe flow velocity profile,[3] $Re = 6000$

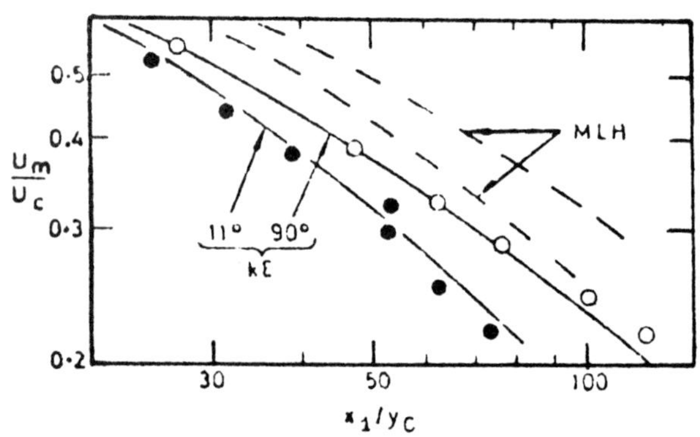

Figure 3.5 Decay of wall jet on conical surfaces[3]
MLH denotes mixing length hypothesis, and the symbols represent the measurements

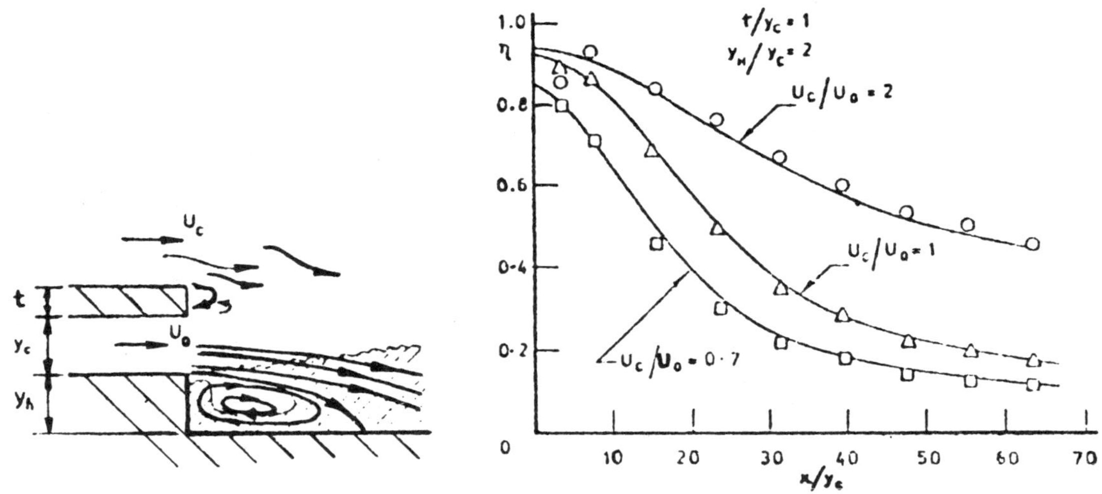

Figure 3.6 Decay of wall jet in terms of adiabatic wall film cooling effectiveness Measurements are represented by symbols and the lines represent the calculations[3,38]

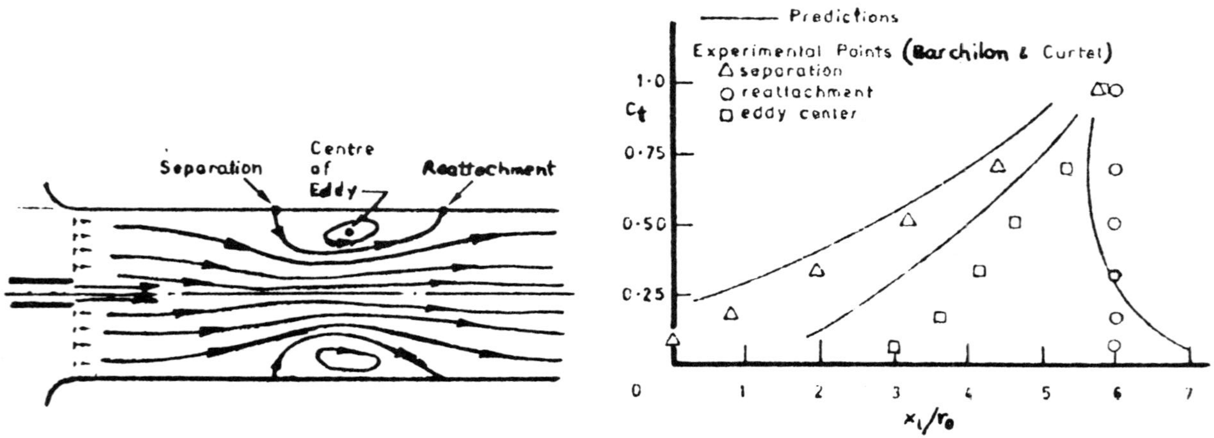

Figure 3.7 Flow reversal in co-flowing jets in confinement and the calculations of the characteristics of reversed flow regions and their experimental verification
△○□ denote measurements of reference 54
—— denotes the calculations of reference 3

form which is a function of the Richardson number, as follows:

$$C_1 = 1.44 + 0.8R_i \frac{\varepsilon}{G}$$

and

$$R_i = \left[\overline{vwr} \frac{\partial}{\partial r} \left(\frac{\bar{W}}{r} \right) \right] / \varepsilon$$

This relation gives the ratio between additional production due to the stress $\overline{vw}$ and the dissipation rate ε; so far this modification has not been extensively tested.

After exhaustive investigation of the above mentioned simple flows, investigators proceeded towards a more complex variety. Gosman *et al.*,[6] for example, turned their attention towards the flow pattern in confined recirculating flows, where the non-similarity and reversal of the flow occur. They computed the flow patterns and turbulence intensity, for a wide variety of flow configurations and these were compared to their corresponding measurements. The tested configurations were the pipe flows,[53] the co-flowing flow of Barchilon *et al.*[54] and of Craya *et al.*,[55] and the swirling co-flowing flow of Craya *et al.*[56] A sample of the predictions of Gosman *et al.*[6] is shown in Figure 3.8 for a co-flowing flow with various degrees of swirl in the central jet. The calculations agree with the experiments at low swirl numbers, but as the swirl number increases to $S \geqslant 0.81$, the agreement deteriorates. Beside these configurations, Gosman *et al.*[6] reported predicted flow pattern and heat transfer variation between the ribs, for the flow past roughness rib elements of Wilkie.[57] The agreement was good for pitch/height ratios

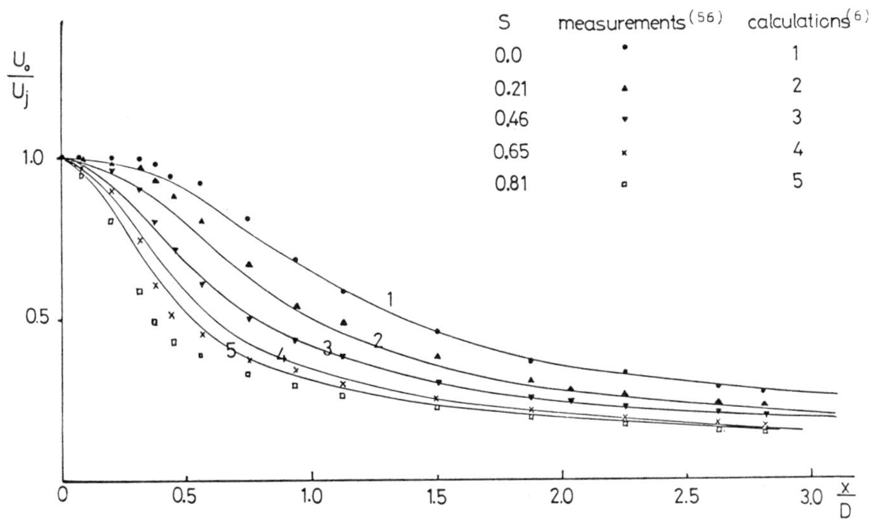

Figure 3.8 Measured and predicted centreline distributions of axial velocity for various swirl numbers in the co-flowing jet arrangement of reference 56

of 7.2 and 15. Although no measurements of velocity and turbulence were reported, these were readily predicted. Also predictions of wall jet flows and flow behind disc were reported and reference should be made to the original paper for more details.

The most relevant flows to this work that Gosman *et al.*[6] investigated are the flows past backward facing step and in sudden enlargement. This latter configuration is the basic feature of the furnace geometries. For sudden enlargement, the predictions were done for the geometry of Abbott *et al.*[58] The predicted length of the wall recirculation zone was in reasonable agreement with the measured values as shown in Figure 3.9. Another group of investigators is Moon *et al.*,[59] who measured and predicted the velocity pattern in an axisymmetric sudden enlargement. Their measurements were obtained by the Laser Doppler Anemometer and the predictions were obtained using a numerical procedure which embodied the $k-\varepsilon$ model as its turbulence model. A sample of their comparisons discussed is shown in Figures 3.10 and 3.11. The rate of decay of centreline velocity, shown in Figure 3.10, is well predicted and the discrepancy is less than 10%. Figure 3.11 shows the measured and predicted radial profiles of mean axial velocity. Discrepancies between measured and predicted mean velocity are shown to be minimum in the central region of the flow. Near the edge of the forward flow region, the velocity decreases sharply and discrepancies of less than 15% are observed.

Because of the wide range of applicability of flows in sudden enlargement geometries, investigations of such flows, using the $k-\varepsilon$ model, are frequently appearing in the current literature; for example, Khalil[60] and Abou Arab.[61] In most cases, reasonable agreement was reported between the calculated results and the corresponding experimental data. Coaxial jets in sudden enlargement have had their share in research, for example Khalil.[12] The importance of this geometry lies in the fact that it resembles boiler furnaces and combustors.

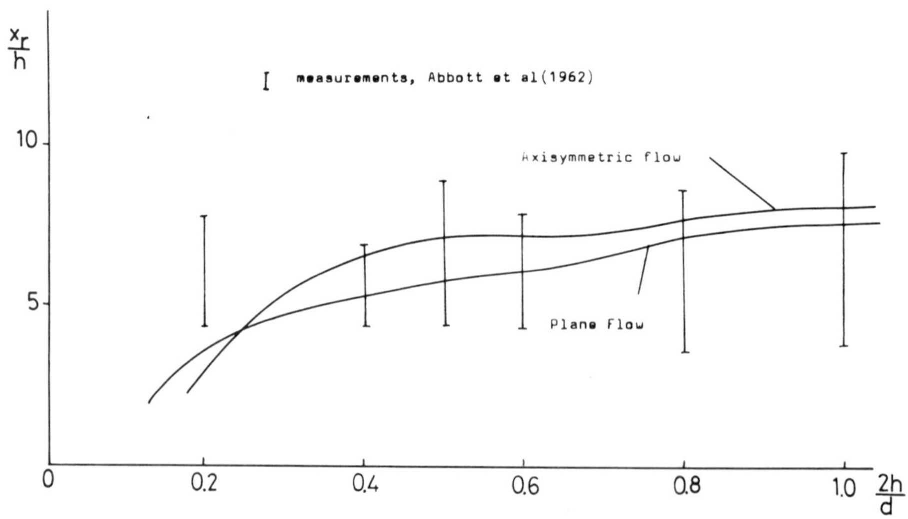

Figure 3.9 Measured and calculated reattachment length in sudden enlargement geometry (measurements of reference 58 and calculations of reference 6)

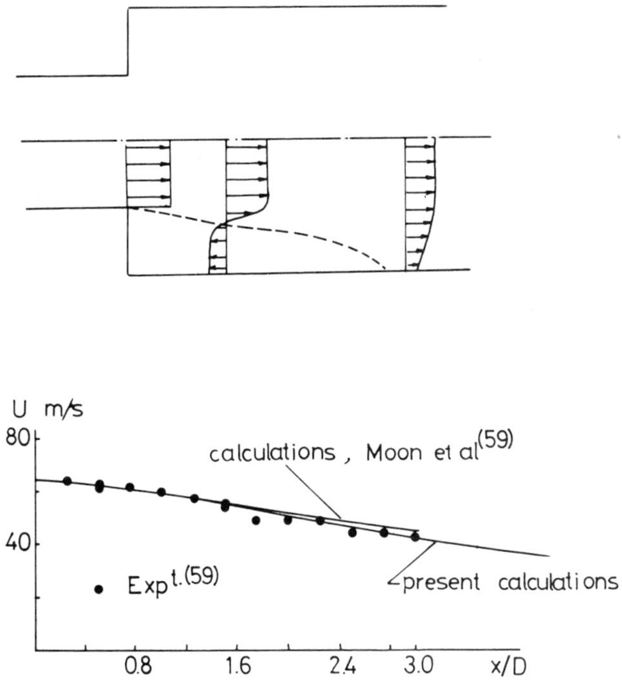

Figure 3.10 Measured and calculated rate of decay of centreline velocity in a sudden enlargement geometry

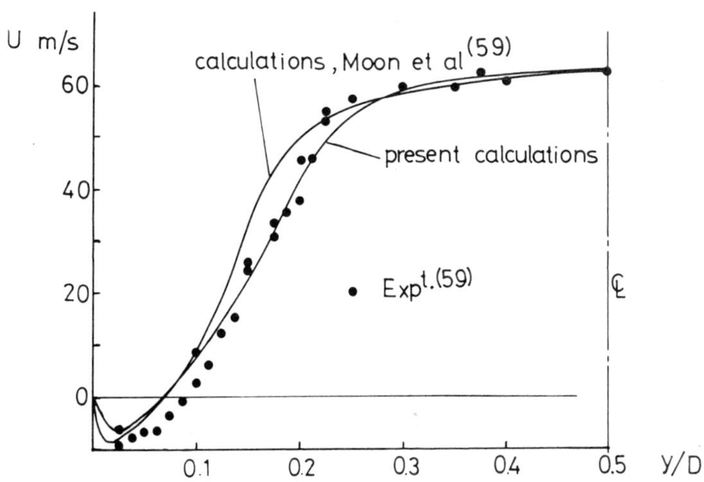

Figure 3.11 Measured and calculated radial profiles of mean axial velocity in a sudden enlargement geometry

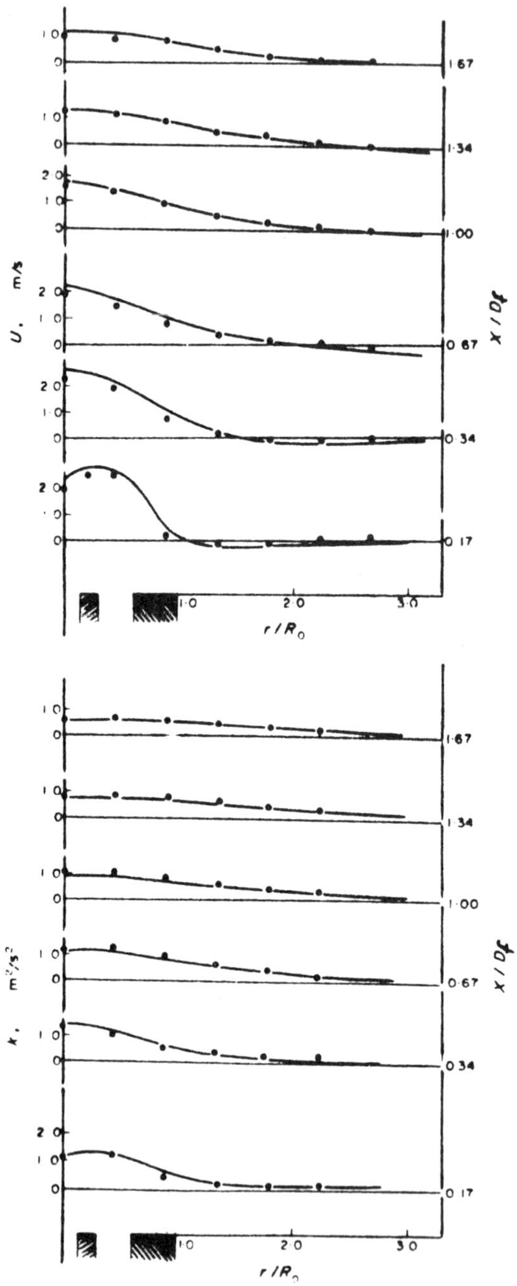

Figure 3.12 Radial profiles of mean axial velocity and kinetic energy of turbulence in the furnace geometry of reference 62
$S = 0$, non-reacting
● measurements of reference 62
— calculations of reference 5

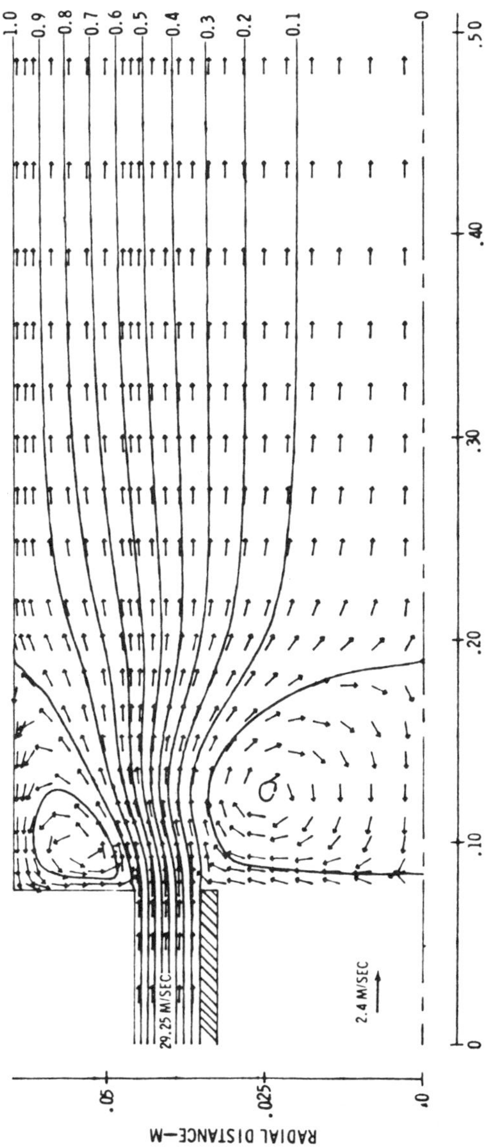

Figure 3.13 Calculated stream lines for confined coaxial jets [64]

The experimental data provided in references 12 and 62 were used by Khalil *et al.*,[5] Hutchinson *et al.*,[16,63] to validate the same model. The experimental data in this case were obtained, in the non-reacting flow conditions, using the Laser Doppler anemometer; the measurements comprised the mean axial, radial and tangential velocity profiles together with their corresponding turbulence intensity distributions. These measurements are shown in Figure 3.12 against the calculations of Khalil *et al.*;[5] on the figure the radial profiles of the mean axial velocity at different axial locations are shown. Under the non-swirling conditions, the calculated results identified the same regions which were identified by the measurements, namely the forward central jet decay and the wall recirculation zone. The size of the wall recirculation zone was well predicted with a maximum discrepancy of less than 5%; the rate of decay of the centreline velocity was also adequately predicted.

The predictive capability of the $k - \varepsilon$ model is further demonstrated by inspecting the results obtained for the flow field resulting from coaxial jets entering a confined sudden expansion by Novick *et al.*[64] Figure 3.13 shows the calculated streamlines, velocity directions and recirculation zones. It can be seen that the wall recirculation zone is provoked by the sudden expansion of the flow area. The central recirculation zone results from the imbalance of mass flux between the inner and outer jets. The mass flux of the outer jet is excessive compared to the central jet, thereby leaving the central jet with insufficient momentum to overcome the adverse static pressure gradient created by the sudden expansion of the outer jet flow. Consequently, a central recirculation zone is formed. Novick *et al.*[64] reported comparisons between predicted and experimentally determined axial velocity variation at two axial locations. These comparisons are shown in Figure 3.14 at x/D_f of 0.25 and 1.8. The first profile exhibits the existence of both the central and wall recirculation zones while at x/D_f of 1.8, both zones disappear. The

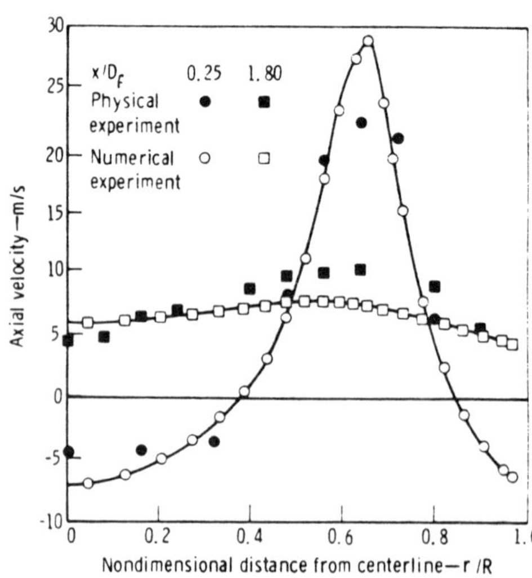

Figure 3.14 Measured and calculated radial profiles of mean axial velocity in confined coaxial jets[64]

obtained agreement is reasonable apart from the observed discrepancies in the vicinity of the central recirculation zone.

The case of swirling coaxial jet flow, which represents the furnace flow configuration with variable degree of swirl in the annular air space, was also investigated[63] (Figure 3.15). Samples of comparisons between the measurements of reference 16 and the predictions are shown in Figures 3.16 and 3.17. The swirl numbers in the annular flows are 0.3 and 0.52. Three main flow regions are identified; namely, the forward central flow region, the wall recirculation zone and the central recirculation zone. The measured and calculated profiles are in reasonable agreement, except near recirculation zones. In this area the size of the recirculation zone is under-predicted by up to 20%. This feature of under-predicting the size of recirculation zone is common when employing the two-equation model to predict recirculating flows; it can be attributed to the assumptions of

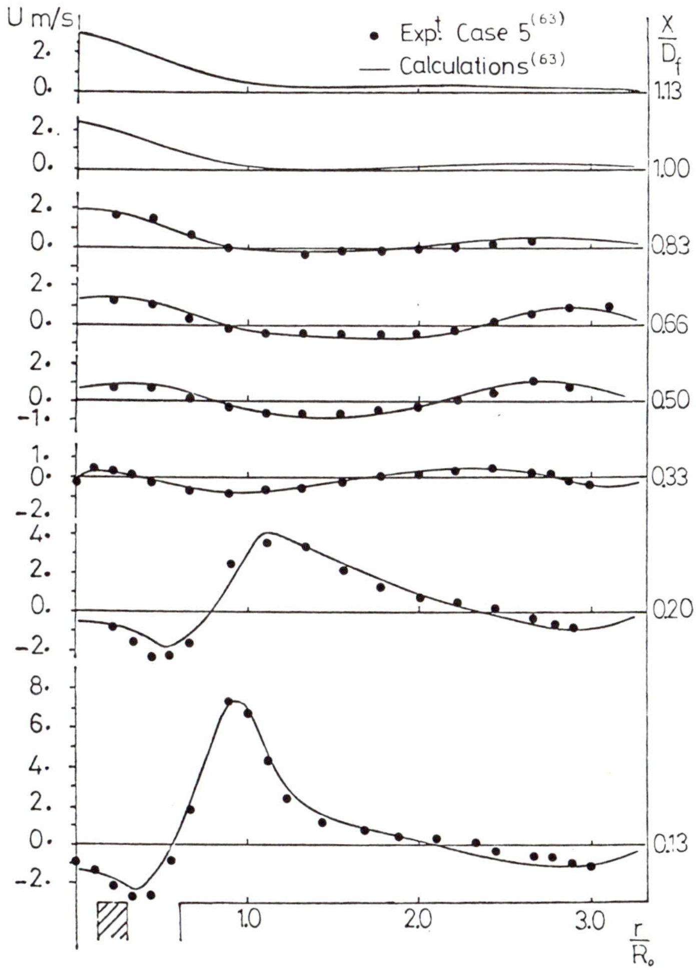

Figure 3.15 Measured and calculated profiles of mean axial velocity

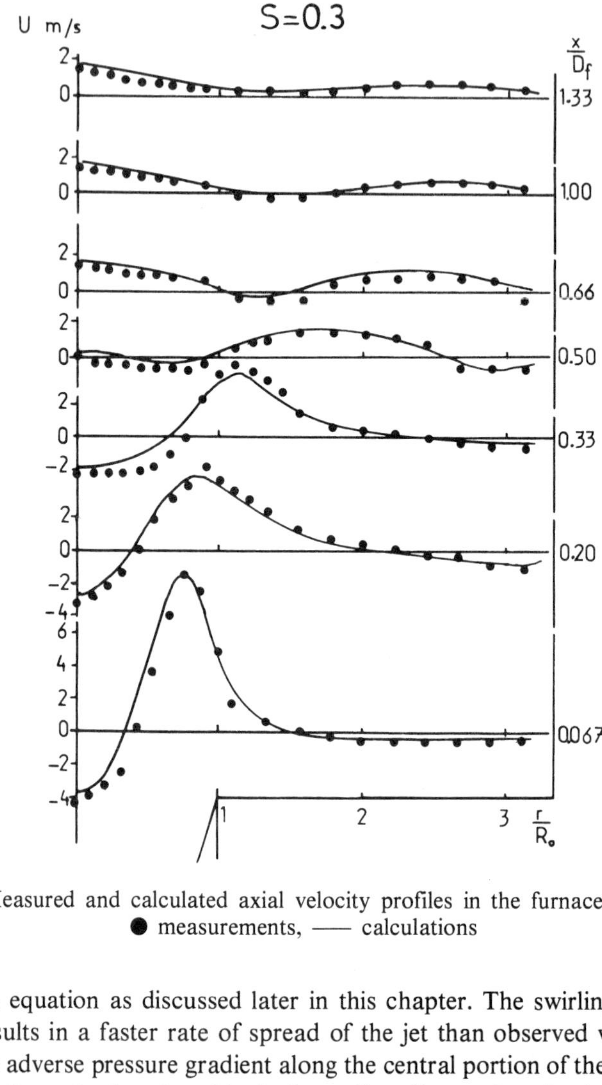

Figure 3.16 Measured and calculated axial velocity profiles in the furnace of reference 16
● measurements, —— calculations

the dissipation equation as discussed later in this chapter. The swirling motion of the annular air results in a faster rate of spread of the jet than observed when no swirl is employed. The adverse pressure gradient along the central portion of the flow causes the reversal of the flow; the lengths of both the wall and central recirculation zones were found to depend strongly on the swirl intensity.[65–67] The predicted effect of swirl intensity on the recirculation zones geometry is illustrated in Figure 3.18. As the swirl intensity increases, the length of the wall recirculation zone decreases due to the faster rate of the turbulent diffusion of momentum. For swirl intensities greater than 0.3, a central recirculation zone occurs; the size of this zone naturally increases with the increase in the swirl intensity. It was observed that, on the whole, the $k-\varepsilon$ model represented adequately swirling furnace flows.

The measurements of the tangential velocity profiles, in the boiler furnace geometry of reference 62, were compared to calculations by Khalil[12] and Hutchinson *et al.*[16] A sample is shown in Figure 3.19; it is clear that the measured and calculated profiles are in reasonable agreement. However, the calculated decay of the tangential momentum

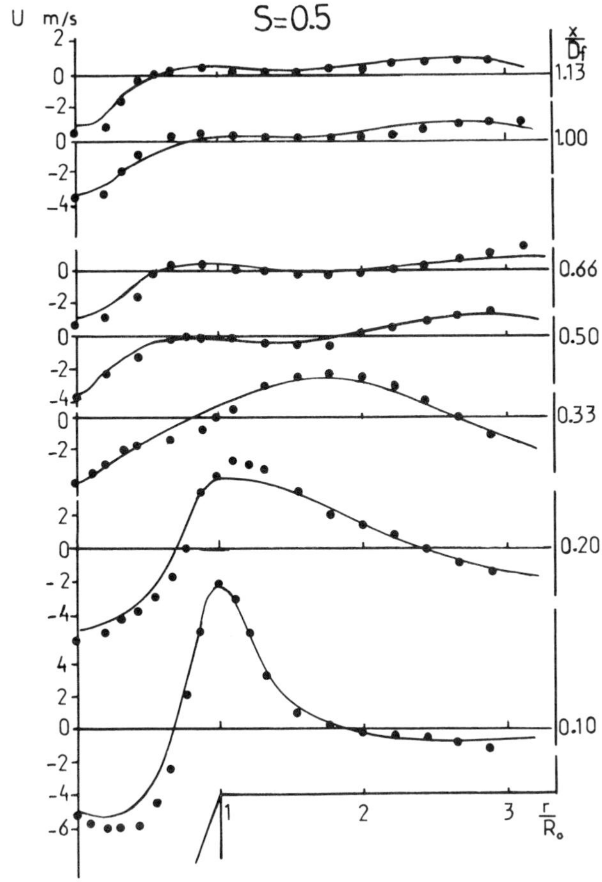

Figure 3.17 Measured and calculated axial velocity profiles in the furnace of reference 16 ● measurements, —— calculations

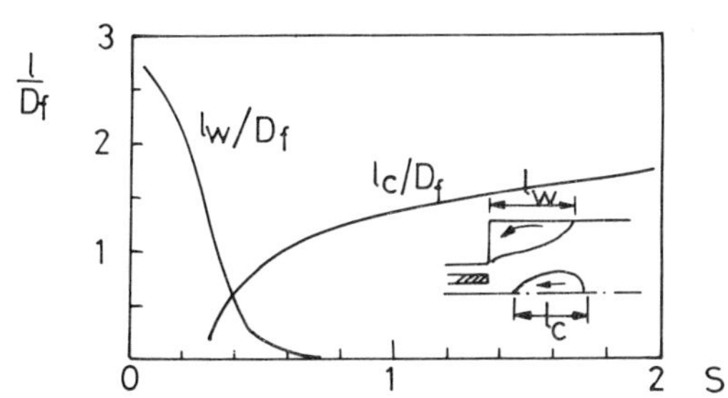

Figure 3.18 Effect of swirl on wall and central recirculation zones

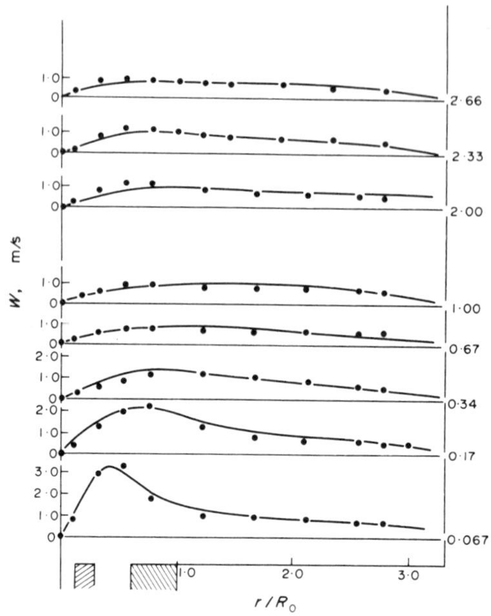

Figure 3.19 Radial profiles of mean tangential velocity in the geometry of references 12 and 62
● measurements of reference 62
—— calculations of reference 12

was faster than the measured one; this feature explains why the length of the central recirculation zone was under-predicted. It is obvious that an over-prediction of the decay of the tangential momentum leads to the prediction of a smaller recirculation zone.

Two defects resulting from the turbulence model, i.e. the under-prediction of the length of the central recirculation zone and the under prediction of the rate at which the wake decays, were observed by other investigators. For example Pope[13] and Assaf[68] reported unsatisfactory agreement between measurements and calculations downstream of bluff bodies in uniform streams.

All these samples of comparisons were meant to indicate that reasonable calculated results can be obtained, using the $k-\varepsilon$ model, in simple and swirling recirculating non-reacting flows. The two major discrepancies which were observed in the predictions of the length of the circulation zone and the spreading rate can be attributed to two things, namely inaccuracy in the measurements and the specification of inlet conditions, and deficiencies in the $k-\varepsilon$ turbulence model.

To overcome the first defect, apart of course from taking strict measures to increase the accuracy of the measurements, the solution domain is increased and correct inlet and boundary conditions are prescribed. This approach will be gone into in more detail when the numerical procedure is discussed. To understand better the deficiencies embodied in the $k-\varepsilon$ model, it is necessary to discuss the model in terms of the constitutive relationship 3.3.1 and the dissipation term. The value of the Reynolds stresses term $\overline{u_i u_j}$ at any point must be uniquely related to the values of k, ε and $(\partial U_i/\partial x_j)$ at the same point. It is clear from Equation 3.3.10, that the Reynolds stresses may be transported by both the mean and fluctuating velocities. To satisfy the above condition, the transport terms must be negligible, which in turn implies that the velocity gradients and Reynolds stresses must be nearly homogeneous. Further, in nearly homogeneous flows, with high Reynolds number, the assumption that all macro-scales of turbulence are proportional can be reasonably made, and hence two scaling parameters k and ε, together with the velocity gradients are sufficient to determine the Reynolds stresses.

The constitutive relation between $\overline{u_i u_j}$, k, ε and $(\partial \bar{U}_i/\partial x_j)$ contains the constant C_μ which is dependent on the rate of strain and the rotation invariant (Pope[69]). A modification was proposed[70] to relate C_μ to the local rate of strain and rotation invariant. Pope[69] defined the following non-dimensional quantities:

$$A_{ij} = \overline{u_i u_j}/k - \frac{2}{3}\delta_{ij}$$

and

$$A_{ij} = A_{ij}[\alpha_{ij} \cdot B_{ij}] \qquad (3.3.27)$$

where

$$\alpha_{ij} = \frac{1}{2}\frac{k}{\varepsilon}\left(\frac{\partial \bar{U}_i}{\partial x_j} + \frac{\partial \bar{U}_j}{\partial x_i}\right)$$

$$B_{ij} = \frac{1}{2}\frac{k}{\varepsilon}\left(\frac{\partial \bar{U}_i}{\partial x_j} - \frac{\partial \bar{U}_j}{\partial x_i}\right)$$

The tensor α_{ij} is non-dimensional and symmetric with zero trace ($\alpha_{ii} = 0$ due to incompressibility), while the tensor B_{ij} is non-dimensional and antisymmetric. The general expression for the tensor A_{ij} is a finite tensor polynomial in terms of all the linearly independent second-order tensors, T_{ij}, and the polynomial coefficients, G_0, which depend upon the invariants of α_{ij} and B_{ij}. Thus

$$A_{ij} = \sum_{n=\lambda}^{n=0} G_0^n T_{ij}^n \qquad (3.3.28)$$

The above modification includes the dependence of C_μ on the rate of strain and rotation invariants and has been applied to swirling flow in cyclone chambers.[70] Samples of the obtained agreement are shown in the original work to which reference should be made for details.

The dissipation Equations 3.3.7 and 3.3.16 of the models considered in this chapter differ only in the transport term. The source term S_ε in the dissipation equation is considered as the most important part of the turbulence model. This source term not only affects the dissipation rate ε but has an important indirect effect on the kinetic energy of turbulence k. Therefore, an error in S_ε may be expected to produce considerable errors in the values of k, and the predicted velocity fields. The source term is related to local quantities by a constant whose value is to be determined. The original ε-equation was developed and tested in two-dimensional, near-equilibrium, boundary layer and free shear flows, where only *one* velocity gradient was modelled in a linear function of the only non-zero independent invariant of the rate of strain. The form of dissipation equation has been criticized by Bradshaw,[71] Priddin[72] and Pope,[13] because it does not contain any terms that relate to the rotation invariants and, therefore, it is applicable only to similar flows. Shear flows, such as boundary layers, jets and wakes, display approximate similarity and are expected to be fairly well represented by turbulence models incorporating this form of the dissipation equation.

Moving over to the Reynolds stresses model, we note that this model has been assessed in free shear flows and boundary layers. The extent to which such model represents the turbulence characteristics of the flow was investigated by Launder *et al.*,[33] Castro,[73] Pope[13] and Launder[74] in mixing layers, plane jet and flow behind bluff bodies. The obtained agreement with experiments indicated a somewhat superior overall universality in two-dimensional boundary layer flows than the standard $k-\varepsilon$ model. In weak shear flows, the obtained agreement with the corresponding measurements was less satisfactory than in the boundary layer flows, but still better than that obtained using the $k-\varepsilon$ model.

The application of the Reynolds stress model to recirculating flow in furnace geometries was reported by Abou Arab.[75] The reported agreement was reasonable in comparison with that obtained with a two-equation turbulence model (the k-vorticity model[3]). Figure 3.20 displays the measured and calculated profiles of mean axial velocity obtained in an axisymmetric sudden enlargement geometry. The diameter and

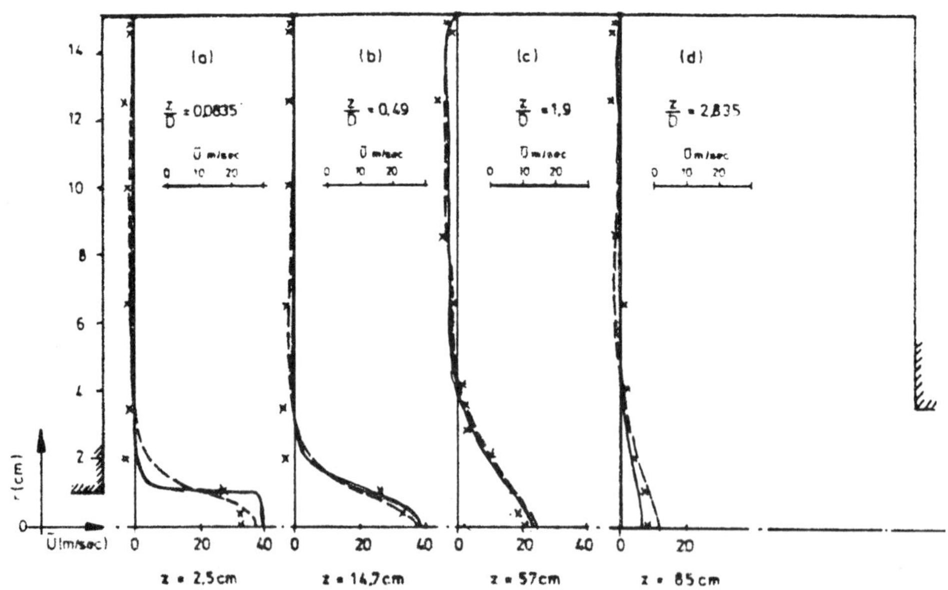

Figure 3.20 Radial distribution of axial flow velocity
---- developed Reynolds Stress model
—— k–w model; x experiments of reference 75

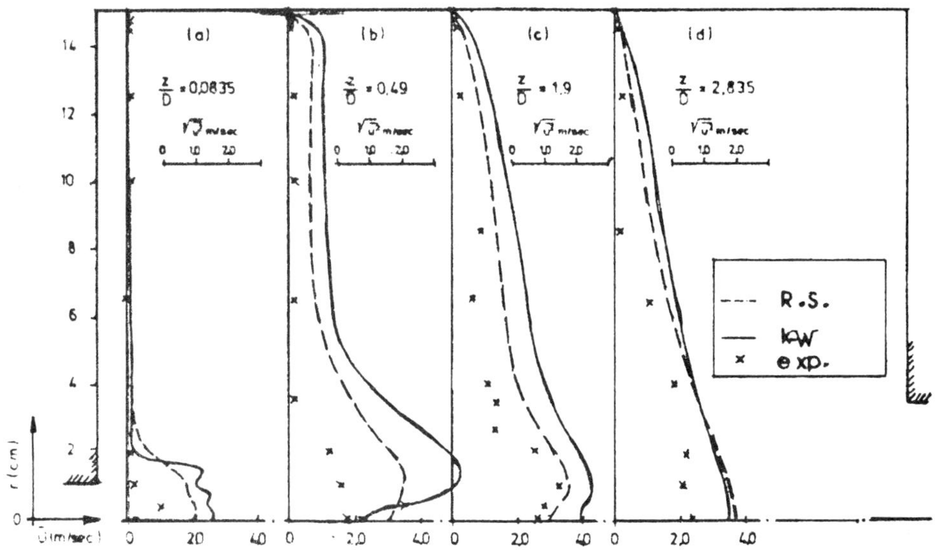

Figure 3.21 Radial distribution of the r.m.s. value of the axial velocity fluctuations[75]

length of the furnace were 0.3 and 1.5 m respectively and it was fitted with a concentric burner of diameter 0.04 m. The measured and calculated profiles were shown[75] to be in reasonable agreement. Discrepancies were observed near the axis of symmetry and particularly near the burner. They can be attributed to the inadequate specification of the inlet velocity and the corresponding turbulence intensity profiles. Measured and calculated profiles of $\sqrt{\bar{u}^2}$ are shown in Figure 3.21. The model predicts the trends correctly but discrepancies of the order of 100% were observed. It should be emphasized that one comparison test for the Reynolds stresses model in such flows cannot lead to concrete conclusions about the model applicability to furnace flow geometries and more comparisons are vitally needed.

3.4 CLOSURE

This chapter has described the partial differential conservation equations of the flow and heat transfer characteristics. These equations are those governing the mass, three components of momentum, chemical species and energy. The equations were expressed in tensor form and methods of averaging the flow entities were discussed. The time-averaged conservation equations and the associated density fluctuation correlations were also described. Favre-averaging the conservation equations, which is one of the methods used, was discussed with special attention to the correlation models and their treatment in reacting flows with strong dominant density variation; the reasons for discarding such a method were mentioned.

The conservation equations were found to contain more unknowns than the number of equations; modelling of some additional terms, therefore, was necessary to close the set of equations. These closures comprise three different aspects of flow, combustion and heat transfer. The turbulent characteristics of the flow were modelled in terms of the 'turbulence models'. Only two turbulence model closure appropriate to furnace flows were considered; the first is the two-equation $k-\varepsilon$ model and the second is the Reynolds stress model. The second closure is that of chemical reaction and flame characteristics; and is arrived at by modelling the rate of chemical reaction and the probability distribution as will be discussed in Chapters 4 and 5. The third modelling direction is that of the heat transfer characteristics of the flame, specially that of the radiative heat transfer, which is discussed in Chapter 6.

In the remainder of the chapter, the two turbulence models were discussed. The wall functions were also considered. The formulation of the two models was expressed in detail to allow examination of their validity in complex furnace geometries. Examples of the application of these models to turbulent non-reacting swirling recirculating flows were displayed and demonstrated the extent to which such models can be applied to complex flow situations.

Although reasonably good agreement was reported in many of these flows, discrepancies were also observed and particularly in strongly swirling and recirculating flows where strong non-similarity is dominant. The deficiencies of the turbulence models were discussed in terms of the form of the Reynolds Stress expression in the mean flow closure and the dissipation equation, and possible modifications were reviewed.

References to Chapter 3

1. Brid, R. B., Stewart, W. E. and Lightfoot, E. N. (1960) *Transport Phenomena* (Wiley, New York).
2. Spalding, D. B. (1972) A novel finite difference formulation for differential expressions involving both first and second derivative. *Int. J. Numerical Methods in Engineering*, **4**, p. 551.
3. Launder, B. E. and Spalding, D. B., (1974) The Numerical computations of turbulent flows. *Computer Methods in Applied Mechanics and Engineering*, **3**, p. 269.
4. Gosman, A. D., Pun, W. M., Runchal, A. K., Spalding, D. B. and Wolfstein, M. (1969) *Heat and Mass Transfer in Recirculating Flows* (Academic Press, London).
5. Khalil, E. E., Spalding D. B. and Whitelaw, J. H. (1975) The calculation of local flow properties in two dimensional furnaces. *Int. J. Heat and Mass Transfer*, **18**, p. 775.
6. Gosman, A. D., Khalil, E. E. and Whitelaw, J. H. (1979) The calculation of two dimensional turbulent recirculating flows, in *Turbulent Shear Flows*, edited by Durst *et al.*, (Springer-Verlag), p. 237.
7. Bergeles, G., Gosman, A. D. and Launder, B. E. (1975) The near field character of a jet discharged through a wall at 90° to a main stream. ASME Paper 75-WA/HT-108.
8. Humphrey, A. J. C. (1978) Numerical calculation of variable property flows in curvilinear orthogonal coordinates. *Can. J. Chem. Eng.* **56**, p. 624.
9. Favre, A. (1969) Statistical equations of turbulent gases, in *Problems of Hydraulics and Continuum Mechanics*, (Society of Industrial and Applied Mechanics, Philadelphia), p. 231.
10. Bilger, R. (1975) A note on Favre averaging in variable density flows. *Combustion Science and Technology*, **1.1**, p. 215.
11. Hutchinson, P., Khalil, E. E. and Whitelaw, J. H. (1976) The calculation of wall heat transfer rate and pollutant formation in axisymmetric furnaces. *Proc. 4th Members Conference of the IFRF.*
12. Khalil, E. E. (1977) Flow and combustion in axisymmetric furnaces, Ph.D. Thesis, London University.

13. Pope, S. B. (1976) The calculation of flow behind bluff bodies with and without combustion, Ph.D. Thesis, London University.
14. Bray, K. N. C. (1978) The interaction between turbulence and combustion. *Proc. 17th Symp. (Int.) on Combustion.*
15. Kolbe, W. and Kollmann, W. (1980) Prediction of turbulent diffusion flames with a four-equation turbulence model. *Acta Astronautica*, **7**, p. 91.
16. Hutchinson, P., Khalil, E. E. and Whitelaw, J. H. (1977) Measurement and calculation of furnace flow properties. *J. Energy*, **1**, p. 212.
17. Hottel, H. C. and Sarofim, A. F. (1967) *Radiative Transfer* (McGraw-Hill, New York).
18. Lockwood, F. C. and Spalding, D. B. (1971) Predictions of turbulent duct flow with significant radiation. Thermodynamics Colloquium, Sociètè Française de Physique, Evian, France.
19. Truelove, J. (1975) Mathematical modelling of radiant heat transfer in furnace. A.E.R.E. Report R.7817.
20. Khalil, E. E. and Truelove, J. S. (1977) Calculation of radiative heat transfer in a large gas fired furnace. *Letters in Heat and Mass Transfer*, **4**, p. 353.
21. Johnson, T. R. and Beer, J. M. (1973) Radiative heat transfer in furnaces; further development of the zone method of Analysis. *Proc. 14th Symp. (Int.) on Combustion*, p. 639.
22. Selcuk, N. (1975) Mathematical modelling of radiative heat transfer in enclosures. Ph.D. Thesis, University of Sheffield.
23. Bousinisque, V. J. (1877) Thèorie de l'ècoulement tourbillant. *Mem. Pres. Acad. Sci.* **23**, p. 46.
24. Prandtl, L. (1925) Bericht uber Untersuchung zur ausgebildeten Turbulenz. *ZAMM 5*, p. 136.
25. Launder, B. E. and Spalding, D. B. (1972) *Mathematical Models of Turbulence* (Academic Press, London).
26. Launder, B. E., Morse, A. P., Rodi, W. and Spalding, D. B. (1972) The prediction of free shear flows, a comparison of six turbulence models. NASA Report SP-311, p. 361.
27. Bradshaw, P., Ferriss, D. H. and Atwell, N. P. (1967) Calculation of boundary layer development using turbulent energy equation. *J. Fluid Mechanics*, **28**, p. 593.
28. Kolmogorov, A. N. (1942) Equations of turbulent motion of an incompressible fluid. *Itr. AK. Nauk, SSSR* (Translation from Russian by D. B. Spalding, 1968).
29. Chou, P. Y. (1945) On velocity correlations and the solution of the equations of turbulent fluctuations, *Quart. J. Applied Mathematics*, **3**, p. 38.
30. Harlow, F. H. and Nakayama, P. I. (1967) Turbulent transport equations. *Physics of Fluids*, **10**, p. 323.
31. Jones, W. P. and Launder, B. E. (1973) Predictions of low Reynolds number phenomena with a two equation model of turbulence. *Int. J. Heat and Mass Transfer*, **16**, p. 1119.
32. Daly, B. J. and Harlow, F. H. (1970) Transport equations of turbulence. *Physics of Fluids*, **13**, p. 2634.
33. Launder, B. E., Reece, G. J. and Rodi, W. (1975) Progress in the development of Reynolds stress turbulence closure. *J. Fluid Mechanics*, **68**, p. 537.
34. Rebiero, M. M. M. (1976) The turbulence structure of free jet flows with and without swirl. Ph.D. Thesis, London University.
35. Pope, S. B. and Whitelaw, J. H. (1976) The calculation of near wake flows. *J. Fluid Mechanics*, **73**, p. 9.
36. Lumley, J. L. (1970) Towards a turbulent constitutive relation. *J. Fluid Mechanics*, **41**, p. 413.
37. Spalding, D. B. (1970) Numerical methods in fluid mechanics and heat transfer, Imperial College, Dept. of Mech. Engineering, Report EF/TN/A/34.
38. Pai, B. R. and Whitelaw, J. H. (1969) Simplification of the razor-blade technique and its application to the measurements of wall shear stress in wall jet flows. *Aeronautical Quart.* **20**, p. 355. See also: The prediction of wall jet flows with particular reference to film cooling, *Progress in Heat and Mass transfer*, **2**, p. 163.

39. Rodi, W. (1972) The prediction of free turbulent boundary layers by use of a two equation model of turbulence, Ph.D. Thesis, London University.
40. Hanjalic, K. and Launder, B. E. (1972) Asymmetric flow in a plane channel. *J. Fluid Mechanics*, **51**, p. 301.
41. El Ghobashi, S. E. (1974) Characteristics of gaseous turbulent diffusion flames in cylindrical chambers, a theoretical and experimental investigation. Ph.D. Thesis, London University.
42. Bray, K. N. C. (1974) Kinetic energy of turbulence in flames, *Proc. AGARD Panel Meeting, April 1974.* See also: *AASU Report 332. 1974.*
43. Khalil, E. E. (1981) Flow and combustion characteristics of turbulent reacting flames, *Proc. Symp. on Fluid Mechanics of Combustion Systems.*
44. Rotta, J. C. (1951) Statistische Theorie Nichthomogenerturbulenz. *Zeitsch F. Physik*, **129**, p. 541.
45. McDonald, H. (1969) The effect of pressure gradient on the law of the wall in turbulent flow, *J. Fluid Mechanics*, **35**, p. 311.
46. Mellor, G. L. (1966) The effects of pressure gradients on turbulent flow near a smooth wall. *J. Fluid Mechanics*, **24**, p. 255.
47. Borgnakke, C. (1977) Turbulence and combustion. Tech. Univ. of Denmark, Dept. of Fluid Mechanics Report.
48. Cebeci, T., Khalil, E. E. and Whitelaw J. H. (1979) Calculation of separated boundary-layer flows. *AIAA Journal*, **17**, p. 1291.
49. Lilley, D. G. (1980) Prospects for Computer Modeling in Ramjet Combustors. AIAA/SAE/ASME 16th Joint Propulsion Conference, Paper AIAA-80-1189.
50. Mathews, L. and Whitelaw, J. H. (1971) The prediction of film cooling in the presence of recirculating flows with a two-equation model of turbulence. Imperial College Report HTS/71/3.
51. El Ghobashi, S. E., Pun, W. M. and Spalding, D. B. (1975) Concentration fluctuations in isothermal turbulent confined coaxial jets. Imperial College, Dept. of Mech. Eng., Report HTS/75/23.
52. Morse, A. P. (1977) Axisymmetric turbulent shear flows with and without swirl. Ph.D. Thesis, London University.
53. Weske, D. R. and Sturov, G. Ye (1974) Experimental study of Turbulent swirled flows in cylindrical tube. *Fluid Mechanics, Soviet research*, **3**, p. 77.
54. Barchilon, M. and Curtet, R. (1964) Some details of structure of an axisymmetric confined jet flow. Symp on Fully Separated Flow, Philadelphia. See also: *J. Basic Engineering*, **86**, p. 777.
55. Craya, A. and Curtet, R. (1953) Sur l'èvaluation d'un jet en espace confine. *C.R.A.S.* **241**, p. 621.
56. Craya, A. and Utrysko, B. (1967) Jets tournants en espace confine. Service de Documentation Scientifique et Technique de l'Armement, Paris.
57. Wilkie, D. (1966) Forced convection heat transfer from surfaces roughened by ribs. *Proc. 3rd Int. Heat Transfer Conference*, p. 1.
58. Abbott, D. E. and Kline, S. J. (1962) Experimental investigations of subsonic turbulent flow over single and double backward facing steps. *J. Basic Engineering*, **84**, p. 317.
59. Moon, L. F., Rudinger, G. and Salter, G. R. (1975) High speed turbulent mixing and combustion. AFOSR-TR-75, 1648.
60. Khalil, E. E. (1978) Numerical computations of turbulent swirling flames in axisymmetric combustors, in *Flow, Mixing and Heat Transfer in Furnaces*, edited by Khalil *et al.* (Pergamon Press), p. 231.
61. Abou Arab, T. W. (1978) Zur Modellierung der Turbulenz in eingeschlossenen drallfreien und verdrallten Strömungen. Ph.D. Thesis, Stuttgart University.
62. Baker, R. J., Hutchinson, P., Khalil, E. E. and Whitelaw, J. H. (1974) Measurements of three velocity components in a model furnace with and without combustion, *Proc. 15th Symp. (Int.) on Combustion*, p. 553.
63. Hutchinson, P., Khalil, E. E., Whitelaw, J. H. and Wigley G. (1976) The calculation of furnace flow properties and their experimental verification. *J. Heat Transfer*, **98**, p. 276.

64. Novick, A. S., Miles, G. A., and Lilley, D. G. (1979) Numerical simulation of combustor flowfields: a primitive variable design capability. *J. Energy*, **3**, p. 95.
65. Syred, N. (1979) Private communication.
66. Khalil, E. E. (1978) Aerodynamic and heat transfer characteristics of axisymmetric confined gaseous flames, *Proc. 6th Int. Heat Transfer Conference*, p. 25.
67. Maccallum, N. R. L. (1975) Private communication.
68. Assaf, H. (1975) Measurements and calculations of flow properties downstream of blunt body flame stabilisers. M.Sc. Thesis, London University.
69. Pope, S. B. (1975) A more general effective viscosity hypothesis. *J. Fluid Mechanics*, **72**, p. 331.
70. Khalil, E. E. (1979) Numerical computation of turbulent flow structure in a cyclone chamber. ASME Paper 79-HT-31.
71. Bradshaw, P. (1973) Effects of streamline curvature on turbulent flow. AGARDO-graph 169.
72. Priddin, C. H. (1975) The behaviour of the turbulent boundary layer on curved, porous walls. Ph.D. Thesis, London University.
73. Castro, I. P. (1977) Numerical difficulties in the calculation of complex turbulent flows. *Proc. Symp. on turbulent shear flows*, **1**.
74. Launder, B. E. (1976) *Topics in Applied Physics Turbulence*, edited by Bradshaw (Springer Verlag).
75. Abou Arab, T. W. (1980) Calculation of swirling flow with a general model of turbulence and its experimental verification using LDA. *Proc. 3rd, Int. Conference on Mechanical Power Engineering, Cairo*, Paper VIII.18.

CHAPTER 4

SINGLE-PHASE TURBULENT COMBUSTION MODELS

4.1 INTRODUCTORY REMARKS

4.1.1 Scope

In the previous chapter it was mentioned that the three major features of the flows discussed in this book are the turbulence characteristics, the combustion characteristics and the heat transfer characteristics. The turbulence characteristics were dealt with in the last chapter; this chapter deals with the combustion characteristics. Naturally the whole aim of combustion modelling is to arrive at the calculation of the temperature $\bar{T}$, density $\bar{\rho}$, mass fraction of species $\bar{M}_a$ and the rate of reaction $\bar{R}_a$.

Practical combustion systems in general, such as those considered in this book, are full of complex features, e.g. the fuel may take the form of liquid d·oplets or solid particles, while the chemical reaction may proceed through literally hundreds of radicals and intermediates. Formation of soot and pollutants, and significant heat transfer by radiation are common. Even in laminar flows, the formation of the reaction rates is subject to large uncertainties, and therefore, the closure of the species conservation equation is, indeed, a formidable task.

In an attempt to understand such complex systems, the cases of single phase, gaseous reacting flows, in which the Reynolds number is high and the Mach number is low, are investigated in this chapter. Chapter 5, however, deals with the case of the two-phase combustion phenomenon of liquid and solid fuels. Engineering combustion systems such as furnace flames, flames stabilized by the presence of bluff bodies, gas turbine combustors and after-burners are, to varying degrees of approximation, encompassed by the above restrictions.

In some circumstances the overall features of turbulent flames are independent of the detailed chemistry involved. This statement was discussed and verified by Spalding,[1] Khalil *et al.*[2] Hutchinson *et al.*,[3] Pope,[4] Khalil,[5] and others. There is reason to believe, therefore, that a theory of turbulent combustion, based on instantaneous equations which contain some uncertainties, may be successful. Such a simple approach was applied to furnace and combustor flows; reasonably good agreement with experiments were reported.[3–10]

The two major differences between a turbulent reacting flow and an inert isothermal flow are the effect of turbulence on combustion and that of combustion on the local turbulence. The effect of turbulence on reaction is of prime importance in the present discussion; it is the most serious obstacle to the formulation of a closed set of time-averaged equations that describes reacting flows. The larger portion of this chapter is devoted to the formulation of closures for reaction equations. Various modelling assumptions appropriate to reacting flows are discussed and their merits and limitations are assessed. The effect of reaction on turbulence is due to changes in the viscosity and more especially in the local density due to reaction; this aspect has been considered by Favre,[11] Bilger,[12] Bray,[13] Pope[4] and Khalil.[14,15] Later in the chapter, the influence of density variation is considered and its effect on local flow properties is assessed.

4.1.2 Outline of the Chapter

The chapter starts by a general description of turbulent combustion theory identifying the various parameters involved in the process of reaction and reaction rate expressions. The first subsection describes the combustion models appropriate to diffusion flames. Three different models of that type are considered and assessed; these models vary in their complexity in order to include the effects of concentration fluctuations. Models appropriate to pre-mixed flame situations are then described and examples of their application to furnace flows are displayed. Multi-step chemical reaction mechanisms are briefly described and the appropriate model is applied to turbulent reacting furnace flames.

A third category of models is considered, where the reaction rate dependency on turbulence is accounted for by species and temperature fluctuations correlations. The validity of such models to furnace and combustor flows is assessed by comparisons with measured data.

A model that solves a transport equation for the probability density function is described and its applicability to furnace flows is investigated.

The chapter ends with a critical review of the models and the possible approaches that make account for molecular transport and detailed chemistry.

4.2 THEORY OF TURBULENT COMBUSTION

This section discusses the theories of turbulent combustion and the assumptions made therein in order to produce a simple set of equations. A reacting system can be characterized by the mass fraction of each chemical species, the temperature and the pressure; the equation of state relates all of these entities to the density. The local time-averaged value of chemical species mass fraction $\bar{M}_a$ is presented as

$$\bar{M}_a = \int_{Ma} M_a P(M_a)\, dM_a \tag{4.2.1}$$

where $P(M_a)$ is the probability density function; the corresponding variance of M_a is expressed as:

$$\overline{m_a^2} = \int_{Ma} (M_a - \bar{M}_a)^2 P(M_a)\, dM_a \tag{4.2.2}$$

and

$$M_a = \bar{M}_a + m_a$$

The conservation equation of fuel mass fraction $\bar{M}_{fu}$ can be expressed in the form of Equation 3.2.9, where $\bar{S}_\Phi$ represents the rate of fuel consumption. The principal unknown term in Equation 3.2.9 is the average reaction rate $\bar{S}_a(\Phi)$; the average density $\bar{\rho}(\Phi)$ is also unknown.

The chemical reaction rate in homogeneous reactions is defined as the rate at which one of the reactants disappears to form products, in other words, the rate at which the products are formed. When the reaction is an oxidation, the fuel concentrations decrease and the rate of formation of the products of combustion is positive. The reaction rate can be expressed (Kondratiev)[16] as

$$R_{fu} = f^n \ (T, P, \text{concentrations}) \tag{4.2.3}$$

and the reaction mechanism expressed as

$$\text{Fuel} + \text{Oxidant} \rightarrow \text{Products}$$

in a single step fashion. The rate controlling mechanism can be postulated to involve the collision or interaction of a single molecule of fuel with a single molecule of oxygen. In this case, the rate of reaction is proportional to the collision of fuel and oxygen. This is because, at a given temperature, the number of collisions is proportional to the local concentrations of reactants in the mixture, i.e.

$$R_{fu} = K_f M_{fu} M_{ox} \tag{4.2.4}$$

The influence of temperature on the rate of reaction may be expressed in terms of the exponential of $(-E/RT)$, suggested by Arrhenius[17] as

$$K_f = A_0 \exp(-E/RT) \tag{4.2.5}$$

The values of A_0 and E/R are determined from experiments and were reported for many elementary reactions by Baulch *et al.*[18] and Waldman *et al.*[19]

Samples of the values of A_0 and E/R for various elementary reactions relevant to the present application are given in Table 4.1.

It should be noted that there is a great difference between the time-averaged reaction rate and the rate of reaction based on averaged entities, i.e.

$$\overline{S(\phi)} \neq S(\bar{\phi})$$

The instantaneous rate of reaction can be generally expressed as

$$R_{fu} = (\bar{M}_{fu} + m_{fu})(\bar{M}_{ox} + m_{ox})(\bar{K}_f + k_f') \tag{4.2.6}$$

The term $(\bar{k}_f + k_f')$ is highly non-linear as it contains exponential terms. If $\exp(T'/T)$ and $(1 + T'/T)^{-\frac{1}{2}}$ are expressed in terms of series expansions, the following equality is obtained:

$$\overline{(\bar{k}_f + k_f')} = A_0 \exp(-E/R\bar{T}) \left[1 + \sum_{n=2}^{\infty} P_n \left(\frac{\overline{T'^n}}{\bar{T}^n} \right) \right]$$

TABLE 4.1

Reaction	A_0	E/R
$CH_4 + OH \rightarrow CH_3 + H_2O$	3×10^{13}	2500
$CH_3 + O_2 \rightarrow CH_2O + OH$	3×10^{13}	15 000
$CH_2O + OH \rightarrow CHO + H_2O$	$3 \times 10^{9} T^{-1}$	500
$CHO + OH \rightarrow CO + H_2O$	$3 \times 10^{10} T^{-1}$	0
$CO + OH \rightarrow CO_2 + H$	$4 \times 10^{9} T^{-\frac{1}{2}}$	0
$H + O_2 \rightarrow HO_2$	1.5×10^{15}	500
$HO_2 + H \rightarrow OH + OH$	2.5×10^{14}	950
$O_2 + M \rightarrow O + O + M$	$2.5 \times 10^{19} T^{-1}$	59 335
$OH + OH \rightarrow H_2 + O$	6×10^{12}	500

Units: mole, cm³, second, °K.

where

$$P_n = \sum_{z=1}^{n} (-1)^{n-z} \frac{n-1!}{(n-z)!((z-1)!)^2} \frac{1}{z} \left(\frac{E}{R\bar{T}}\right)^z$$

and

$$\bar{R}_{fu} = A_0 \bar{\rho}^2 \bar{M}_{fu} \bar{M}_{ox} \exp(-E/R\bar{T}) \left[1 + a_0 \frac{\overline{m_{fu} m_{ox}}}{M_{fu} M_{ox}} + a_1 \left(\frac{\sqrt{\overline{T'^2}}}{T}\right)^2 + a_2 \left(\frac{\overline{T' m_{ox}}}{\bar{T}\bar{M}_{ox}} + \frac{\overline{T' m_{fu}}}{\bar{T}\bar{M}_{fu}}\right)\right] \tag{4.2.7}$$

with the constants a_0, a_1 and a_2 given as:

$$a_0 = 1, \quad a_1 = \frac{1}{2}(E/R\bar{T})^2 - E/R\bar{T}$$

$$a_2 = E/R\bar{T}$$

Equation 4.2.7 can be rewritten as

$$\bar{R}_{fu} = A_0 \bar{\rho}^2 \bar{M}_{fu} \bar{M}_{ox} \exp(-E/R\bar{T})[1 + F] \tag{4.2.8}$$

where the term F includes the influence of turbulence on time-averaged reaction rate. The main objective of the present theory is to overcome the difficulty of modelling such non-linear terms and to discuss the possible alternative approaches to modelling.

In the remainder of this chapter, various models of combustion, used in the calculations of furnace flames, are discussed in the light of the assumptions imposed, the model

TABLE 4.2

Model	Basic assumptions	Scalar temporal distributions	Reactions	Application
1	fast chemical reactions reactants do not coexist ($\tau_s \gg \tau_k$)	—	single step irreversible	general, *qualitative*, physically controlled diffusion flames
2	fast chemical reactions reactants can exist at same place but different times ($\tau_s \gg \tau_k$)	square wave	single step irreversible	general, *quantitative*, physically controlled diffusion flames
3	fast chemical reactions, reactants can exist at same place but different times ($\tau_s \gg \tau_k$)	clipped Gaussian probability distribution	single step irreversible	*quantitative*, physically controlled diffusion flames
4	finite reaction rate eddy break up expression ($\tau_k \gg \tau_s$)	clipped Gaussian probability distribution	single step irreversible	*quantitative*, kinetically influenced premixed flames and mixing zones
5	finite reaction rate eddy break up expression ($\tau_k \gg \tau_s$)	clipped Gaussian probability distribution	multi-intermediate-step reversible	*quantitative*, kinetically influenced, detailed chemistry. Premixed flames and mixing zones
6	finite reaction rate correlation expression $\tau_k \simeq \tau_s$	clipped Gaussian probability distribution	single step irreversible	*quantitative*, general, diffusion/premixed flames and arbitrary fuelled system
7	finite reaction rate correlation expression cross correlations $\tau_k \simeq \tau_s$	clipped Gaussian probability distribution	single step irreversible	*quantitative*, general, diffusion/premixed flames and arbitrary fuelled suitable for pollutant calculations

formulation, their physical implications and their validity to the various flame flow configurations. The models considered in this book include models assuming infinitely fast chemical reactions, appropriate to diffusion flames, and others which assume finite reaction rate more suited to premixed flame situations. Models appropriate to arbitrary fired flames are also considered. The various equations solved in each model and the basic assumptions involved are shown in Table 4.2, which also indicates the type of flame to which a particular model is applicable. The validity of each of these models, in turbulent furnace flows, is assessed by comparisons between measurements and calculations, as demonstrated later in the chapter.

4.2.1 Fast Chemical Reaction Models for Diffusion Flames

Early attempts to analyse diffusion flames evolved from the original work of Burke and Schumann,[20] who considered the oxygen as a negative fuel, and so obtained a scalar quantity which is conserved under the chemical reaction. They considered that the reaction possessed the features of being single step, irreversible and infinitely fast. This approach led to the result that the unpremixed fuel and oxidant concentrations could be derived from the model, which assumed that the reaction was so fast that fuel and oxidant did not co-exist anywhere, except within an infinitely thin flame sheet.

In applying this model approximation, it was inherently assumed that the chemical reaction is frozen everywhere outside the thin reaction zone (flame sheet). This assumption was found to be unrealistic in many diffusion flames, and hence modifications to account for co-presence of the reactants, although at different times, were made, Naguib,[21] El Ghobashi[22] and Bilger.[23]

From the above main assumptions and the work of Shvab,[24] Zeldovich[25] and Williams,[26] it can be concluded that a linear combination of species conservation equations for the unpremixed reactants, fuel and oxidant, yields an equation whose form is identical to the one governing the conservation of chemically inert species, i.e. neither equation has a source term. This statement is only true when the following two assumptions are made:

(a) That the turbulent transport coefficients for the reactants and the products, at each point in the field, are equal.
(b) The fuel and oxygen combine always in a stoichiometric ratio i to produce $(i + 1)$ kg of products.

The resultant variable from the combination is the mixture fraction $\bar{f}$, which is defined as

$$\bar{f} = \frac{\bar{\xi} - \bar{\xi}_A}{\bar{\xi}_F - \bar{\xi}_A} \tag{4.2.9}$$

where

$$\bar{\xi} = \bar{M}_{\mathrm{fu}} - \bar{M}_{\mathrm{ox}}/i$$

or

$$\xi = \bar{M}_{\mathrm{fu}} + \bar{M}_{\mathrm{pr}}/(1 + i)$$

The subscripts A and F denote the air and the fuel stream conditions at inlet. The conservation equation of $\bar{f}$ is expressed, in the steady state, and neglecting density fluctuations, as:

$$\frac{\partial}{\partial x_j} \bar{\rho}\bar{U}_j\bar{f} = \frac{\partial}{\partial x_j} \Gamma_f \frac{\partial \bar{f}}{\partial x_j} \tag{4.2.10}$$

and

$$\Gamma_f = \mu_{\text{eff}}/\sigma_f \quad \text{and} \quad \sigma_f = 0.9$$

The three models described here are based on the very same assumptions described above but differ in the way in which the turbulence fluctuations are presented. These various models are described in detail below. The same procedure will be followed for the three models, first the model assumptions will be mentioned, then the model will be formulated and finally it will be assessed.

4.2.1.1 *Model 1*

Model Assumptions

The fuel and oxygen cannot exist at the same place at any time.
The reaction is infinitely fast.
Equilibrium is attained.
The reaction is single step and irreversible.
All the exchange coefficients are equal.

Model Formulation

The flame is actually divided into two regions:

1. The first region of the flame is the one where the oxidant and the products co-exist, i.e.

$$0 < \bar{f} < \bar{f}_{\text{st}}$$

$$\bar{M}_{\text{fu}} = 0$$

$$\bar{M}_{\text{ox}} = \bar{M}_{\text{ox}A} \left[1 - \frac{\bar{f}}{\bar{f}_{\text{st}}} \right] \leqslant \bar{M}_{\text{ox}A} \tag{4.2.11}$$

2. The second region is the one where the fuel and the products co-exist, i.e.

$$1 > \bar{f} > \bar{f}_{\text{st}}$$

$$\bar{M}_{\text{fu}} = \bar{M}_{\text{fu}F} \, [(\bar{f} - \bar{f}_{\text{st}})/(1 - \bar{f}_{\text{st}})] \tag{4.2.12}$$

$$\bar{M}_{\text{ox}} = 0$$

where $\bar{f}_{\text{st}}$ is the stoichiometric value of the mixture fraction and is given by:

$$\bar{f}_{\text{st}} = [(i\bar{M}_{\text{fu}F}/\bar{M}_{\text{ox}A}) + 1]^{-1} \tag{4.2.13}$$

and

$$\bar{M}_{pr} = 1 - \bar{M}_{ox} - \bar{M}_{fu}$$

The linear relationship between the instantaneous mass fractions of fuel and oxidant and $\bar{f}$ is shown in Figure 4.1. The enthalpy at each point in the flame is obtained in a similar manner to $\bar{f}$ by solving an equation for $\bar{H}$:

$$\bar{H} = \int_0^T \bar{M}_j \bar{C}_{pj} d\bar{T} + H_{fu}\bar{M}_{fu} + \tfrac{1}{2}(U^2 + V^2 + W^2) \qquad (4.2.14)$$

and that

$$\bar{C}_{p,j} = \sum_{n=0}^{m} a_n \bar{T}^n \quad \text{with} \quad m = 3 \qquad (4.2.15)$$

The mass fractions and the enthalpy are known functions of f, and are the temperature and density.

In Model 1, where the fluctuations of f and H are neglected, this omission allowed the solutions to be obtained with a fast rate of convergence. This model has been previously discussed by Williams,[26] Spalding,[27] Bilger[23] and Khalil;[5] its assumptions were found to be appropriate for diffusion flames and where excessive air is used. The model is an over-simplification of the features of real flame situations; some available experimental evidence is contrary to this model, for example Bilger[28,29] showed that in the mixing zones of diffusion flames, oxygen and fuel concentrations have finite values.

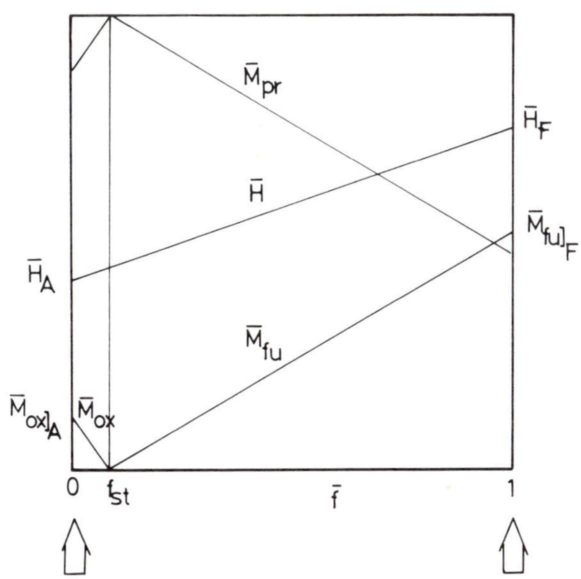

Figure 4.1 Relation between f and scalar properties

Assessment of Model

In the early stages of combustion modelling, Equations 4.2.9 to 4.2.15 were used to obtain the local values of the mixture fractions and enthalpies. They were applied to the calculation of furnace flows; the work of Pun and Spalding[30] reported the calculations of local flow characteristics in an axisymmetric furnace with zero and finite swirl intensities. Unfortunately no comparisons with experimental data were reported. The investigations of Gosman *et al.*[31] were reported for another axisymmetric furnace at zero and finite swirl of combustion air; yet again no comparisons with experiments were done. Such comparisons appear in the work of Khalil *et al.*,[2] where the present model was incorporated to calculate the local flow and heat transfer characteristics in a furnace model; a sample of the obtained comparisons of velocity field and recirculation zones is shown in Figures 4.2 and 4.3. The team of Pai *et al.*[32] at the International Flame Research Foundation investigated the flow and combustion in a large furnace firing gaseous fuels. The turbulence model used was that of a mixing length and the combustion model was the flame sheet model. Comparison with the experimental data was reported and samples of the obtained prediction of velocities and temperatures are

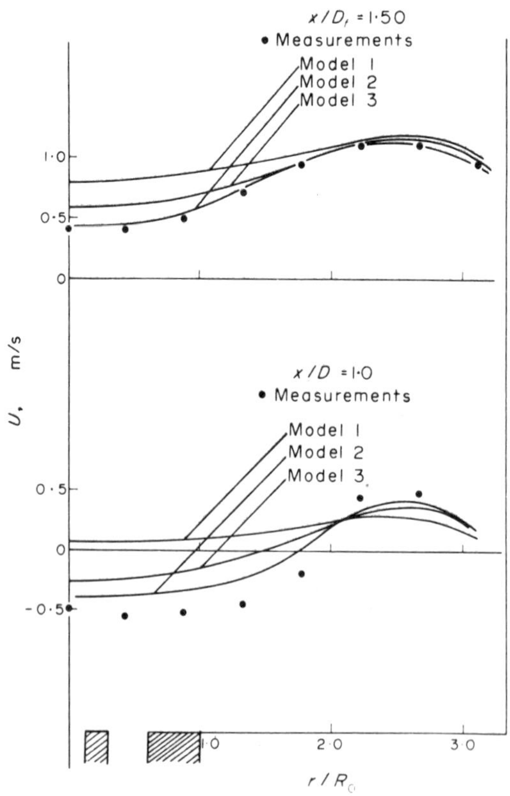

Figure 4.2 Radial profiles of mean axial velocity, $S = 0$
● Experiments of reference 2 —— Predictions of reference 2

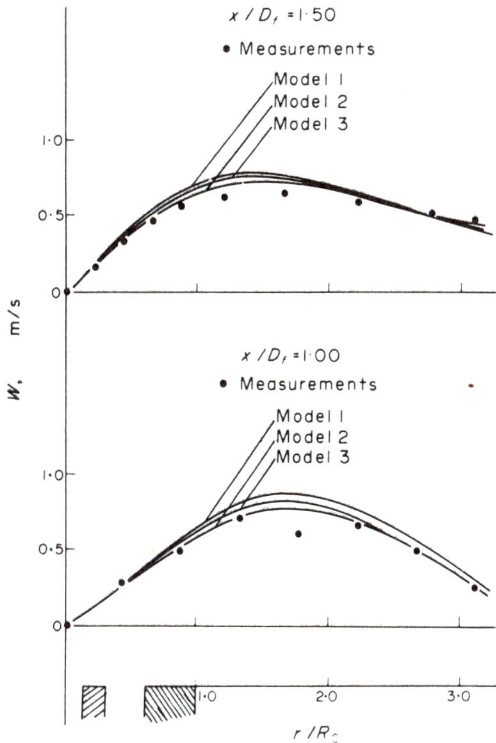

Figure 4.3 Radial profiles of mean tangential velocity, $S = 0$
● Experiments of reference 2 —— Predictions of reference 2

shown in Figures 4.4 and 4.5. Naturally, discrepancies were observed in the vicinity of the reaction zone; they are attributed to the inadequacy of both the turbulence and combustion models to perform in such complex flow configuration.

These comparisons make evident the deficiencies of such a model approximation which prevents it from predicting adequately the local gas temperatures and the reaction zone, and hence the local densities and heat release rates. Consequently the local velocities were not accurately predicted. This discrepancy can certainly be produced by the neglect of concentration fluctuations.

Now that the defects of Model 1 have been exposed, we shall proceed to discuss how to remedy such defects. To begin with, the value of any scalar entity is assumed as a function of the mixture fraction f and hence

$$\bar{\Phi} = \int_0^f \Phi(f)P(f)\, df$$

with f representing either the fuel or the oxidant streams. Generally, in turbulent reacting flows, the fuel and oxygen concentration fluctuations have a measurable value which varies from one location to another in the flame zone. Due to these fluctuations

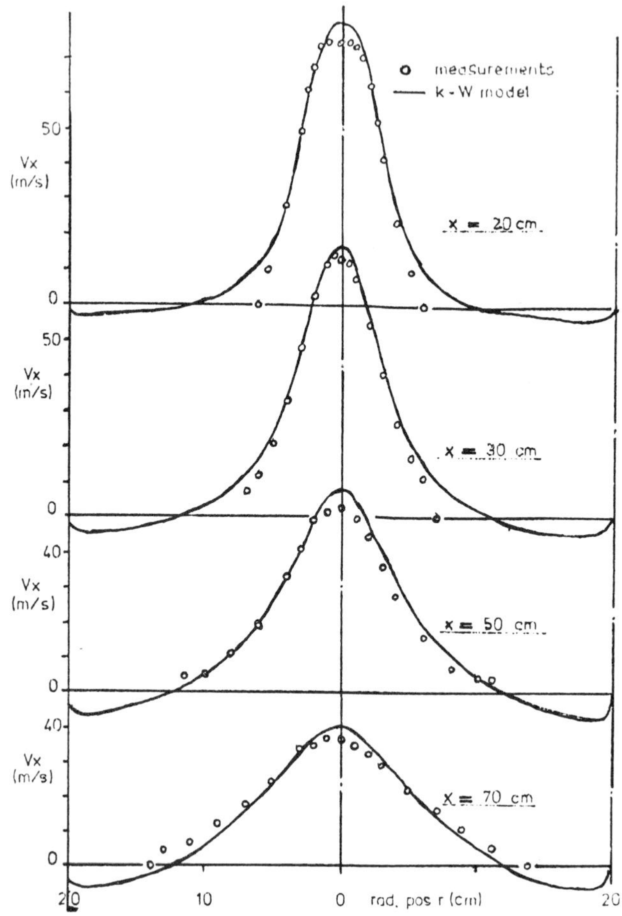

Figure 4.4 Measured and predicted axial velocity profiles in a confined jet[32]

observed by, for example, Bilger *et al.*[28] and El Ghobashi,[22] the fuel and the oxidant may exist at the same location but at different times. The definition of the probability density function $P(\Phi)$ of the scalar entity Φ, which takes the form

$$P(\Phi)\,\delta\Phi = \lim_{t\to\infty} \frac{1}{t}\,\Sigma\delta t \tag{4.2.16}$$

is shown in Figures 4.6 and 4.7 for a continuous distribution and that of two delta functions.

From Equation 4.2.16 it is implied that $P(\Phi)\,\delta\Phi$ is the fraction of time that $\Phi(t)$ spends between Φ and $(\Phi + \delta\Phi)$. It also implies that, over a long period of time, the sum of the values of $P(\Phi)\,\delta\Phi$ must be equal to unity: i.e.

$$\int_{-\infty}^{\infty} P(\Phi)\,d\Phi = 1 \tag{4.2.17}$$

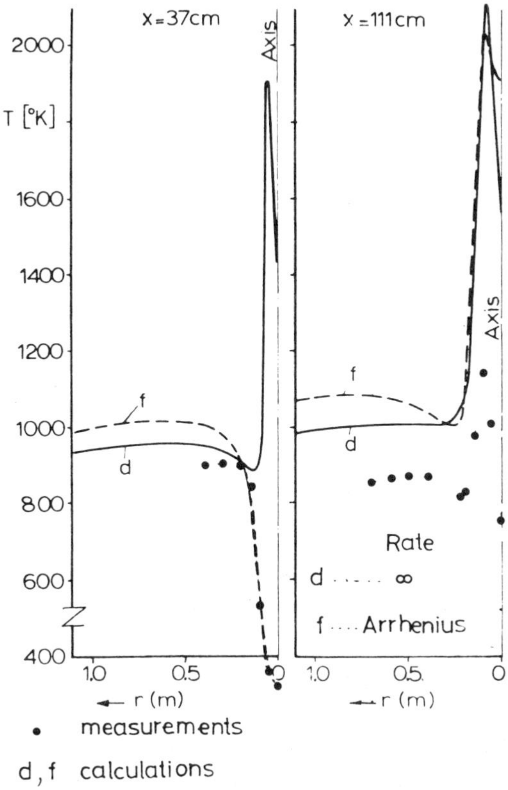

Figure 4.5 Predicted and measured temperature profiles, M–2 flame 29, (reference 32)

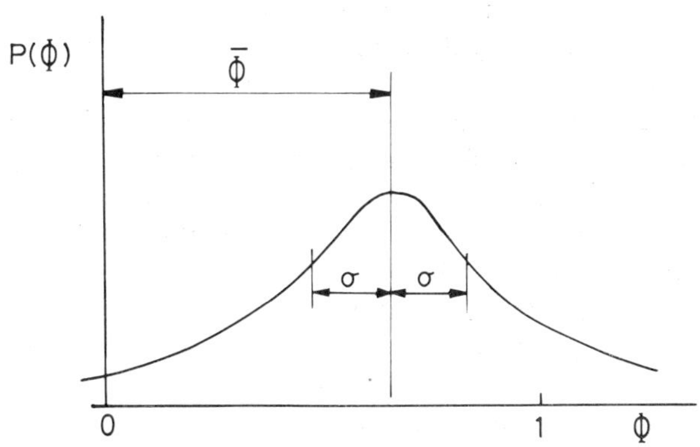

Figure 4.6 Probability distribution

Figure 4.7 Delta function distribution

and hence, the time-averaged value of $\bar{\Phi}$ can be expressed as,

$$\bar{\Phi} = \int_{-\infty}^{\infty} \Phi P(\Phi)\, d\Phi \tag{4.2.18}$$

which represents the first moment of $P(\Phi)$ about the origin $\Phi = 0$; for any other scalar variable $\gamma(\Phi)$ its time-averaged value is expressed as:

$$\bar{\gamma} = \int_{-\infty}^{\infty} \gamma(\Phi)\, P(\Phi)\, d\Phi \tag{4.2.19}$$

The effect of turbulent fluctuations on the local flow properties can be introduced by including second and higher-order correlations of the concentration. Introducing the concentration fluctuation 'g' whose definition is

$$g = (f - \bar{f})^2 \tag{4.2.20}$$

constitutes a second-order closure. The values of $\bar{f}$ and g are used to calculate the probability distribution.

The time-averaged value of any scalar $\Phi(f)$ can be obtained from the knowledge of $P(f)$, which in turn can be calculated by different approaches which vary in complexity. The simplest form of $P(f)$ is the two delta functions at $f = 0$ and $f = 1$, with a top-hat random fluctuation as suggested by Spalding.[1] The assumption of a Gaussian distribution which extends from $-\infty$ to ∞, reported by Hawthorne *et al.*[33] and Thring *et al.*,[34] is unrealistic when Φ represents the mixture fraction f, fuel mass fraction, temperature, density, etc. Such an assumption allows the passive scalars to attain negative values or excessively large ones. Another alternative, namely a beta function in the physical limits between $f = 0$ and $f = 1$, was proposed by Richardson *et al.*[35] Many other shapes of $P(\Phi)$ were proposed, for example by Rhodes *et al.*;[36] these were reviewed and assessed by El Ghobashi *et al.*[37] Clipped Gaussian distributions were first suggested by Naguib[21] and El Ghobashi.[22] To adequately represent $P(f)$ in reacting flames, they also proposed that such a distribution can be applied to furnace flow calculations. The Gaussian distribution with limits of $-\infty$ and ∞ was clipped at $f = 0$ and 1 on the basis of the impossibility of having negative f or f greater than unity. The clipped parts were replaced by two delta functions at the two physical limits of $f = 0$ and 1 representing oxidant and fuel streams.

The assumed shapes of the probability distributions were based on a mathematical formulation of the problem. Experimental evidence indicated that the form of $P(\Phi)$ depended on the local turbulence, time scales, reaction rate and location in the flame, Pope,[4] Moss[38] and Lewis *et al.*[39] Pope[4,40] proposed that a transport equation for $P(\Phi)$ should be solved, and the obtained values used in the model afterwards. In this approach, the values of $P(\Phi)$ are dependent on the rate of reaction, time scales of turbulence and reaction and on the flow properties.

In this chapter, two approaches which are used to arrive at the probability density distribution are discussed:

A. The first approach is to assume a shape for the probability density functions, and here two types are examined;

1. The delta functions at $f = 0$ and $f = 1$, i.e. square wave distribution of f with time.
2. A Gaussian distribution between $f = 0$ and $f = 1$ together with two delta functions at $f = 0$ and 1.

B. The second approach is to calculate the probability distribution functions, by solving the transport equation for $P(\Phi)$. The discussion of this approach comes much later in the chapter.

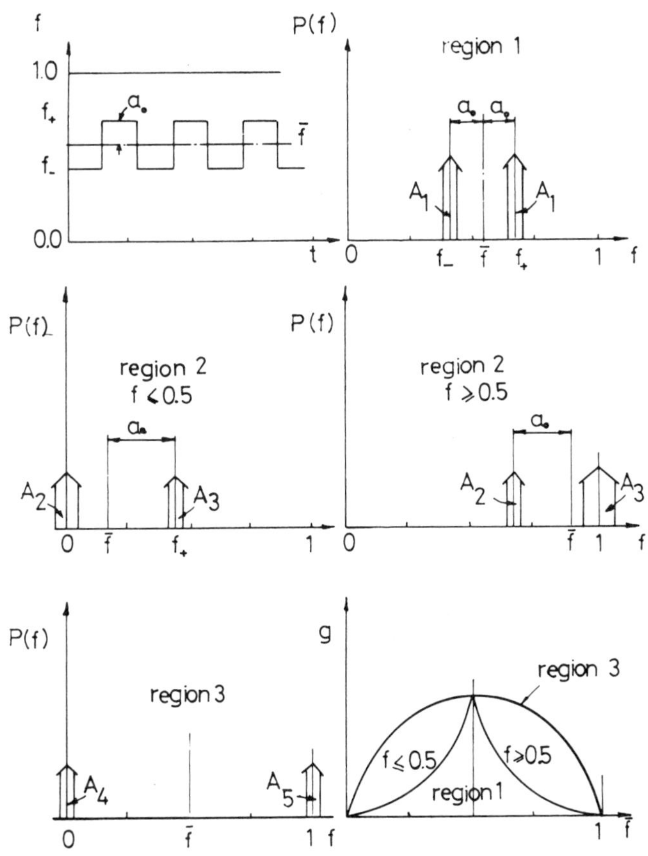

Figure 4.8 Probability distributions for a square wave variation of mixture fraction

4.2.1.2 *Model 2*

Model Assumptions

The fuel and oxygen may coexist at the same place but at different times
The reaction is infinitely fast
Equilibrium is attained
The reaction is single step and irreversible
The exchange coefficients are equal
The probability density function is represented by double delta functions at $f = 0$ and 1 and by a corresponding square wave temporal distribution.

Model Formulation

The assumed probability distribution, in this case, and the corresponding temporal distribution are divided into four separate regions, as shown in Figure 4.8.

1. Regions where $1 > f > 0$

$$P(f) = A_1[\delta(f_t) + \delta(f)] \text{ and } A_1 = \tfrac{1}{2} \qquad (4.2.21)$$
$$f_+ = \bar{f} + g^{\frac{1}{2}} \text{ and } f_- = \bar{f} - g^{\frac{1}{2}}$$

2a. Regions where $\bar{f} < 0.5$ and $\bar{f} - g^{\frac{1}{2}} < 0$

$$P(f) = A_2 \cdot \delta(0) + A_3 \cdot \delta(f_+)$$
$$A_2 = g/(\bar{f}(\bar{f} + g/\bar{f})) \qquad (4.2.22)$$
$$A_3 = \bar{f}(\bar{f} + g/\bar{f})$$

2b. Regions where $\bar{f} > 0.5$ and $\bar{f} + g^{\frac{1}{2}} > 1.0$

$$P(f) = A_2 \cdot \delta(f_-) + A_3 \cdot \delta(1)$$
$$A_2 = (1 - \bar{f})/(1 - \bar{f} + g/(1 - \bar{f})) \qquad (4.2.23)$$

and

$$A_3 = g/((1 - \bar{f})^2 + g)$$

3. Regions where large oscillations of f are imposed, such that either the fuel or the oxidant is present, the probability density function, in this instance, is given by:

$$P(f) = (1 - \bar{f}) \cdot \delta(0) + \bar{f} \cdot \delta(1)$$

and

$$g_{\max} = (1 - \bar{f})(\bar{f}) \qquad (4.2.24)$$

The local values of the concentration fluctuations g, are obtained from the corresponding conservation equation, expressed in the following steady state form

$$\underset{\text{I}}{\frac{\partial}{\partial x_j} \bar{\rho} \bar{U}_i g} = \underset{\text{II}}{\frac{\partial}{\partial x_j} \Gamma_g \frac{\partial g}{\partial x_i}} + \underset{\text{III}}{C_{g_1} \mu_{\text{eff}} \left(\frac{\partial \bar{f}}{\partial x_j} \right)^2} - \underset{\text{IV}}{C_{g_2} \bar{\rho} \frac{\varepsilon}{k} g} \qquad (4.2.25)$$

In this equation, term I represents the net rate of convection, and term II denotes the turbulent diffusion of the entity g. Term III denotes the production of concentration fluctuations due to non-uniformity of mixture fraction. The dissipations of the fluctuations due to the molecular diffusion is represented by term IV. The values of the constants C_{g_1} and C_{g_2} are 2.8 and 2.0 respectively and were obtained by Naguib[21] and El Ghobashi.[22] The exchange coefficient Γ_g is expressed as $\Gamma_g = \mu_{eff}/\sigma_g$, where σ_g is 0.9.

The validity of Equation 4.2.25 together with its embodied assumptions was examined by Naguib,[21] El Ghobashi,[22] Khalil[5] and Hutchinson *et al.*[8] As f is a passive scalar g, its concentration fluctuation equation does not contain any terms of a chemical reaction nature.

Model Assessment

The local values of $\bar{T}$, $\bar{\rho}$, $\bar{M}_{fu}$ and $\bar{M}_{ox}$ were obtained from Equation 4.2.18 and the probability distributions of Equations 4.2.21 to 4.2.24. This model was previously used by Spalding,[1] El Ghobashi *et al.*,[41] Khalil *et al.*[2] and Richter[42] for turbulent diffusion flames. Spalding[1] investigated the effects of concentration fluctuations on the characteristics of round jets and concluded that it provides better agreement with experiments than Model 1. Khalil *et al.*[1] reported comparisons between measured and computed flow pattern in a furnace firing natural gas. Figures 4.9 and 4.10 exhibit the level of agreement obtained in this furnace. It is also clear from these figures that the predictions obtained using this model are better than those obtained using Model 1.

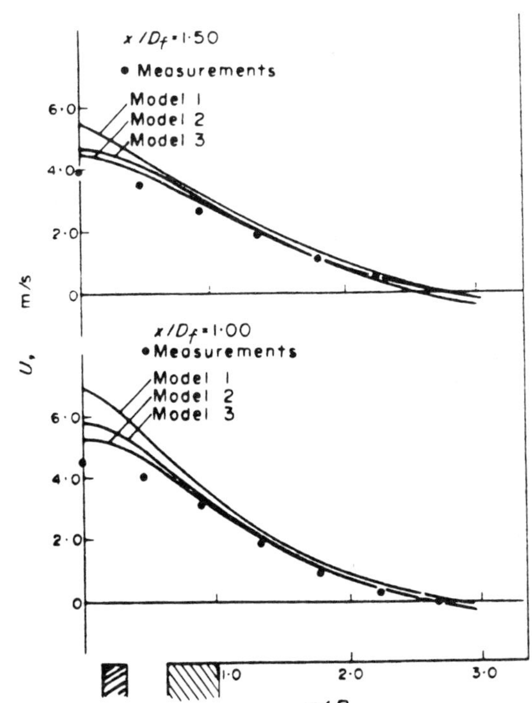

Figure 4.9 Radial profiles of mean axial velocity, $S = 0$
● Experiments of reference 2 —— Predictions of reference 2

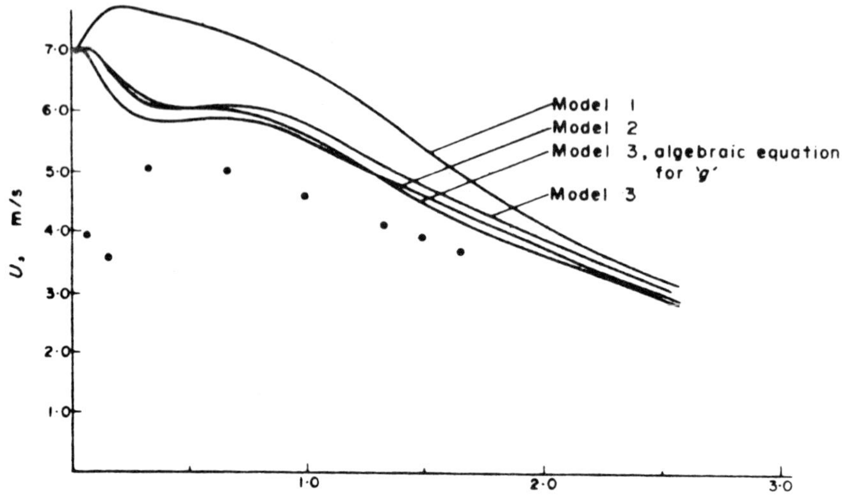

Figure 4.10 Centreline distribution of mean axial velocity; $S = 0$
● Experiments of reference 2 —— Predictions of reference 2

The assumption implied in the model of a square wave temporal distribution is primitive and only approximate. Other assumed shapes such as saw teeth and trapizoidal form are of the same nature; their application in predicting diffusion flames yields results similar to those of the square wave form.

4.2.1.3 *Model 3*

Model Assumptions

The fuel and oxygen may coexist at the same place but at different times
The reaction is infinitely fast
Equilibrium is attained
The reaction is single step and irreversible
The exchange coefficients are equal
Probability distribution of mixture fraction is a clipped Gaussian.

Model Formulation

The difficulty which stems from the presence of the unwanted tails of the conventional Gaussian distribution at $f < 0$ and $f > 1$, which are physically unrealistic, is taken care of by the presence of the two Dirac delta functions at $f = 0$ and $f = 1$. The resultant clipped Gaussian probability distribution is expressed, as given by Naguib,[21] in the form

$$P(f) = \frac{1}{\sigma\sqrt{2\pi}} \exp\left(-\frac{1}{2}\left(\frac{f-\mu}{\sigma}\right)^2\right)[D(f) - D(f-1)] + A\delta(0) + B\delta(1) \quad (4.2.26)$$

where μ is the value of f giving maximum probability; σ is the variance; $D(f)$ is the

Heaviside step function defined as, $D(\xi) = 0$, $\xi < 0$ and $D(\xi) = 1$, $\xi > 0$; and $\delta(f)$ is the Dirac delta function. The probability distribution is shown in Figure 4.11.

The time-averaged mixture fraction f, obtained from Equation 4.2.10, is expressed in the form 4.2.18 as:

$$\bar{f} = B + \int_0^1 \frac{f}{\sigma\sqrt{2\pi}} \exp\left(-\frac{1}{2}\left(\frac{f-\mu}{\sigma}\right)^2\right) df \qquad (4.2.27)$$

and

$$B = \int_1^\infty P_0(f)\, df$$

$$A = \int_{-\infty}^0 P_0(f)\, df \qquad (4.2.28)$$

$$P_0(f) = \frac{1}{\sigma\sqrt{2\pi}} \exp\left(-\frac{1}{2}\left(\frac{f-\mu}{\sigma}\right)^2\right)$$

The corresponding value of g, the square of concentration fluctuations, obtained from Equation 4.2.25, is given as:

$$g = B + \int_0^1 \frac{f^2}{\sigma\sqrt{2\pi}} \exp\left(-\frac{1}{2}\left(\frac{f-\mu}{\sigma}\right)^2\right) df - \bar{f}^2 \qquad (4.2.29)$$

Values of $\bar{f}$ and g were calculated from their conservation Equations 4.2.10 and 4.2.25 and μ and σ determined from a table in terms of $\bar{f}$ and g by a searching technique.[21] The time-averaged value of any property Φ and its variance can be obtained through the

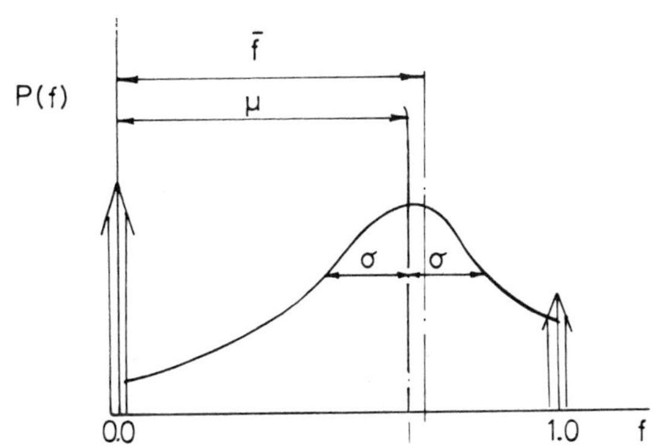

Figure 4.11 Clipped Gaussian probability distribution of mixture fraction

equations:

$$\bar{\Phi} = A\bar{\Phi}_A + B\bar{\Phi}_F + \int_0^1 \frac{\Phi(f)}{\sigma\sqrt{2\pi}} \exp\left(-\frac{1}{2}\left(\frac{f-\mu}{\sigma}\right)^2\right) df \qquad (4.2.30)$$

and

$$\overline{\phi^2} = B\bar{\phi}_F^2 + \int_0^1 \frac{(\Phi(f))^2}{\sigma\sqrt{2\pi}} \exp\left(-\frac{1}{2}\left(\frac{f-\mu}{\sigma}\right)^2\right) df - \bar{\Phi}^2 \qquad (4.2.31)$$

As the instantaneous values of Γ, ρ, M_{fu} and M_{ox} were assumed to be a function of f, the corresponding time-averaged values can be obtained from Equation 4.2.30 and the square of the fluctuations obtained from Equation 4.2.31.

Model 3 requires the solution of the conservation equations for $\bar{f}$ and g and uses, in addition, tabulated values of μ and σ for given f and g.

Assessment of Model

This present model has been widely applied to furnace and combustion chamber flows. El Ghobashi's investigation was devoted to combustion modelling and its verification. The combustion Model 3, incorporated in his calculations, resulted in reasonably good agreement with experiments and suggested the application of this model to complex geometries. Hutchinson *et al.*,[8] used Model 3 to calculate the local flow properties in confined furnace flows. An example of results obtained in such flows, is the calculation of temperature distribution in the hydrogen flame of reference 43. For this geometry Khalil[5] and Hutchinson *et al.*[7] calculated the profiles of mean gas temperatures and compared then with the experiments; the comparisons are shown on Figure 4.12. These measured and calculated profiles were in reasonable agreement; the centreline gas temperature distribution was predicted rather well. Khalil[5] applied this model to various furnace configurations, such as that of Cernansky *et al.*,[44] Wu *et al.*[45] for swirl intensities of zero and 0.84, Gunther *et al.*[46] and the configuration of Bilger *et al.*[28,29] The reported agreement was satisfactory for engineering purposes of representing the flow and flame characteristics.

The measured and calculated gas temperature profiles in the furnace of Wu *et al.*,[45] reported by Khalil[5] are shown in Figure 4.13 for two swirl intensities. The predictions were, generally, in agreement with the measurements. As expected, a longer flame was identified, by the calculations, for the low swirl intensity. Figure 4.14 identifies the centreline temperature distribution and illustrates the extent to which the present model represents the flow and combustion characteristics.

In situations where the assumption of rapid reaction is valid, the only uncertainty which hinders present models is the determination of $P(f)$. Experimental research (Moss)[38] gives evidence that the probability distribution does not have a clipped Gaussian form, but depends on rate of reaction and time scales. However, for diffusion flames, the comparisons shown here, and those of Khalil[15] and Hutchinson *et al.*,[8] are favourable and suggested that the model presents a good description of the diffusion flames.

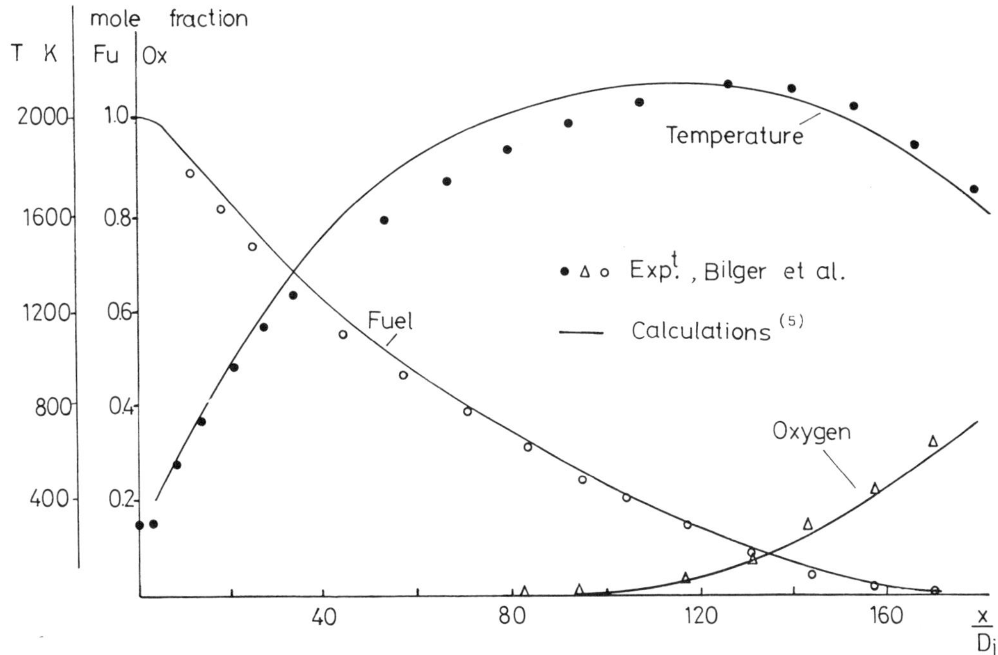

Figure 4.12 Measured and predicted centreline distributions of temperature and species concentrations in a confined hydrogen flame, reference 43

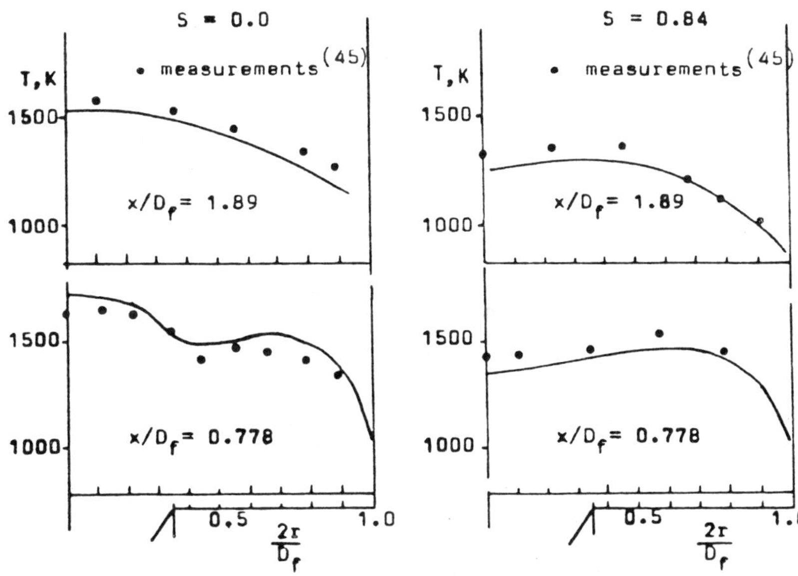

Figure 4.13 Measured and calculated gas temperature profiles in the furnace of reference 45; calculations of reference 5

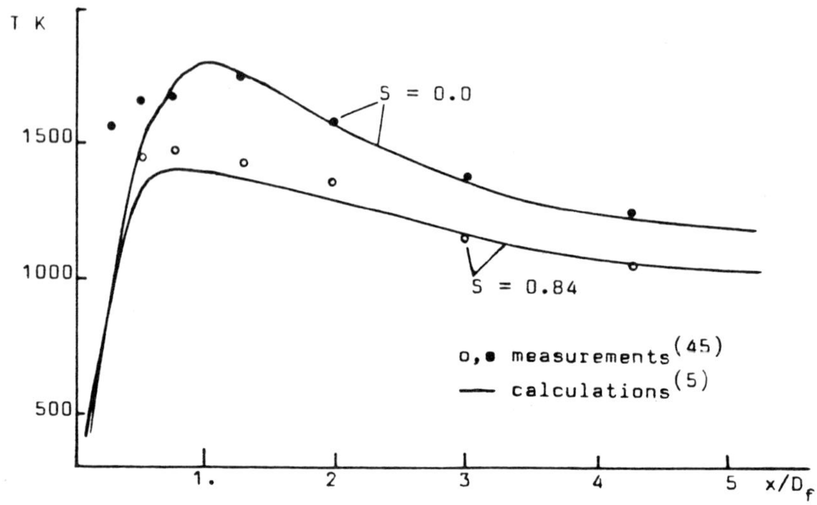

Figure 4.14 Measured and calculated centreline temperature distributions in the furnace of reference 45

4.2.2 Finite Reaction Models for Premixed Flames

In premixed flame situations in gas turbine after burners and baffle stabilized burners, i.e. in the flow situations where the fuel and oxidant are mixed prior to the combustion chamber, the assumption of infinitely fast chemical reaction is invalid. It is, therefore, necessary to provide for a finite rate of reaction. A source term is required in the conservation equation of species to account for the disappearance of species at a finite rate.

The reaction mechanism between fuel and oxygen can take place in intermediate steps, producing radicals in the initiation reactions to yield carbon dioxide and water vapour for hydrocarbon fuels. The oxidation of hydrogen and hydrocarbons takes place in multi-step reactions; each of these steps is governed by a particular reaction rate. This rate appears in the corresponding species equation to account for the rate of formation or disappearance of the species. It is expressed in the Arrhenius form, and is strongly temperature dependent as indicated by the exponential term; it is also dependent, albeit to a lesser degree, on the local concentrations of reactants.

In turbulent reacting flows, the effect of turbulence on reaction rate is significant as demonstrated by Equation 4.2.8. The effect of turbulence–chemistry interaction is represented in that equation by the term (F).

Alternative models were proposed; Spalding[47] proposed his eddy break-up model which accounted for the time scales of turbulence and combustion. The Borghi[48,49,50] approach was founded on the basis of Equation 4.2.7 and the solution of transport equations for the correlations $\overline{m_{fu}m_{ox}}$, $\overline{m^2_{fu}}$, $\overline{m^2_{ox}}$ and other correlations. In this section, a finite reaction rate of a single step mechanism is considered. Two different models, suitable for premixed flames, are discussed here. The first model is based on the eddy break-up assumptions of Spalding.[47] The fuel conservation equation is solved with the appropriate reaction rate term, thus yielding the fuel mass fraction, which together with the value of $\overline{m^2_{fu}}$ obtained from its conservation equation, are used to obtain $P(M_{fu})$.

Consequently other scalar properties are obtained. The second model is similar to the first one but is more appropriate to situations where intermediate steps are incorporated. The same solution procedure is used for each species, as described later in the chapter.

4.2.2.1 *Model 4*

Model Assumptions

The fuel and oxidant streams are premixed
The reaction rate is finite
The reaction is single step and irreversible
Probability distribution of fuel mass fraction is clipped Gaussian.

Model Formulation

In the mixing zone of fuel and oxidant, the rate of single-step chemical reaction can be expressed as

$$R_{fu} = A_0 \rho^2 M_{fu} M_{ox} \exp(-E/RT) \tag{4.2.32}$$

Time averaging this Arrhenius expression and neglecting turbulent fluctuations yields

$$\bar{R}_{fu} = A_0 \bar{\rho}^2 \bar{M}_{fu} \bar{M}_{ox} \exp(-E/R\bar{T}) \tag{4.2.33}$$

This expression is similar to the expression 4.2.8 when F equals zero, i.e. when the influence of turbulence on the reaction rate is neglected.

The difficulty with using expression 4.2.8 to evaluate the reaction rate is that it is necessary to solve additional equations for the conservation of $\overline{m^2_{ox}}$, $\overline{m_{fu} m_{ox}}$, $\overline{T'^2}$, $\overline{m_{ox} T'}$, $\overline{m_{fu} T'}, \ldots$, thus large computer storage is required. To avoid solving these additional equations, Spalding[47] proposed a reaction rate based on the species concentration fluctuations and the rate of break-up of the eddies. The eddy break-up reaction rate is expressed as:

$$\bar{R}_{E.B.U} = C_R \bar{\rho} \frac{\varepsilon}{k} (\overline{m^2_{fu}})^{\frac{1}{2}} \tag{4.2.34}$$

where C_R is a constant of the model. The term ε/k has the dimensions of s^{-1}, and the diffusion time scale $\tau_{s_0} = k/\varepsilon$ should be compared to the chemical time scale τ_k given by Borghi[48] as

$$\tau_k = [A_0 \bar{\rho}(\bar{M}_{ox} + i\bar{M}_{fu}) \exp(-E/R\bar{T})]^{-1} \tag{4.2.35}$$

This is done to determine the flame type. These two reaction rates are compared together, in the computational solution procedure, and the smaller of them is taken to represent the effective controlling rate. Evidently, the smaller rate of reaction corresponds to the larger of the time scales (τ_{s_0} and τ_k); it is natural that the reaction is controlled by the process that takes the larger time.

In the regions where τ_{s_9} is larger than τ_k, the mixing of the reactants is slow and the reactants are at a suitable temperature and concentrations to react as soon as they intimately mix. Therefore, the reaction is diffusion controlled; i.e. it is controlled by the rate given by Equation 4.2.34. On the other hand, when τ_{s_0} is very small, which corresponds to a large dissipation rate of eddies and rapid mixing, the reaction is kinetically influenced. This implies that the reactants are in intimate contact but their temperature and concentrations are not suitable for the reaction to proceed; the smaller rate, in this case, is that given by Equation 4.2.33.

To determine the eddy break-up rate of reaction, it is necessary to evaluate $\overline{m^2_{fu}}$. The conservation equation of $\overline{m^2_{fu}}$, reported by Borghi,[49] Bray,[51] Bilger,[12] Hutchinson *et al.*[8] and Khalil[5] was expressed in the general steady state form as

$$\underset{\text{(I)}}{\frac{\partial}{\partial x_j}\bar{\rho}\bar{U}_j\overline{m^2_{fu}}} = \underset{\text{(II)}}{D_l \operatorname{grad} \overline{m^2_{fu}}}$$

$$\underset{\text{(III)}}{-2D_l\,\overline{(\operatorname{grad} m_{fu}\cdot \operatorname{grad} m_{fu})}}$$

$$\underset{\text{(IV)}}{+\,2D_t \operatorname{grad} \bar{M}_{fu}\cdot \operatorname{grad} \bar{M}_{fu}}$$

$$\underset{\text{(V)}}{-\operatorname{grad} \overline{u_j m_{fu} m_{fu}}}$$

$$\underset{\text{(VI)}}{-\,2K_f[\bar{M}_{fu}\overline{m_{fu}m_{ox}} + \bar{M}_{ox}\overline{m^2_{fu}}]} \tag{4.2.36}$$

Term I represents the net rate of convection; term II denotes the laminar diffusion of $\overline{m^2_{fu}}$. Term III represents the dissipation of the fluctuations due to molecular diffusion. The net generation rate due to the non-uniform distribution of local mean mass fraction is expressed in term IV. Terms V and VI contribute to the turbulent diffusion of the entity $\overline{m^2_{fu}}$ and its production due to chemical reaction respectively.

The value of the turbulent diffusion coefficient D_t was given by Monin *et al.*[52] as

$$D_t = \mu_t/\sigma_f$$

It should be noted that the term which represents the dissipation of the entity $\overline{m^2_{fu}}$, term III, is proportional to the quantity that is being destroyed by viscous action, i.e. $\overline{m^2_{fu}}$ times the local density times ε/k. The latter term has the dimensions of $(\text{time})^{-1}$. The term ε/k can be regarded as the rate of decay multiplier for the turbulence energy k. This means that the dissipation of the mean square velocity fluctuations (i.e. of k) and the dissipation of the entity $\overline{m^2_{fu}}$ proceed at proportional rates (Spalding).[1] Term III is expressed as

$$2D_l\,\overline{\frac{\partial m_{fu}}{\partial x_j}\cdot\frac{\partial m_{fu}}{\partial x_j}} = 2\bar{\rho}\,\frac{\varepsilon}{k}\,\overline{m^2_{fu}} \tag{4.2.37}$$

This modelled form of the dissipation term does not explicitly include the laminar Schmidt number (i.e. the ratio of kinematic viscosity to diffusivity). This means that its use is limited to Schmidt numbers whose values are close to unity, and consequently can be applied to gaseous flame configurations. The turbulent diffusion term V is modelled, Borghi,[49] Hutchinson *et al.*[8] and Khalil[5] in the form

$$\frac{\partial}{\partial x_j}\overline{u_j m_{fu} m_{fu}} = -\frac{\partial}{\partial x_j}\frac{\mu_t}{\sigma_f}\frac{\partial}{\partial x_j}\overline{m^2_{fu}} \tag{4.2.38}$$

The generation term of $\overline{m^2_{fu}}$ due to the non-uniform spatial distribution of $\bar{M}_{fu}$ is expressed as:

$$2D_t \text{ grad } \bar{M}_{fu} \cdot \text{grad } \bar{M}_{fu} = C_{g1} G_{gfu} \tag{4.2.39}$$

The modelled equation for the transport of $\overline{m^2_{fu}}$ is expressed as:

$$\frac{\partial}{\partial x_j}\bar{\rho}\bar{U}_j\overline{m^2_{fu}} = \frac{\partial}{\partial x_j}\frac{\mu_t}{\sigma_f}\frac{\partial\overline{m^2_{fu}}}{\partial x_j} + C_{g_1}G_{fu} - C_{g_2}\bar{\rho}\frac{\varepsilon}{k}\overline{m^2_{fu}}$$

$$-2A_0\bar{\rho}^2\bar{M}_{fu}\bar{M}_{ox}\exp(-E/R\bar{T})\left[\frac{\overline{m_{ox}m_{fu}}}{\bar{M}_{ox}} + \frac{\overline{m^2_{fu}}}{\bar{M}_{fu}}\right] \tag{4.2.40}$$

and

$$G_{gfu} = \mu_{eff}\left(\frac{\partial\bar{M}_{fu}}{\partial x_j}\right)^2$$

The last component of the source term appears due to the presence of the finite source term in the conservation equation of $\bar{M}_{fu}$, in contrast to the sourceless mixture fraction conservation equation. Previous investigations using a similar model, e.g. by Pope,[4] indicated that the reaction rate effect on the fluctuations of $\overline{m^2_{fu}}$ is significant and that the term should be included in the model. The entity $m_{fu}m_{ox}$, which appears in Equation 4.2.40, is obtained from the solution of the corresponding conservation equation as discussed later in the chapter.

The reader's attention is drawn to the fact that previous investigations, by Mason *et al.*[53] and Pope,[4] neglected the effect of reaction rate on $\overline{m^2_{fu}}$ in their modelling of Equation 4.2.36. Later in the work of Khalil[5] and Khalil *et al.*,[9] account was taken of this effect. Early on, the value of the eddy break-up constant C_R was proposed by Spalding to be 0.53. Khalil[5] however, obtained a value of C_R of 1.07 through using computer optimization. This value is in agreement with the value of 1.1 reported by Pope[4] for turbulent confined flames.

Assessment of the Model

The present model requires the solution of the conservation equations for $\bar{M}_{fu}$, $\overline{m^2_{fu}}$ and $m_{ox}m_{fu}$. The temporal distribution of M_{fu} corresponds to a clipped Gaussian probability distribution whose characteristics were obtained from local values of $\bar{M}_{fu}$ and $\overline{m^2_{fu}}$, in a similar manner to Model 3. Model 4 is suitable for premixed flames where fuel and oxidant can coexist and was successfully used by Mason *et al.*[53] Pope,[4] Khalil[5] and Khalil *et al.*[9] Mason and Spalding[53] investigated the flow of premixed stoichiometric propane-air flame in a duct whose dimensions were 76.2 mm × 25.4 mm, with various inlet velocities. The flame was stabilized on a cylindrical flame stabilizer. The calculations were obtained with the aid of the eddy break-up model of Equation 4.2.34. A

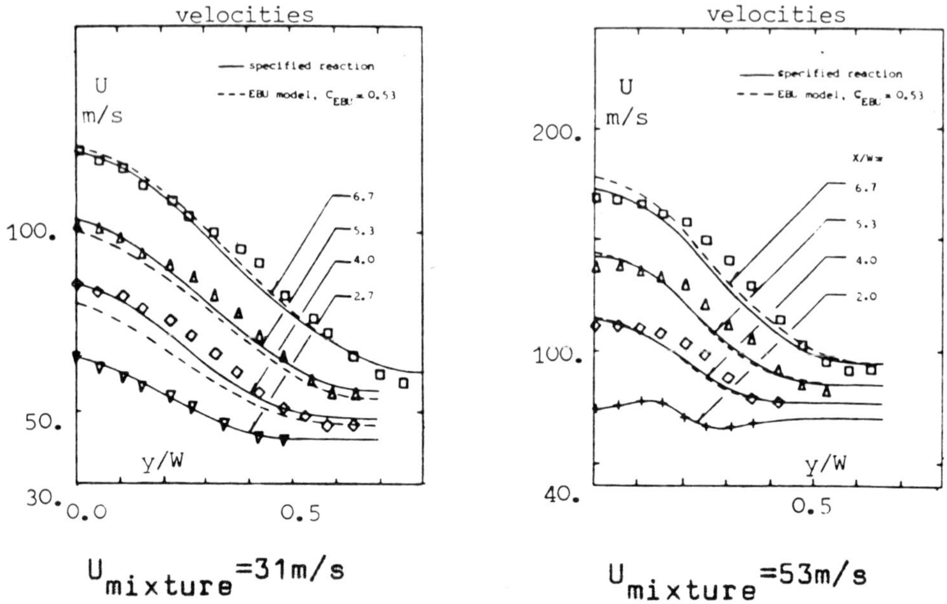

Figure 4.15 Comparisons between measured and calculated axial velocities in a propane–air flame, in a duct[53]

sample of their comparisons is shown in Figure 4.15. The measured and calculated axial velocity profiles are shown in the figure, and are in reasonable agreement. Figure 4.16 displays the comparison between the measured and the computed mean oxygen concentrations; a general agreement is observed at down-stream locations from the duct inlet. This supports the use of the model in the calculation of furnaces and combustors. It is worth noting that the value of $(\overline{m_{fu}^2})^{\frac{1}{2}}$ was obtained from algebraic expressions.[5] Further applications of the eddy break-up model were reported by Khalil *et al.*[2] and Hutchinson *et al.*,[3,7] and the calculations well represented the flow and flame characteristics.

The present Model 4 with the solution of a differential equation for the transport of $\overline{m_{fu}^2}$, was previously tested for premixed flame situations as described by Khalil.[5] A sample of the predicted velocity and temperature profiles, obtained for the furnace of Beltagui *et al.*,[54] is shown in Figures 4.17 and 4.18. Figure 4.17 demonstrates the ability of the model to predict axial velocity pattern and to identify the forward and recirculation zones, in a town gas premixed flame. The measured and the predicted local gas temperature profiles are as shown in Figure 4.18; the predictions lie within ± 100 °K from the experimental data.

Another example to be found in Khalil[5] is the prediction of the centreline temperature distribution in a cylindrical furnace due to Steward *et al.*[55] The comparisons are shown in Figure 4.19. It is clear from the figure that the centreline temperature distribution is well represented.

Generally, the predictions obtained with the aid of Model 4 demonstrated that this model is appropriate to premixed flame situations where the rate of reaction is both finite and very high. In other words, the model is restricted to situations where the

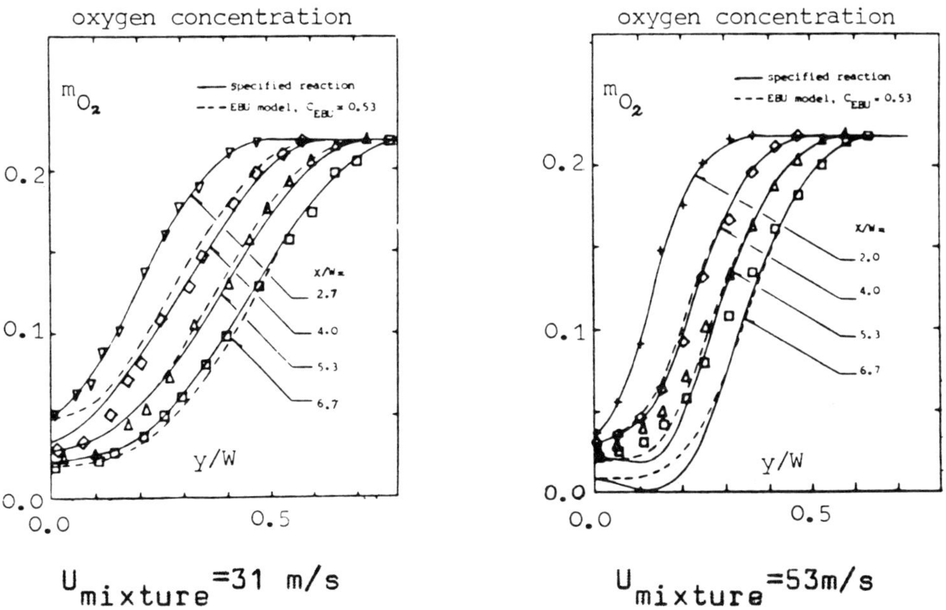

Figure 4.16 Comparison between measured and calculated oxygen concentrations in a propane–air flame in a duct[53]

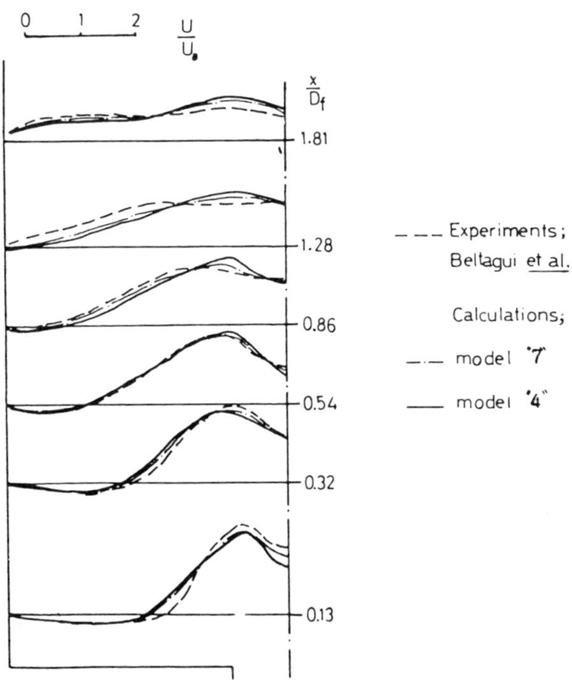

Figure 4.17 Measured and calculated axial velocity profiles in a premixed confined flame

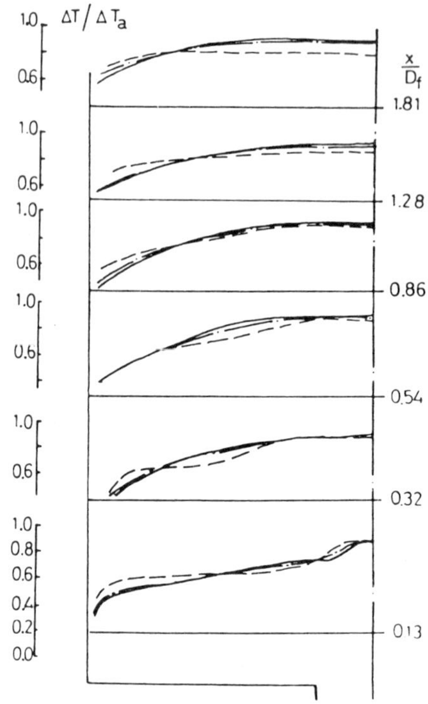

Figure 4.18 Measured and calculated gas temperature profiles in a premixed confined flame

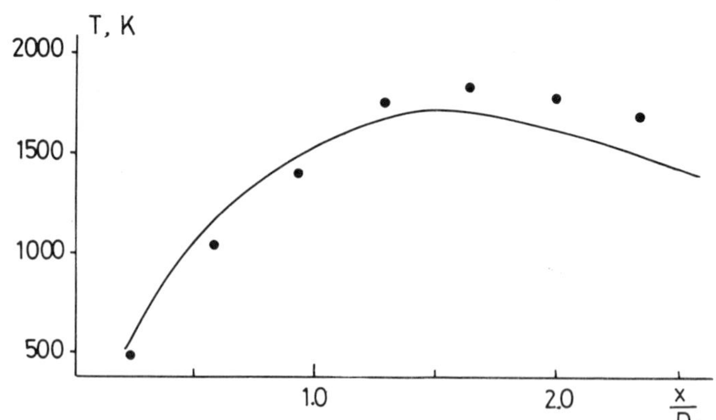

Figure 4.19 Centreline temperature distribution
● Experiments of reference 55 —— Calculations of model 4

Domkohler number $N_D = \tau_s/\tau_k$ is very small. The assumptions of single step is not, however, appropriate for many gaseous flames; especially when the radical concentrations are required. A model that allows for multi-step reaction is therefore needed.

4.2.2.2 *Model 5*

Model Assumptions

The fuel and oxidant streams are premixed
The reaction rate is finite
The reaction is multi-step and reversible
The probability distribution of fuel mass fraction is clipped Gaussian.

Model Formulation

The model is similar to Model 4, except for one difference and that is the chemical reaction takes place in intermediate steps. The net reaction rate is evaluated from the forward and backward reaction rates. Similar to Model 4, the influence of turbulence was included, through an eddy break-up rate. A clipped Gaussian probability distribution was also assumed to represent the temporal distribution of the various species. Simple reaction mechanisms for the oxidation of gaseous fuels were suggested for methane by, for example, Waldman *et al.*,[19] Enikolopyan,[56] Baulch *et al.*[18] and Bowman.[57] Simple reaction mechanisms for the oxidation of other gaseous fuels are also available (Waldman *et al.*).[19] The methane reaction mechanism of Waldman *et al.*[19] is shown in Figure 4.20; only eight major reaction steps of this mechanism are considered. They were selected on the basis of their major contribution to the formation of the intermediate radicals, these selected reactions are:

$$
\begin{aligned}
CH_4 + OH &\rightleftharpoons CH_3 + H_2O \\
CH_3 + O_2 &\rightleftharpoons CH_2O + OH \\
CH_2O + OH &\rightleftharpoons CHO + H_2O \\
CHO + OH &\rightleftharpoons CO + H_2O \\
CO + OH &\rightleftharpoons CO_2 + H \\
H_2 + O &\rightleftharpoons OH + H \\
H_2O + O &\rightleftharpoons OH + OH \\
O_2 + M &\rightleftharpoons O + O + M
\end{aligned}
\qquad (4.2.41)
$$

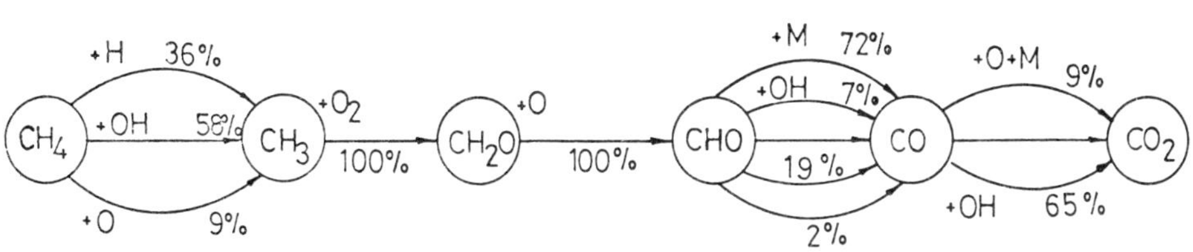

Figure 4.20 Methane oxidation paths

These reaction steps are concerned only with the oxdation path of CH_4. The NO_x concentration is always too small to affect the main reaction mechanism or to influence the energy based properties; its formation path is discussed separately in Chapter 6. The constants of each rate of the forward and backward reactions were calculated according to the data of Waldman *et al.*[19]

Despite the computational complex involved in using the mechanism of 4.2.41, it was nevertheless applied to furnace flow computations. The procedure in this model demands that a conservation equation of the chemical species, for each of the reaction steps, is solved. These equations are expressed in a similar form to 3.2.3, with the mass fraction M_a denoting any species, and the rate of formation or disappearance as the source or sink term. The forward reaction rate coefficient K_f is expressed as

$$K_f = B\bar{T}^n \exp(-E/R\bar{T}) \qquad (4.2.42)$$

where B is the pre-exponential coefficient. The backward reaction rate coefficient K_b is obtained from the equilibrium constant K_c as

$$K_b = K_f/K_c \qquad (4.2.43)$$

This equilibrium constant K_c is expressed, at atmospheric conditions, as

$$K_c = \exp(-\Delta G^\circ/R\bar{T}) \qquad (4.2.44)$$

where

$$\Delta G^\circ = \sum_{J=1}^{N} (\nu_J - \nu_j)\, g^\circ_J$$

$$g^\circ_J = \bar{H}_J - \bar{T}\bar{S}_J = \text{Gibbs free energy} \qquad (4.2.45)$$

$$S_J = \int_T C_{p,J}\frac{d\bar{T}}{\bar{T}} + C_{s,J} = \text{species entropy}$$

$$\bar{H}_J = \int_T \bar{C}_{p,J}\, d\bar{T} + (C_H - a_{-1})_J$$

and

$$\sum_{J=1}^{N} \nu_j \mathring{W}_J \rightleftharpoons \sum_{J=1}^{N} \nu_{j}'' \mathring{W}_J \qquad (4.2.46)$$

where ν_j and ν_J are the stoichiometric coefficients of the reactants and products, and $(C_H - a_{-1})_J$ and $C_{s,J}$ are constants related to the heat of formation and entropy of the species.

As mentioned before, the complexity of this model arises from the great number of conservation equations that are to be solved, namely conservation equations for all the species considered together with the aerodynamic conservation equations. In the species conservation equations, the square of the concentration fluctuation g_J is obtained from an algebraic form of Equation 4.2.40 by equating the generation of g_J to its dissipation, as reported by Hutchinson *et al.*[7] and Khalil.[5]

Assessment of Model

Several approaches which are similar to this present model, i.e. they solve the conservation equations for the individual reaction steps, were reported by, among others, Borghi[50] for the process of carbon monoxide oxidation in a mixing layer. Similar computations also performed by Leuchter[58] for hydrogen–oxygen reactions in the mixing region of two adjacent streams, were only concerned with boundary layer flows and did not apply their method to the flow situations where substantial recirculation zones occur. The work of Richter *et al.*[42] reported computations of furnace flames; his model depended basically on the simple reaction mechanism for methane oxidation:

$$\begin{aligned} CH_4 + \tfrac{3}{2}O_2 &\xrightarrow{\infty} CO + 2H_2O \\ CO + \tfrac{1}{2}O_2 &\rightleftharpoons CO_2 \\ CO + H_2O &\underset{\infty}{\overset{\infty}{\rightleftharpoons}} CO_2 + H_2 \end{aligned} \tag{4.2.47}$$

The reaction rate of the second reaction step was obtained from the expression suggested by Kozlov.[59] Richter[60] computed the flow in the furnace of International Flame Research Foundation. Figure 4.21 shows a sample of the comparison between measured and calculated volumetric concentrations of H_2O, CO_2, CO and H_2 along the furnace. The agreement obtained for the CO_2 distribution is satisfactory along the furnace centreline and away from the burner. Predictions of H_2 are also shown on the figure and are in reasonable agreement with the corresponding measurements.

Khalil[5] and Hutchinson *et al.*[7] predicted the centreline distribution of CH_4 and CO concentrations in town and natural gas flames in the furnace of Lenze *et al.*[61] Both measurememts and calculations are in satisfactory agreement, the trend is correctly predicted and the discrepancies are less than 10%, as shown in Figure 4.22. The approximations embodied in this model can be viewed as being an over-simplification of the physical situation. The approximations, such as replacing the *g*-equation by an algebraic expression and reducing the reaction mechanism to a limited number of steps,

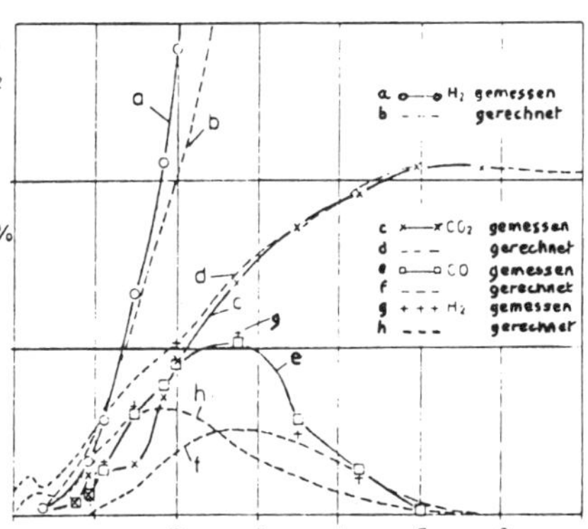

Figure 4.21 Axial distributions of species concentrations in the Ijmuiden Furnace, IFRF (reference 42)

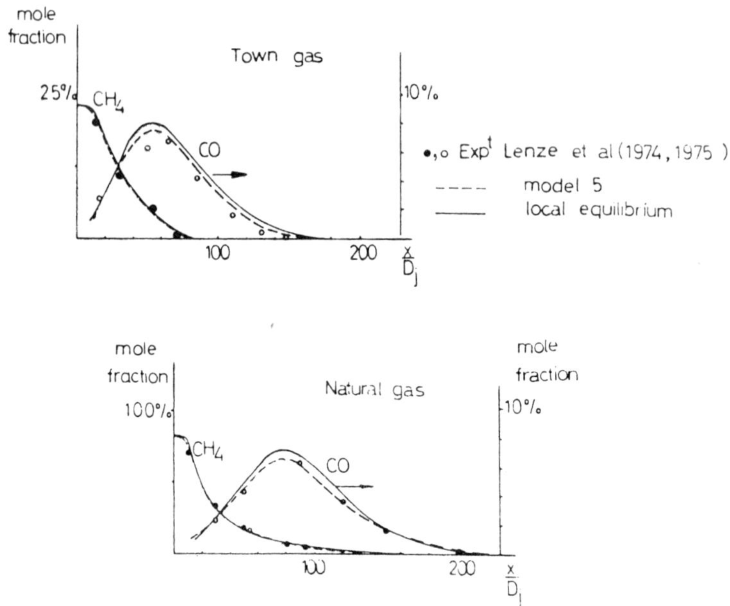

Figure 4.22 Centreline distributions of CH_4 and CO in the furnace of reference 61; predictions are those of reference 5

contribute to the observed discrepancies. Despite these discrepancies, the model adequately represents the global rate of reactions and consequently the heat release rates. A major disadvantage of the model is the excessively large computation time involved. However, the use of this model is needed most when the intermediate species concentrations are important and should be known.

4.2.3 Turbulence–Chemistry Interactions Models

The effect of turbulence fluctuations on the reaction rate is determined by decomposing the instantaneous reaction rate, given by Equation 4.2.4, into mean and fluctuating components and evaluating them through their relation with the mean flow properties. Many previous attempts to formulate a statistical theory for turbulent reacting flows were concerned with isothermal reactions, e.g. Donaldson *et al.*[62,63]

Donaldson and Varma[63] presented an attempt at illustrating how second-order closure techniques may be used in obtaining detailed calculations of turbulent flames. They considered most of the second-order correlations, and defined expressions for the decay of the concentration fluctuation correlations, such as $\overline{m_a m_b}$, in terms of the initial value $(\overline{m_a m_b})_0$, kinematic viscosity, velocity and the characteristic length scale. Two typical time-scales for the diffusion and chemistry were defined; their ratio, a sort of Damkohler parameter, was found to be a key factor in the flame behaviour. The major criticism which can be made of this approach is that it is applicable only to isothermal reactions, with K_f = constant. Such closure is useful for chemical reactors where relatively small reaction-temperature dependence is observed.

Donaldson *et al.*[63] employed simple shapes of the probability density functions of the Dirac delta functions. Imposing such a restriction on the probability distribution is undesirable and may lead to erroneous conclusions (Bush *et al.*[64]). The closure of Patterson[65] comes along the same line as Donaldson *et al.*[63] Other investigations neglected the effect of density changes on the velocity field, Dopazo *et al.*[66] When the chemical reactions take place in the mixing zone, the time-averaged reaction rate depends on the mixing pattern and chemical kinetics.

Up to now, the models that were discussed are appropriate only to particular flow situations. For example, models 1 to 3 are appropriate to diffusion flames. Otherwise, i.e. when the fuel and oxidant are intimately premixed prior to the combustion chamber, Models 1, 2 and 3 are no longer suitable. Then it is clear that a model incorporating an explicit finite rate of reaction is to be employed. The finite reaction rate models incorporated a single-step reaction mechanism (Model 4) or a multi-step reaction mechanism (Model 5) and were found appropriate to premixed flames.

The present subsection, however, incorporates general models which can be applied to arbitrarily fired flames where none of the above models is appropriate by itself.

4.2.3.1 *Model 6*

Model Assumptions

The reaction rate is finite
Equilibrium is attained
The reaction is single step and irreversible
Temperature fluctuations correlations are neglected
Triple order correlations are neglected
Probability distribution of mass fractions is clipped Gaussian
The turbulent diffusion coefficients of various species are equal.

Model Formulation

The rate of chemical reaction given by Equation 4.2.4 was decomposed to the mean and fluctuating entities, as shown by Equation 4.2.6. Time averaging the rate of reaction with series expansion to the exponential term, yields Equations 4.2.7 and 4.2.8. Equation 4.2.7 contains second-order correlations; the magnitude and sign of the entity $\overline{m_{fu}m_{ox}}$ depends entirely on the manner in which the reactants are introduced into the reaction domain (Borghi *et al*).[50] For non-premixed flames, the correlation $\overline{m_{fu}m_{ox}}$ is always negative or zero since fuel and oxygen concentrations diminish towards each other. Other correlations such as $\overline{m_{ox}T'}$, $\overline{m_{fu}T'}$ and $\overline{T'^2}$ have to be modelled and will be considered later in this chapter. Several expressions for the conservation equation of this entity $\overline{m_{fu}m_{ox}}$ were suggested.[5,49,51] The transport equation for the correlation $\overline{m_{fu}m_{ox}}$ is expressed, for steady state flows, as:

$$\underset{\text{(I)}}{\frac{\partial}{\partial x_j}\bar{\rho}\bar{U}_j\overline{m_{fu}m_{ox}}} = \underset{\text{(II)}}{D_l\,\text{grad}\,\overline{m_{fu}m_{ox}}} - \underset{\text{(III)}}{2D_l\,\overline{\text{grad}\,m_{fu}\cdot\text{grad}\,m_{ox}}}$$

$$+\underset{\text{(IV)}}{\frac{2u_t}{\sigma_f}\,\text{grad}\,\bar{M}_{fu}\cdot\text{grad}\,\bar{M}_{ox}} - \underset{\text{(V)}}{\text{grad}\,\overline{u_j m_{fu}m_{ox}}}$$

$$\underset{\text{(VI)}}{-K_f[(\bar{M}_{ox} + i\bar{M}_{fu})\,\overline{m_{fu}m_{ox}} + i\bar{M}_{ox}\overline{m^2_{fu}} + \bar{M}_{fu}\overline{m^2_{ox}}}$$

$$+\,\overline{(m_{ox} + im_{fu})m_{fu}m_{ox}}] \qquad (4.2.48)$$

where: D_l is the laminar exchange coefficient. Term I represents the net rate of convection of $\overline{m_{fu}m_{ox}}$; term II is the laminar diffusion of the entity in question; and term III denotes the dissipation of the fluctuations due to molecular diffusion. The net generation rate due to the non-uniform distribution of local mass fractions is expressed term IV. Terms V and VI denote the turbulent diffusion of $\overline{m_{fu}m_{ox}}$ and its generation due to chemical reaction respectively. The dissipation term of the entity $\overline{m_{fu}m_{ox}}$, term III, is proportional to the quantity that is being destroyed by viscous action, i.e. $\overline{m_{fu}m_{ox}}$, times the local density times ε/k, in other words, term III is proportional to $\bar{\rho}\ \varepsilon/k\ (\overline{m_{fu}m_{ox}})$. The term ε/k, denoting reciprocal of turbulence time scale, is regarded as the rate of decay multiplier for the turbulence energy k; therefore

$$2D_l\,\overline{\frac{\partial m_{fu}}{\partial x_j}\frac{\partial m_{ox}}{\partial x_j}} = 2\bar{\rho}\,\frac{\varepsilon}{k}\,\overline{m_{fu}m_{ox}} \qquad (4.2.49)$$

The use of the above expression is limited to Schmidt numbers with values close to unity. The turbulent diffusion term V, is modelled,[5] in a similar manner to that for $\overline{m^2_{fu}}$, in the form

$$\frac{\partial}{\partial x_j}\overline{u_j m_{fu}m_{ox}} = \frac{\partial}{\partial x_j}\frac{\mu_t}{\sigma_f}\frac{\partial}{\partial x_j}\overline{m_{fu}m_{ox}} \qquad (4.2.50)$$

The modelling of this term was investigated by, e.g. Khalil,[5] Bilger[12] and Bray.[51] The generation term of the entity $\overline{m_{fu}m_{ox}}$, due to the non-uniform spatial distribution of the entities $\bar{M}_{fu}$ and $\bar{M}_{ox}$, can be either positive or negative according to the distribution of the reactants and is expressed as

$$2\,\frac{\mu_t}{\sigma_f}\,\text{grad}\,\bar{M}_{fu}\,\text{grad}\,\bar{M}_{ox} = C_{g_1}G_{g_2} \qquad (4.2.51)$$

where C_{g_1} is a constant and G_{g_2} is expressed as

$$G_{g_2} = \mu_{\text{eff}} \frac{\partial \bar{M}_{\text{ox}}}{\partial x_j} \cdot \frac{\partial \bar{M}_{\text{fu}}}{\partial x_j}$$

The triple-order correlations such as $\overline{m_{\text{fu}} m^2_{\text{ox}}}$ and $\overline{m_{\text{ox}} m^2_{\text{fu}}}$ were generally neglected,[5,48] on the grounds that their influence was negligibly small at low activation energy, which is true for most of the hydrocarbon oxidation reactions. The modelled form of the conservation equation of $\overline{m_{\text{fu}} m_{\text{ox}}}$ is expressed[5] as

$$\frac{\partial}{\partial x_j} \bar{\rho} \bar{U}_j \overline{m_{\text{fu}} m_{\text{ox}}} = \frac{\partial}{\partial x_j} \frac{\mu_{\text{eff}}}{\sigma_f} \frac{\partial \overline{m_{\text{fu}} m_{\text{ox}}}}{\partial x_j} + C_{g_1} G_{g_2} - 2\bar{\rho} \frac{\varepsilon}{k} \overline{m_{\text{fu}} m_{\text{ox}}}$$

$$-A_0 \bar{\rho} \bar{M}_{\text{ox}} \bar{M}_{\text{fu}} \exp(-E/R\bar{T}) \left[(\bar{M}_{\text{ox}} + i\bar{M}_{\text{fu}}) \frac{\overline{m_{\text{fu}} m_{\text{ox}}}}{\bar{M}_{\text{fu}} \bar{M}_{\text{ox}}} + \frac{\bar{m}^2_{\text{ox}}}{\bar{M}_{\text{ox}}} + i \frac{\bar{m}^2_{\text{fu}}}{\bar{M}_{\text{fu}}} \right] \quad (4.2.52)$$

The Damkohler number N_D defined by Damkohler,[67] as the ratio of a mixing distance (or time) to a reaction distance (or time) which equals unity, when the two distances (or times) are equal. An expression of this number N_D is given[5] as

$$N_D = \tau_s / \tau_k \quad (4.2.53)$$

where τ_k is the time scale of chemistry given by equation 4.2.35, while τ_s is defined as the stretching time of the flame eddies:

$$\tau_s = \left(\left| \frac{\partial \bar{U}_i}{\partial x_i} \right| + \frac{\varepsilon}{k} \right)^{-1} \quad (4.2.54)$$

this accounts for the mixing time in connection with the mean motion of the fluid and the random turbulence (Spalding).[68,69] Introducing these definitions in the last three terms of Equation 4.2.52 makes it feasible to estimate the contributions of chemical kinetics and turbulent mixing to the generation of $\overline{m_{\text{fu}} m_{\text{ox}}}$. In flame situations where the reactants are in intimate contact, due to rapid mixing which is characterized by large values of ε/k, the reaction is kinetically influenced if $N_D \ll 1.0$. The influence of chemical kinetics on the formation of the entity $\overline{m_{\text{fu}} m_{\text{ox}}}$ is hence negligible and only the terms $G_{g_1} G_{g_2} - 2\bar{\rho}\, \varepsilon/k\, \overline{m_{\text{fu}} m_{\text{ox}}}$ have significant roles in generation/destruction of $\overline{m_{\text{fu}} m_{\text{ox}}}$. In the flame situations where the temperature is high enough for the reaction to proceed, but the reactants are not in intimate contact, the reaction is controlled by mixing; the chemical time τ_k is very small relative to τ_s, i.e. $N_D \gg 1.0$.

The variation of time scales τ_s, τ_{s_0} and τ_k and N_D along a mixing zone of a co-flowing methane–air flame, reported by Khalil,[5,70] is demonstrated in Figures 4.23 and 4.24 respectively. In Figure 4.25, the variation of N_D with temperature is shown; as the temperature increases, N_D increases and the reaction is more dependent on mixing.

The present model requires the solution of the same equations as Model 4, in addition to two conservation equations for the transport of $\overline{m_{\text{fu}} m_{\text{ox}}}$ and $\overline{m^2_{\text{ox}}}$. The transport equation for the entity $\overline{m^2_{\text{ox}}}$ is similar to Equation 4.2.40 but with appropriate source terms.

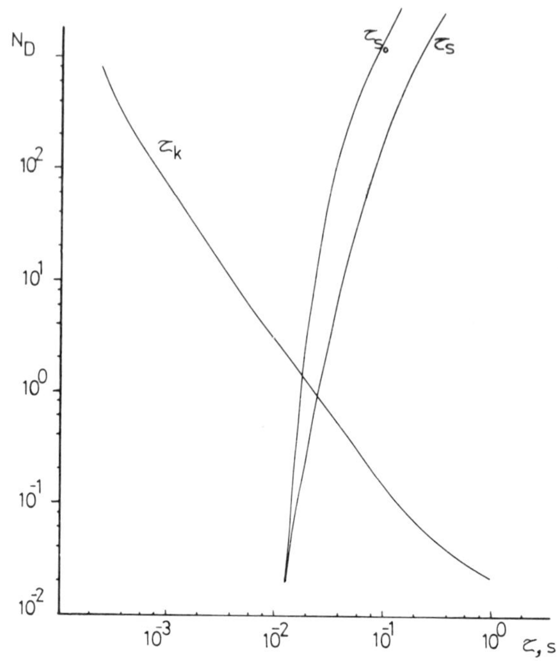

Figure 4.23 Relation between various time scales and N_D

The local flow properties were also obtained in similar manner through the specification of $P(M_{fu})$.

Investigations of Hutchinson *et al.*[7] reported an approximation of the conservation equations of $\overline{m_{fu}^2}$, $\overline{m_{fu}m_{ox}}$ and $\overline{m_{ox}^2}$. This was accomplished by neglecting the convection and diffusion fluxes, thus yielding algebraic forms for the balance of production and dissipation of the scalar entity. This approximation when applied to furnace flames[7] yielded results which are different by 3–8% from those obtained from the solution of the full differential equations.

Assessment of Model

This model was applied[48] to turbulent reacting premixed ducted flames, by Hutchinson *et al.*[7] to axisymmetric furnace flows, and by Hutchinson *et al.*[8] to turbulent arbitrarily fired flame configurations. The reported agreement was satisfactory for engineering design purposes and suggested the application of the model to furnace and combustor design. A systematic study of the model in laboratory experimental furnaces, was reported[5] for the furnaces of references 8, 28 and 54.

Measured and calculated velocity profiles were reported,[8] in an axisymmetric furnace with and without swirl in the combustion air. Figure 4.26 demonstrates the capability of Model 6 in predicting the axial velocity profiles at various axial locations and for a swirl number of 0. Figure 4.27 shows the predicted centreline temperature distribution for a swirling flame, with $S = 0.52$.

Model 6 was also used,[5] to predict the centreline temperature distribution for the flow

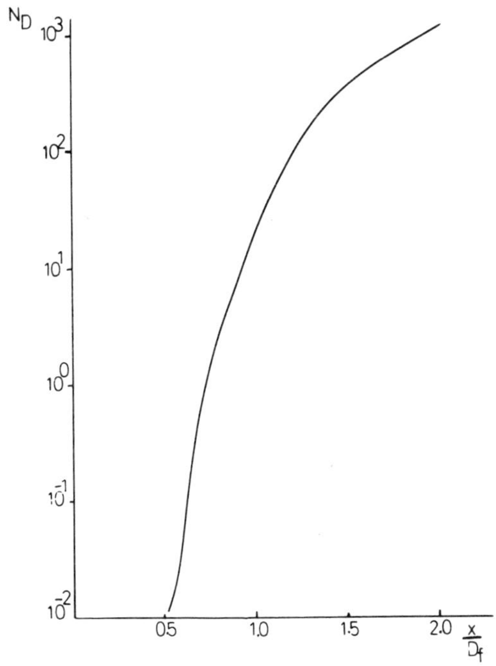

Figure 4.24 Distribution of N_D along the flame

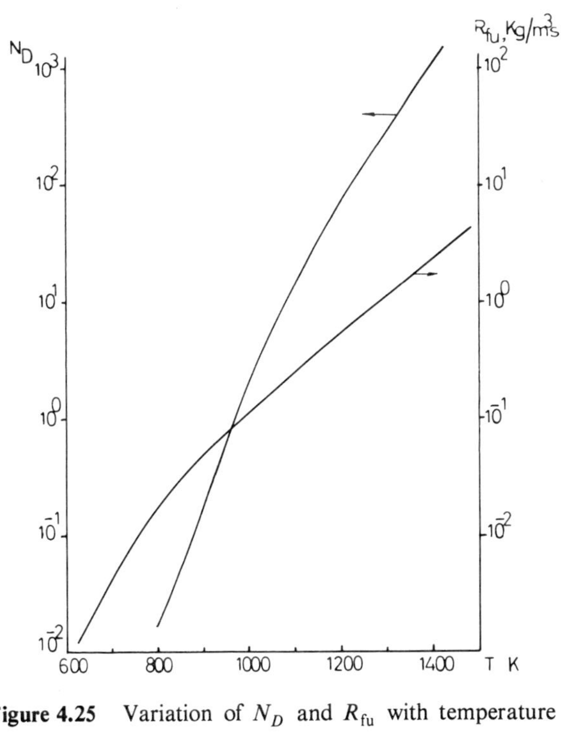

Figure 4.25 Variation of N_D and R_{fu} with temperature

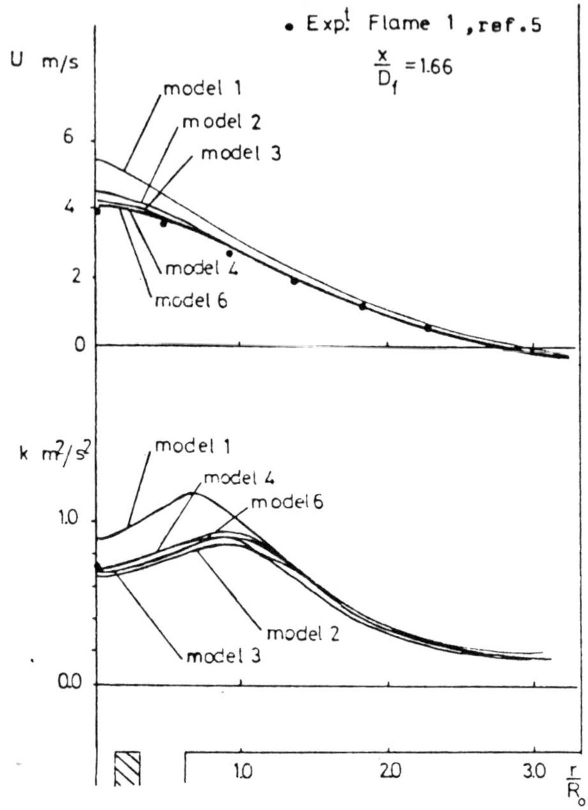

Figure 4.26 Radial profiles of mean velocity and kinetic energy in the furnace of reference 5: a comparison between various combustion models

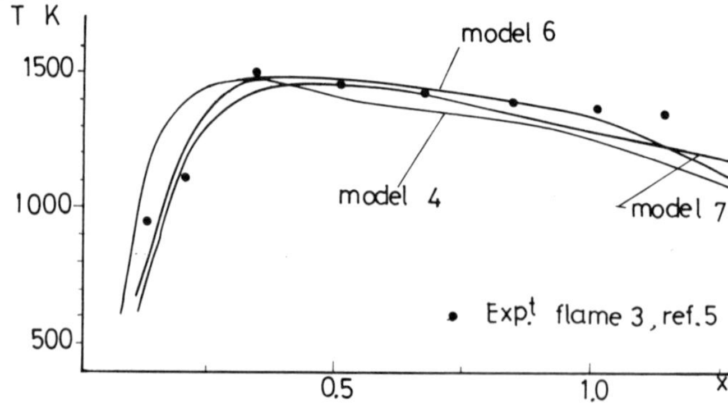

Figure 4.27 Centreline temperature distribution in the furnace of reference 5, a comparison between combustion models

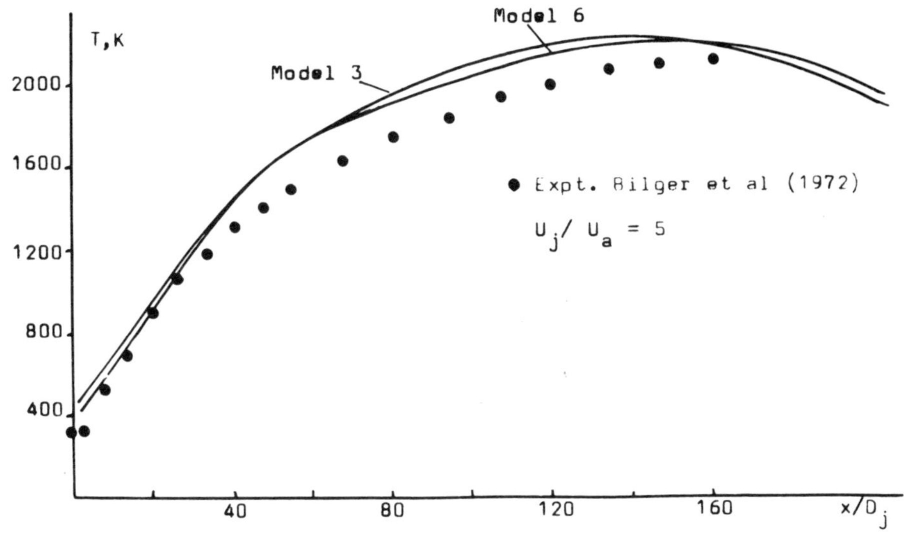

Figure 4.28 Centreline distribution of mean gas temperature in a hydrogen diffusion flame: a comparison between combustion models

of Bilger *et al.*[28] The flow configuration comprised a co-flowing hydrogen-air flame with a jet/air velocity ratio of 5. The calculated temperature distribution and the corresponding measurements are shown in Figure 4.28. In the same figure the corresponding calculations obtained with the aid of Model 3 are shown. It can be concluded from the figures that the flow is well represented by the models' assumptions. As N_D, in the present case, is greater than unity, Model 3 can be applied and is recommended for this type of flows as it demands the solution of fewer equations than Model 6.

Model 6 faces criticism on two counts. The first is that it involves the solution of a relatively large number of differential equations. The second and more important defect is that the model ignores the effect of temperature fluctuations on the reaction rate expression. In many flow situations, this omission can lead to serious errors especially at large activation energies. The next model, 7, accounts for the temperature fluctuation correlations, and further comments to Model 6 will be given after the description of Model 7 so as to allow a general assessment.

4.2.3.2 *Model 7*

Model Assumptions

The reaction rate is finite
Equilibrium is attained
The reaction is single step and irreversible
The triple-order correlations are neglected
Probability distribution of mass fractions is clipped Gaussian
The turbulent diffusion coefficients of the various species are equal.

Model Formulation

In many situations, the activation energy is both very high and temperature dependent, for example in the pollutant reactions. As explained earlier, the time mean rate of reaction is expressed by Equation 4.2.7 and higher-order correlations $\overline{m_{fu}^2 T'}$, $\overline{m_{ox}^2 T'}$ and $\overline{T'^3}$ were neglected. The truncation errors due to the series expansion, in Equation 4.2.7, to the second term only, i.e. Model 6, were discussed by Borghi.[48] He also found out that the approximation of the rate of reaction by a series expansion deteriorated as a representation when $E/R\bar{T}$ was increased, i.e. for high activation energy and a low mean gas temperature for a specified level of temperature fluctuations. A temperature fluctuation intensity of 10% and $E/R\bar{T}$ of 10, produced an error of 1.5% in the rate expression; this error increased as the activation energy increased. Hence, in Model 7, which is used for reactions where the activation energy is large, e.g. pollutant formation $E/R\bar{T} \simeq 67$, the second-order correlations $\overline{m_{fu}^2}$, $\overline{m_{ox}^2}$, $\overline{m_{fu}T'}$, $\overline{m_{ox}T'}$ and $\overline{T'^2}$ are incorporated in the rate expression. Higher order correlations such as $\overline{T'^3}$, $\overline{m_{fu}m_{ox}T'}$, $\overline{m_{ox}T'^2}$, $\overline{m_{fu}T'^2}$, $\overline{m_{fu}m_{ox}^2}$ and $\overline{m_{ox}m_{fu}^2}$ were neglected.[5] Borghi[48] and Donaldson *et al.*[63] evaluated algebraically the magnitude of these terms and they were found to be negligible. For example for $E/R\bar{T}$ of 20 and $\sqrt{\overline{T'^2}}/\bar{T}$ of 0.15, the truncation errors were less than 0.5%.

Model 7 solves the conservation equations for the entities $\overline{m_{fu}^2}$, $\overline{m_{ox}^2}$, $\overline{m_{fu}m_{ox}}$, $\overline{m_{fu}T'}$, $\overline{m_{ox}T'}$ and $\overline{T'^2}$. From the values of $\bar{M}_{fu}$ and $\overline{m_{fu}^2}$, the characteristics of the assumed clipped Gaussian probability distribution of M_{fu} were obtained in a similar manner to that of Model 4. The local values of $\bar{T}$, $\bar{\rho}$ and $\bar{M}_{ox}$ were consequently obtained from Equation 4.2.30 and 4.2.31 with M_{fu} replacing the mixture fraction f.

Assessment of Model

Model 7 represents an attempt to calculate the time-averaged reaction rate, taking into account the effects of turbulence and chemistry. The effects of turbulence are characterized explicitly by ε/k and implicitly by certain terms such as $\overline{m_{fu}m_{ox}}$, $\overline{T'^2}$, $\overline{m_{fu}^2}$, . . . etc. The effects of chemistry are characterized by the time scale τ_k, E/R and A_0. The great number of equations, to be solved for this model, demands large computing facilities. Assessment of the model was reported by Hutchinson *et al.*[8] for an axisymmetric furnace flow at different swirl intensities. Khalil[10] reported many comparisons for test cases to examine the validity of Model 7, e.g. the comparisons done for the flow of Bowman *et al.*[71] using natural gas, Paauw[72] and Lockwood *et al.*[73] Further comparisons were reported by Khalil *et al.*[2] and Khalil[5] for small gas-fired experimental furnaces. A few comparisons in large scale furnaces were carried out by Khalil[5] and Khalil *et al.*[74] to demonstrate the capability of the Model 7 to predict adequately real furnace flows. The examples given below were selected for demonstration purposes, and the reader is advised to refer to the original papers for more detail.

The model was used to predict the local flow pattern in the axisymmetric furnace of Bowman *et al.*[71] The furnace had a central fuel injection and annular air, the methane jet/air velocity ratio was 1/22, which allowed for 10% excess air; the furnace was water cooled. Figure 4.29 shows measured and calculated mean axial velocity profiles, for $X/D_f = 0.157$. The calculated velocity was in good agreement with the measurements

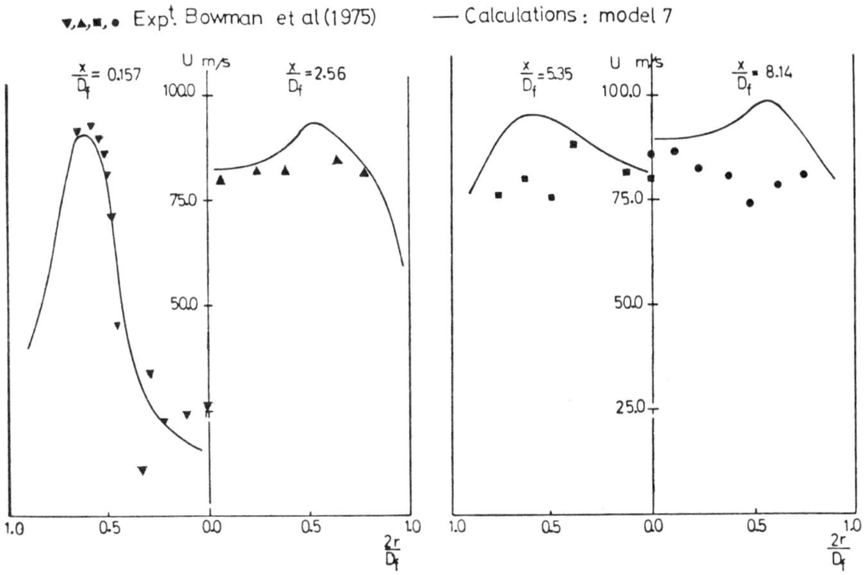

Figure 4.29 Measured and calculated profiles of mean axial velocity in the furnace of Bowman *et al.* (1975)

except near the injector wall, where discrepancies were observed and were attributed to the two equation turbulence model deficiencies.[5] The local gas temperature profiles are shown in Figure 4.30; it can be observed that the rate of flame spread was under-predicted, the flat profiles of mean gas temperature suggested that only a small portion of the total available chemical heat release occurred beyond the initial stirred reactor region of 2.5 D_f.

Radial profiles of oxygen concentrations are shown in Figure 4.31; these concentrations were over-predicted in the wall vicinity and better agreement is observed at further downstream locations. The over-prediction of oxygen concentration at the first measuring locations is consistent with an under-predicted rate of flame spread.

To ensure better mixing in the combustion chamber, in many furnace flows the burner is made of a divergent quarl of air nozzle with central fuel jet. Comparisons were given[5] for the flow in the conical furnace of Paauw;[72] the conical angle was 22° and the small diameter of the chamber and its length were 0.3 and 1.4 m respectively. Figure 4.32, reported the comparison between measured and calculated mean axial velocity profiles, in the natural gas–air flame with 5% excess air. Model 7 was selected (Khalil)[5] on the grounds that the local values of N_D were of the order of unity, and hence, the assumptions of Models 3 and 4 were not appropriate. The agreement was good and the discrepancies were less than 15% at further downstream locations. Profiles of mean gas temperature are shown in Figure 4.33, which indicates general agreement. The model correctly predicted the trends and particularly so for the locations of peak temperature associated with reaction zones.

Another example is the prediction of mixture fractions in a diffusion flame which was reported by Khalil[5] for the furnace of Lockwood *et al.*[73] The results are shown in

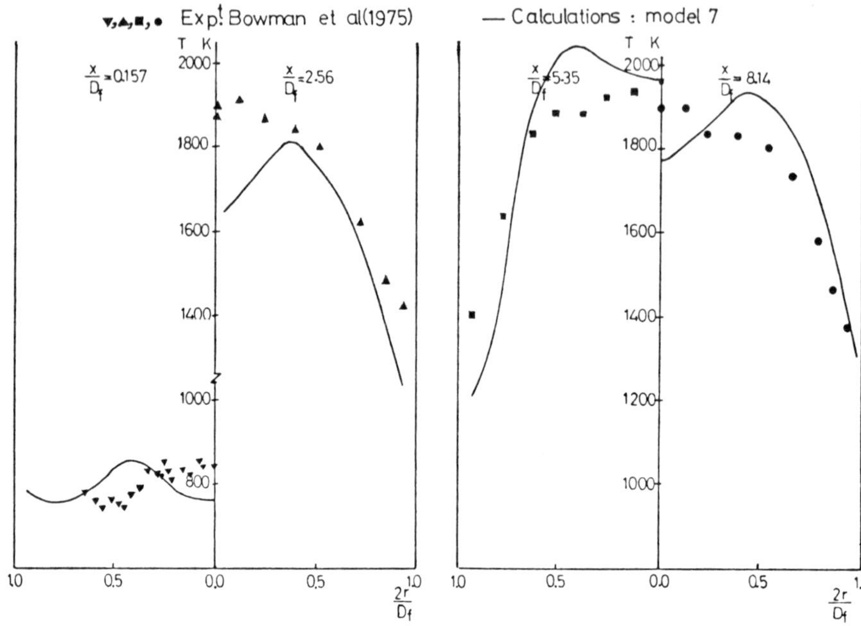

Figure 4.30 Measured and calculated profiles of mean gas temperature in the furnace of Bowman *et al.* (1975)

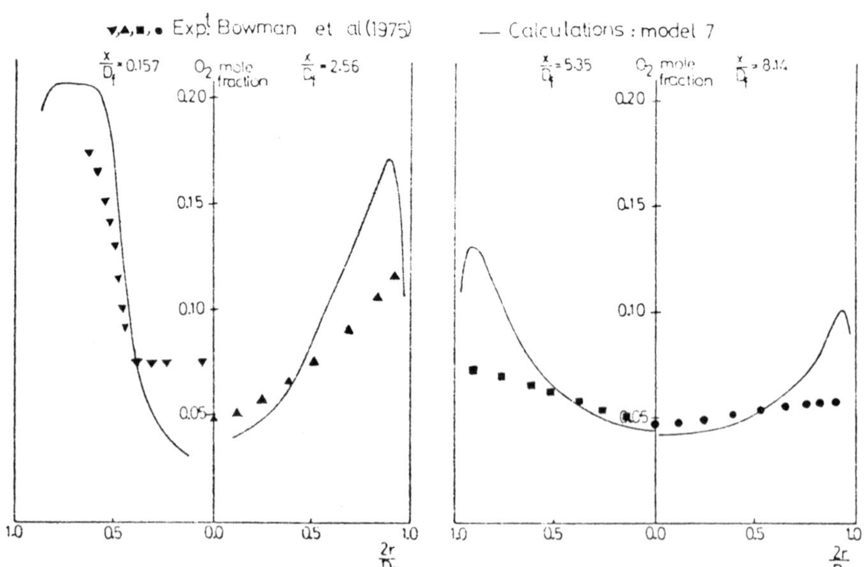

Figure 4.31 Measured and calculated profiles of oxygen mole fraction in the furnace of Bowman *et al.* (1975)

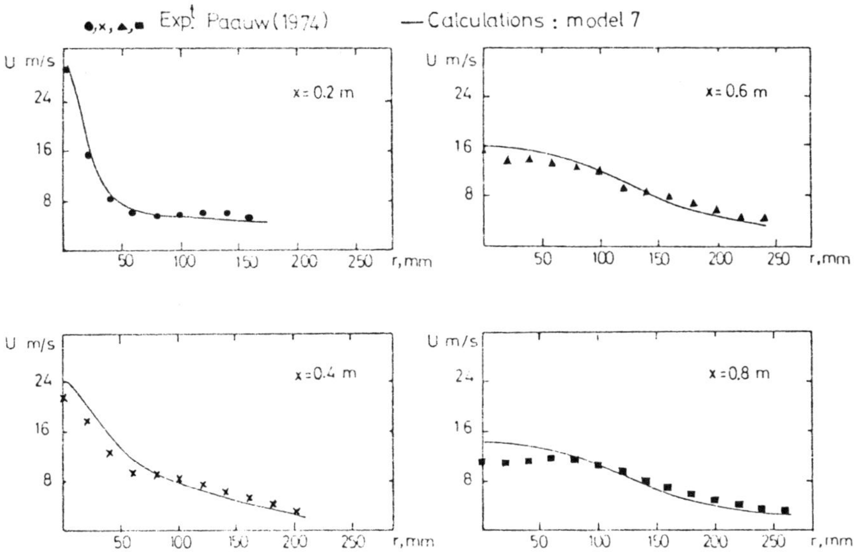

Figure 4.32 Measured and calculated profiles of mean axial velocity in the conical furnace of Paauw (1974)

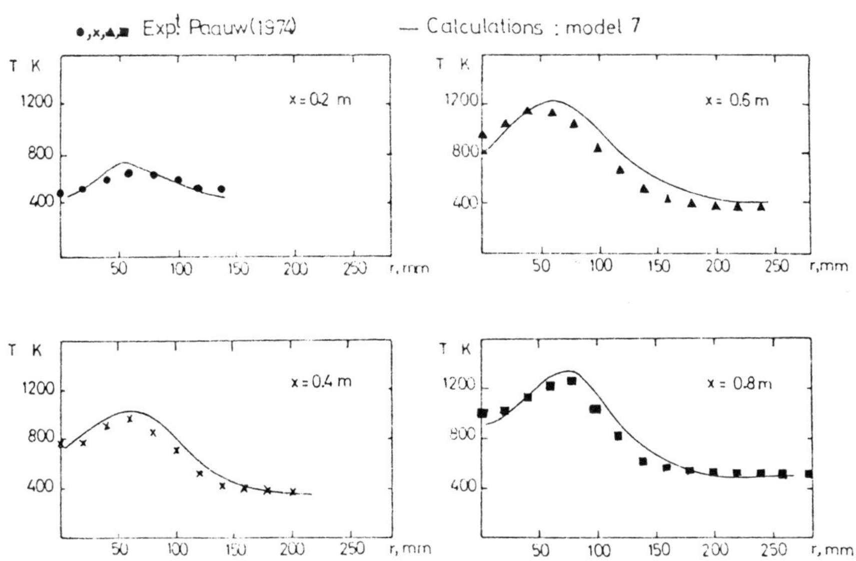

Figure 4.33 Measured and calculated profiles of mean temperature in the conical furnace of Paauw (1974)

Figure 4.34 where reasonable agreement is observed. On this figure, the centreline distributions of *f* are shown; two predicted distributions are presented; one was obtained with Model 3 and the other with Model 7. Both models resulted in reasonable agreement, with Model 7 yielding better agreement near the burner.

Calculations of flow pattern and gas temperature profiles in a confined axisymmetric natural gas flame, are shown in Figures 4.35, 4.36 and 4.37. The furnace diameter and length were 0.3 and 0.9 m respectively and the furnace was divided into five water-cooled segements which were used for the wall heat flux measurements. Measured and calculated profiles of mean axial velocity are shown in Figure 4.35 for a swirl number of 0.52. The calculations adequately represented the measured data, and particularly the decay of the jet. The burner geometry was of the diverging quarl type which enables

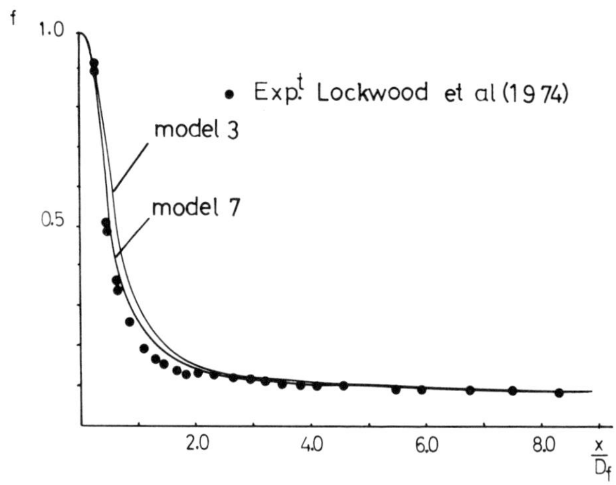

Figure 4.34a Centreline distribution of mixture fraction in a diffusion flame: a comparison between combustion models

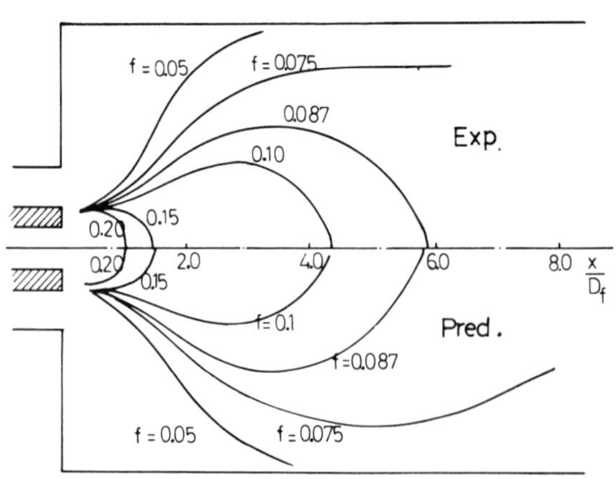

Figure 4.34b Contours of isomixture fraction in a diffusion flame: a comparison between combustion models

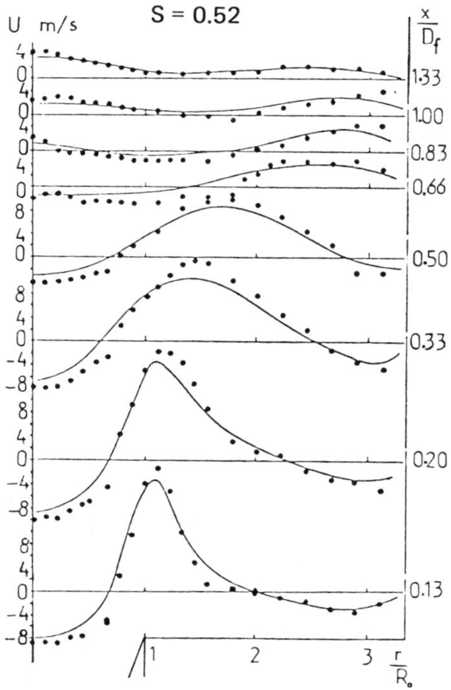

Figure 4.35 Axial velocity profiles in furnace of reference 8
● Experiment —— Predictions

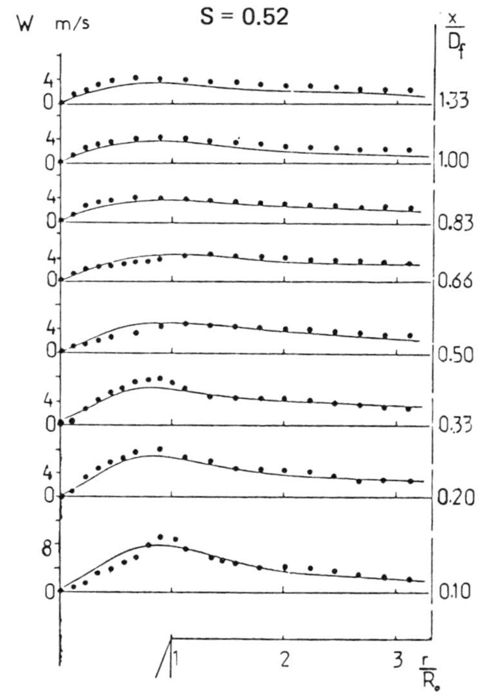

Figure 4.36 Tangential velocity profiles in furnace of reference 8
● Experiment —— Predictions

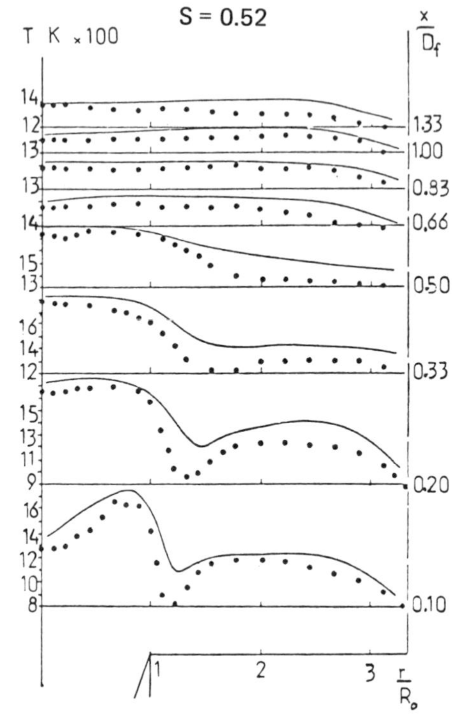

Figure 4.37 Gas temperature profiles in furnace of reference 8
● Experiment —— Predictions

mixing of reactants prior to the furnace itself. The wall recirculation zone was well predicted, but the size of the central recirculation zone was underpredicted and is attributed to the turbulence model assumptions as was discussed earlier.

The corresponding calculations of the tangential velocity component are shown for the same flow parameters, in Figure 4.36. Reasonable agreement was observed, and the rate of decay of swirl velocity was adequately predicted. The measured and calculated gas temperature profiles are shown in Figure 4.37 and the model assumptions represented adequately the various flame regions. The progress of the reaction along the furnace was well predicted as the centreline temperature increased at further downstream locations. The profiles exhibited temperature troughs in the wake of the air stream, and the discrepancies in this region were of the order of 150°K. At further downstream locations, i.e. in the post-flame region, the profiles flattened out and, naturally, the observed agreement became good.

A word about Models 1 to 7: it is evident from the previous discussions that Models 6 and 7 are more general than Models 1 to 5. This is because they include the effects of turbulent fluctuations and chemical kinetics in the reaction rate expressions. They are expensive to run, for they require the solution of many conservation equations, and hence are only most appropriate to arbitrary fired flames, i.e. $N_D \simeq 1.0$, where Models 1 to 5 fail to yield satisfactory results.

Due to the series expansion of the exponential term in the reaction rate expression, Models 6 and 7 are applied only to situations with not too high $E/R\bar{T}$, i.e. of the order of

30; and relatively moderate temperature fluctuation intensities. At higher activation energy reactions, the truncation errors increase to more than 15% and the solution of the conservation equation for the scalar correlations diverges.

4.2.4 Probability Approach to Turbulent Reacting Flows

In all the models discussed earlier, the time-averaged scalar properties were obtained from Equation 4.2.19 with the aid of a presumed probability density function. For Models 1 to 3, the characteristics of such function were calculated from the mean and variance of the passive scalar f, and from the mean and variance of M_{fu} for Models 4 to 7. Various shapes of the probability function $P(\Phi)$ were assumed for reacting flows; square wave form,[1] saw tooth and sinusoidal,[37] clipped Gaussian[21,22] and beta function.[35] The first three forms were an over-simplification of the temporal distribution of Φ and the corresponding $P(\Phi)$. These situations are shown in Figure 4.38 (Spalding).[75]

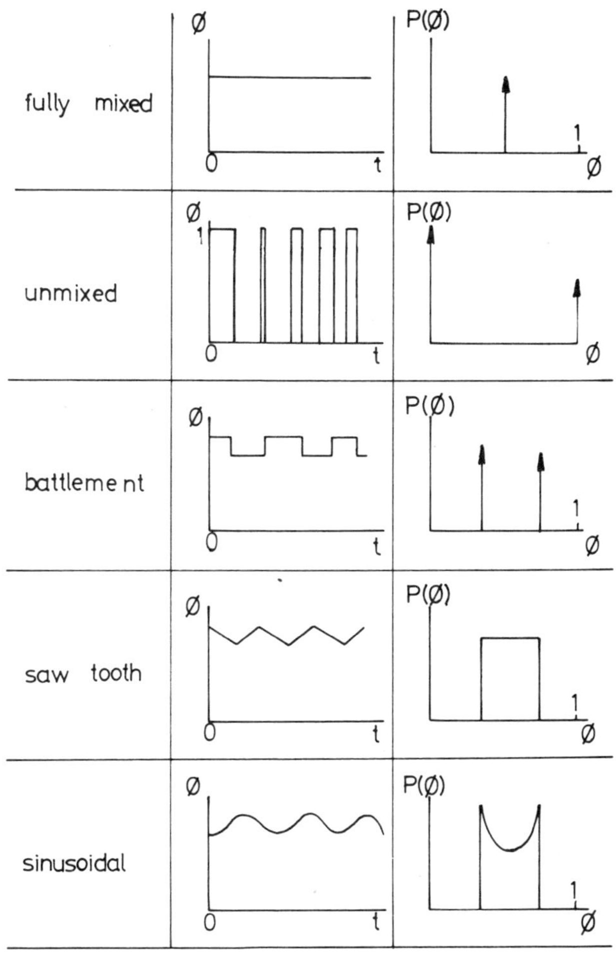

Figure 4.38 Forms of assumed probability distributions

The values of $P(\Phi)$ are expressed in algebraic forms that incorporate $\bar{\Phi}$ and $\overline{\phi^2}$. The more complex forms of clipped Gaussian and beta function were proposed as the probability distribution of the scalar entities in reacting flows by several investigators, among whom are Naguib,[21] El Ghobashi[22] and Richardson.[33] Experimental evidence was scarce at the time these models were proposed; the few measured distributions that were present supported the assumption of Gaussian distribution along the centreline of an inert jet flow. At once this distribution was extended to situations of reacting flows of the physically controlled type (Model 3), where the transport equation for the passive scalar f was similar to that in an inert flow. The beta function proposed by Richardson[33] was also supported by some experimental evidance, Batt.[76] Statistical analysis of turbulent reacting flow in idealized stirred combustors (Krambeck *et al.*),[77] indicated that the probability density for first order reactions is a beta function.

Recent experimental evidence, obtained in reacting flows (Odidi,[78] Kennedy *et al.*[79] and Khalil[80]) showed that the probability distribution is near Gaussian along the centreline. The radial variation of $P(\Phi)$ for a jet diffusion flame at $X/D_j = 40$ for the configuration of Kennedy and Kent[79] indicated an intermittent nature, and the intermittency spike grows. A sample of the probability distribution, obtained by Kennedy *et al.*[79] is shown in Figure 4.39.

Lewis and Moss[39] reported that the temperature probability density functions (Pdfs) are characteristically bi-modal in a confined turbulent premixed flame; they reported that the burnt gas and unburnt mixture temperatures were well defined on the centreline by a comparatively narrow Gaussian distribution. The measurements of Moneib,[81] in a round free flame suggested that the probability distribution of the scalar T is a non-Gaussian function.

Model Assumptions and Formulation

In view of the different observed probability distributions in reacting flames, the shape of the function $P(\Phi)$ must be dependent on the flame characteristics, turbulence characteristics and flow conditions.[80] An equation for the transport of the joint probability distribution was suggested by Pope.[4] A single probability distribution equation can be readily obtained by reducing the number of scalars to one. Consequently, the transport equation proposed by Pope[4] for a single probability distribution is expressed as

$$\bar{\rho}\frac{D}{Dt}\overline{P(\Phi)} = -\frac{\partial}{\partial x_i}\bar{\rho}\overline{P(\Phi)u_i} - \frac{\partial}{\partial \Phi}\left[\overline{P(\Phi)}\,S(\Phi) + \frac{\partial}{\partial \Phi}\bar{\rho}\overline{P(\Phi)\Gamma\left(\frac{\partial \Phi}{\partial x_i}\right)^2}\right] \quad (4.2.55)$$

The term $S(\Phi)$ denotes the instantaneous sink/source term of the entity Φ. The various terms of the equation are modelled; the probability-velocity correlation is modelled in a similar manner to that of $\overline{\phi u_i}$ as

$$\overline{P(\Phi)u_i} = -C_3\frac{K^2}{\varepsilon}\frac{\partial}{\partial x_i}\overline{P(\phi)} \quad (4.2.56)$$

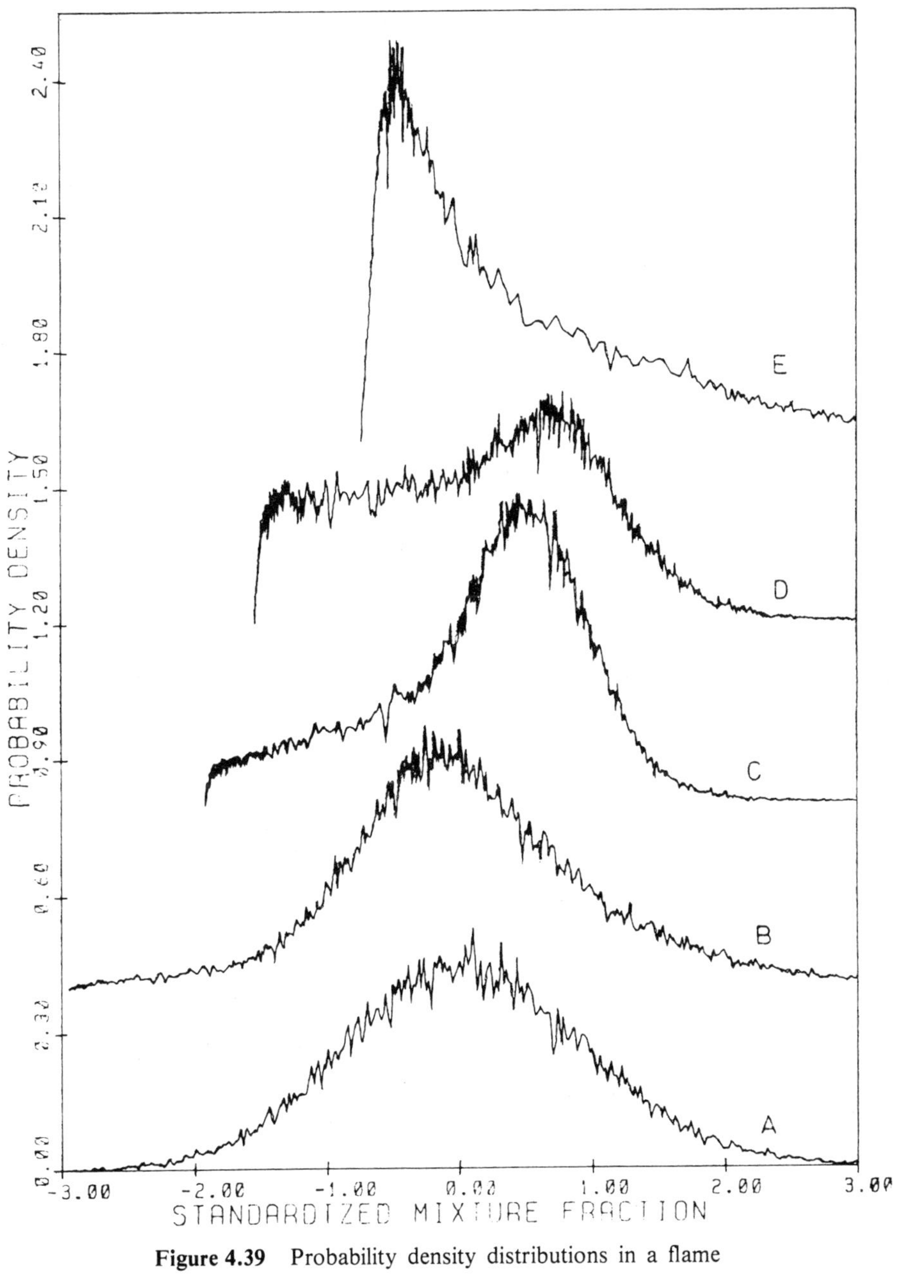

Figure 4.39 Probability density distributions in a flame

This term represents the turbulent diffusion of $P(\Phi)$. The molecular action term $G(\Phi)$ is modelled[4] as

$$G(\Phi) = \overline{P(\Phi)\Gamma\left(\frac{\partial\Phi}{\partial x_i}\right)^2},$$

where $\Gamma(\partial\Phi/\partial x_i)^2$ refers to small fluctuations on the micro-scale, while $P(\Phi)$ is dominated by large fluctuations associated with the macro-scale. Hence, it might be safely concluded that the two quantities are uncorrelated and $G(\Phi)$ can be expressed as

$$G(\Phi) = \overline{P(\Phi)}\,\overline{\Gamma\left(\frac{\partial\Phi}{\partial x_i}\right)^2} \tag{4.2.57}$$

And the term can be readily modelled as

$$G(\Phi) = \overline{P(\Phi)}\, C_4 \frac{\varepsilon}{k}\overline{\phi^2} \tag{4.2.58}$$

The modelled form of Equation 4.2.55 then becomes

$$\bar{\rho}\frac{D}{Dt}\overline{P(\Phi)} = \frac{\partial}{\partial x_i} C_3\bar{\rho}\frac{k^2}{\varepsilon}\frac{\partial}{\partial x_i}\overline{P(\Phi)} - \frac{\partial}{\partial\Phi}\left[\overline{P(\Phi)S(\Phi)} + C_4\bar{\rho}\frac{\varepsilon}{k}\overline{\phi^2}\frac{\partial}{\partial\Phi}\overline{P(\Phi)}\right] \tag{4.2.59}$$

The above equation can be reduced to inert constant density flows by putting $S(\Phi)$ to zero and $\bar{\rho}$ to one. The corresponding conservation equations for $\bar{\Phi}$ and $\overline{\phi^2}$ can be obtained[4] by integrating Equation 4.2.59, and hence producing:

$$\frac{D}{Dt}\bar{\Phi} = \frac{\partial}{\partial x_i} C_3 \frac{k^2}{\varepsilon}\frac{\partial\bar{\Phi}}{\partial x_i} \tag{4.2.60}$$

and

$$\frac{D}{Dt}\overline{\phi^2} = \frac{\partial}{\partial x_i} C_3 \frac{k^2}{\varepsilon}\frac{\partial\overline{\phi^2}}{\partial x_i} + 2C_3\frac{k^2}{\varepsilon}\left(\frac{\partial\bar{\Phi}}{\partial x_i}\right)^2 - 2C_4\frac{\varepsilon}{k}\overline{\phi^2} \tag{4.2.61}$$

Equation 4.2.60 is identical to 4.2.10 for $\bar{f}$ and Equation 4.2.61 is similar to Equation 4.2.25. The values of the constants C_3 and C_4 were taken[4] as 0.127 and 0.925 respectively. These values are consistent with the results of Naguib.[21] The decay of the probability distribution in a stagnant, homogeneous flow is represented as

$$\frac{\partial}{\partial t}\overline{P(\Phi)} = -C_4\frac{\varepsilon}{k}\overline{\phi^2}\frac{\partial^2\overline{P(\Phi)}}{\partial\Phi^2} \tag{4.2.62}$$

Note that there are no turbulent diffusion terms and no chemical kinetic source terms. The solution of Equation 4.2.62 for steady state is of the form:

$$\overline{P(\Phi, x)} = \frac{1}{\sigma\sqrt{2\pi}}\exp\left(-\frac{1}{2}\left(\frac{\Phi-\bar{\Phi}}{\sigma}\right)^2\right) \tag{4.2.63}$$

and

$$\sigma = \left[\frac{C_4}{C_3}\left(\frac{\partial \Phi}{\partial x_i}\right)^2\right]^{\frac{1}{2}} \frac{k^{3/2}}{\varepsilon} \tag{4.2.63}$$

The probability distribution is Gaussian with the standard deviation proportional to the gradient of $\bar{\Phi}$ multiplied by the turbulent length scale. These solutions of the modelled probability distribution equation are in accord with intuition and thus support the modelling assumptions.

An equation similar to that of Pope[4] for the transport of $P(\Phi)$ was discussed by Borghi[82] for constant density flows, except that the equation was in the Favre-averaged form. Borghi modelled the various terms in the probability density transport equation differently from Pope.[4] For constant density flow he expressed the probability equation as

$$\frac{\partial}{\partial t}\overline{P(\Phi)} = -\frac{\partial}{\partial x_i}\overline{\mu\delta(Y-\Phi)} - \frac{\partial}{\partial \Phi}[\overline{P(\Phi)}S(\Phi)]$$
$$-\frac{\partial}{\partial \Phi}\overline{\left[\delta(Y-\Phi)\frac{\partial}{\partial x_i}\Gamma\frac{\partial Y}{\partial x_i}\right]} \tag{4.2.64}$$

where Φ is a random variable, and Y is a physical variable. The term $\delta(Y-\Phi)$ is the Dirac function,[82] non-zero only when Φ equals Y.

The work of Bonniot *et al.*[83] was concerned with the development of a probability equation model and particularly with the joint probability distribution. The modelled equation was in the form:

$$\frac{D}{Dt}\tilde{P}(\alpha,\beta,\gamma,t) = \frac{6D}{\lambda_\alpha^2}\frac{\partial}{\partial\alpha}(\alpha-\tilde{\alpha})\tilde{P}(\alpha,\beta,\gamma,t) + \frac{6D}{\lambda_\beta^2}\frac{\partial}{\partial\beta}(\beta-\tilde{\beta})\tilde{P}(\alpha,\beta,\gamma,t)$$
$$+\frac{6D}{\lambda_\gamma^2}\frac{\partial}{\partial\gamma}(\gamma-\tilde{\gamma})\tilde{P}(\alpha,\beta,\gamma,t) - \frac{\partial}{\partial\alpha}[\tilde{P}(\alpha,\beta,\gamma,t)\,S(\alpha,\beta,\gamma)] \tag{4.2.65}$$

where λ is a characteristic micro-scale.

Assessment of Model

The present approach solves a transport equation for the probability density function. The time-averaged reaction rate is, consequently, obtained from the integrated product of the instantaneous rate and the local probability, in a manner similar to that of the previous models.

This model was applied to predict furnace flows in very few flow configurations, Pope[4] and Khalil.[84,85] The situation considered by Pope[4] was that of flow behind flame holder. A premixed air–propane mixture was made to flow over an annular V-gutter mounted centrally in a circular duct,[86] whose diameter was 0.203 m, and length was 1.05 m. Figure 4.40 shows the measured and the calculated axial velocity profiles for fuel to air ratio of 0.05. The agreement was reasonable and the discrepancies observed

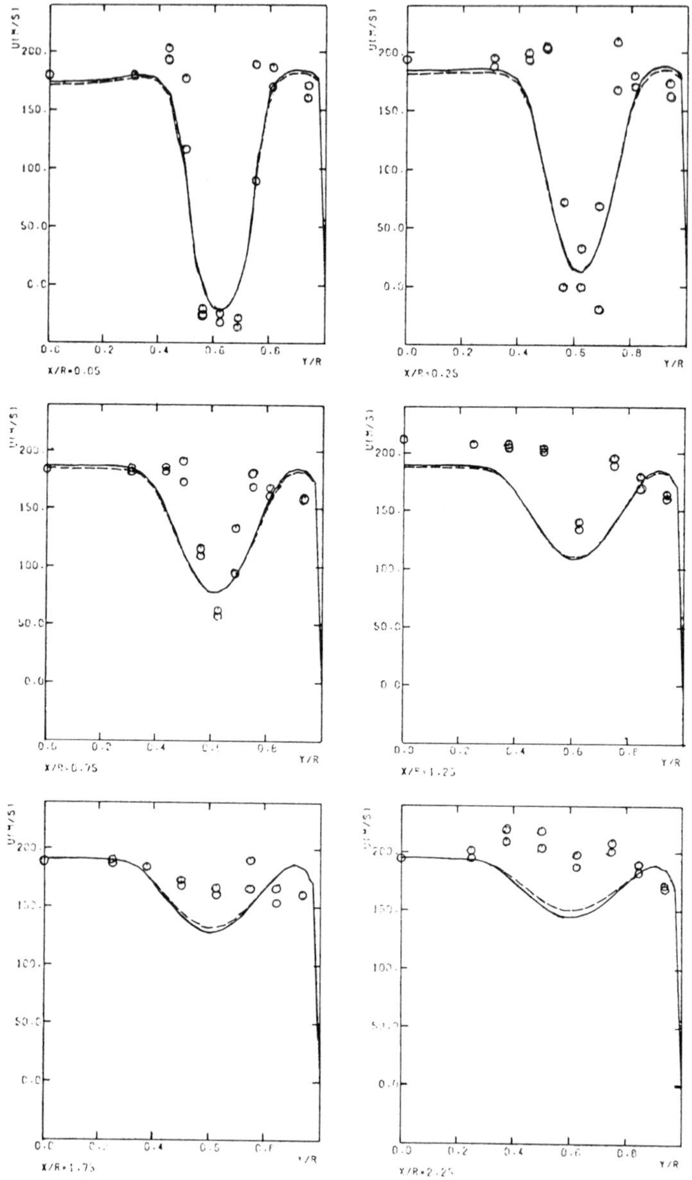

Figure 4.40 Measured and calculated profiles of mean axial velocity
O Experiments of reference 86 —— Predictions of reference 4

were attributed to the scatter in the data, and again to the turbulence model assumptions.

The measured and the calculated local gas temperature distributions are shown in Figure 4.41 for the same geometry and under the same operating conditions. Discrepancies were observed in regions of steep gradient. Pope[4] attributed these observed discrepancies to serious errors in the gas analysis, and as a consequence the temperatures, which were determined from the equation of state, were not correct.

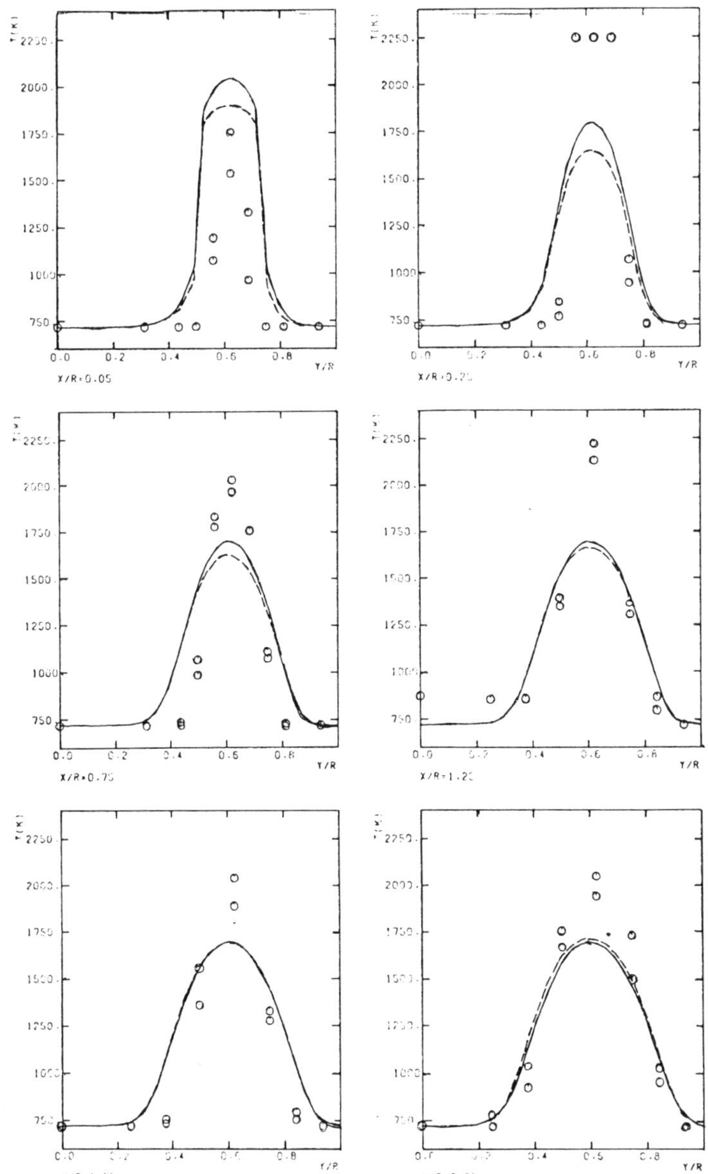

Figure 4.41 Measured and calculated profiles of temperature
O Experiments of reference 86 —— Predictions of reference 4

The probability approach of Pope[4] was applied[84,85] to a turbulent premixed flame of propane–air stoichiometric mixture in a rectangular duct of Cushing *et al.*[87] and Howe *et al.*[88] The flame was stabilized on a cylindrical flame stabilizer. Comparisons between the measured and the calculated mean axial velocity of reference 84 are shown in Figure 4.42. Also on the figure are shown the eddy break-up model results of Mason *et al.*[53] It is clear from the figure that the model predicts correctly the flow velocity field.

Measured and calculated local profiles of mean oxygen concentrations are shown in

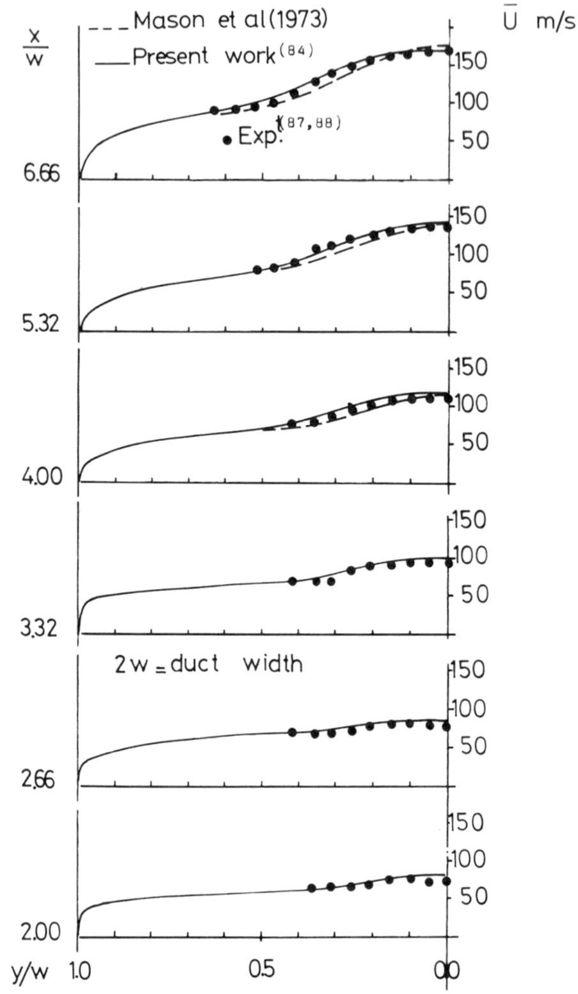

Figure 4.42 Measured and calculated axial velocity profiles

Figure 4.43, and are in satisfactory agreement. The calculations of Mason *et al.*[53] are also shown in the same figure; these calculations indicated discrepancies of the order of 20% away from the duct inlet section. The calculated rates of oxygen consumption were also obtained and are compared with the measured values in Figure 4.44. The rates were in good agreement; the maximum discrepancy was less than 15% and suggested that the model of Pope,[4] represented by Equation 4.2.55, predicts correctly the flame properties.

The present probability approach yields reasonable agreement in simple flame situations. Nevertheless, further modelling refinement is still needed. The joint probability distribution should be incorporated in the modelling of reactive flows but one must be cautious of making additional assumptions that can be a source of increasing uncertainty in the modelled equations.

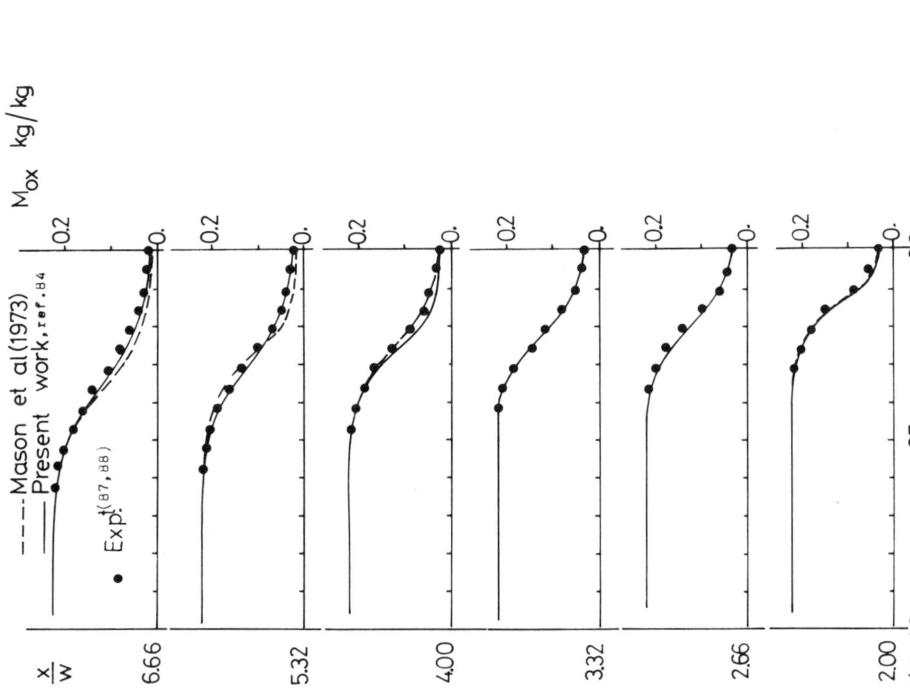

Figure 4.43 Measured and calculated oxygen concentrations

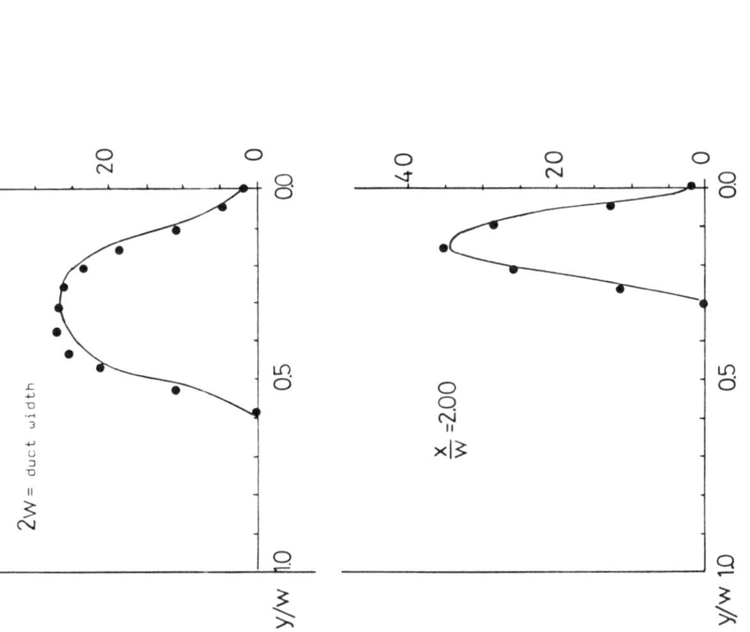

Figure 4.44 Measured and calculated oxygen disappearance rate for the furnace of references 87 and 88

4.2.5 Effect of Density Fluctuations

The chemical and thermodynamic properties of the fluid affect the velocity field through the density and viscosity which appear in the continuity and momentum equations. At high Reynolds numbers, the flow pattern is independent of the viscosity and, consequently, attention is focused on the influence of the density variations. The change in the density may be affected by the gradients of the temperature and species concentration or by the compressibility of the fluid. In the reacting flows, all these three agents may act simultaneously, thus causing the density to vary by as much as an order of magnitude.

In addition to the correlation terms arising in constant density flows, five additional density fluctuation correlation terms result from the decomposition of the convective terms in the conservation equations. These terms are $\overline{\rho' u_j}$, $\overline{\rho' u_i u_j}$, $\overline{\rho' u_i u_j u_k}$, $\overline{\rho' \phi}$ and $\overline{\rho' u_j}$. In reacting flows, the density changes its value by an order of magnitude, and hence none of these terms can be neglected. Bilger[12] investigated the influence of these terms specially in the vicinity of reaction zones.

Khalil[5,70] investigated the influence of the terms $\overline{\rho' u_j}$ and $\overline{\rho' \phi}$ on the reaction zone and flow pattern for a hydrogen–air diffusion flame.[28] This particular flame configuration was selected because of the strong density variation encountered; the density of hydrogen is very much lower than that of air. Figure 4.45 illustrates the effects of including the various terms which appear in the conservation equation; these terms are of the form $\overline{\rho' \phi}$ and $\overline{\rho' u_j}$, the triple and higher-order correlations being neglected. When the conservation equations are expressed in terms of Favre-averaged properties, i.e. $\tilde{\Phi}$ instead of time averaged $\bar{\Phi}$, these correlation terms disappear, but in order to compare the calculations with experiments the correlations terms should be calculated.

Additional conservation equations for $\overline{\rho' u_j}$ and $\overline{\rho' \phi}$ were solved,[70] together with the same equations for mass, momentum, energy and species conservation as in Models 6 and 7. In Figure 4.45, curve 1 shows the radial profiles of mean gas temperature when the terms $\overline{\rho' u_j}$ and $\overline{\rho' \phi}$ are taken into account, while curve 3 shows the same distributions when these correlations were neglected. Calculations obtained with algebraic expressions of the terms $\overline{\rho' \phi}$ and $\overline{\rho' u_j}$ were also reported[70] and are shown in Figure 4.45 by

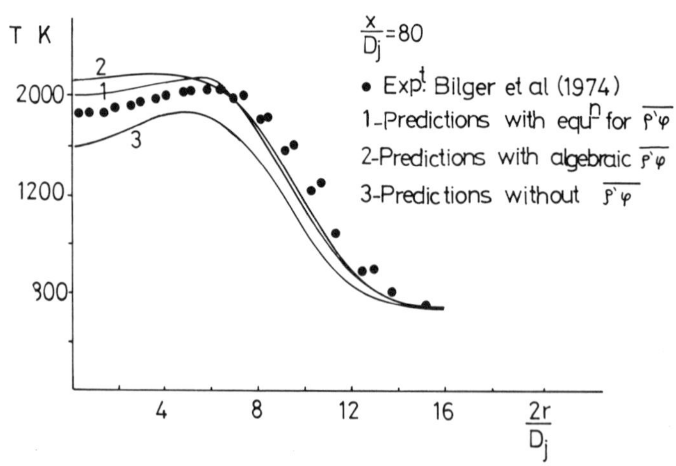

Figure 4.45 Effect of density–velocity terms on the local temperature profiles[70]

curve 2. The effect of these terms on the pollutant concentrations is shown in Figure 4.46 for the same flame. It is clear that the neglect of these terms results in under-predicted NO_x concentrations, thus illustrating the significant role of the density fluctuation correlations.

The entities $\overline{\rho' u_j}$, $\overline{\rho' \phi}$ were calculated from the differential equations for their transport as described by Hutchinson *et al.*[7,8] but can also be obtained from an approximate form given by Khalil[5] as:

$$\overline{\rho' u_j} = -\frac{\mu_{\text{eff}}}{\sigma_\rho}\frac{\partial\bar{\rho}}{\partial x_j}$$

and

$$\overline{\rho' \phi} = -\frac{\mu_{\text{eff}}}{\sigma_\rho}\frac{\partial\bar{\Phi}}{\partial x_j} \tag{4.2.66}$$

In Model 6 the assumption of Equation 4.2.66 was used and the correlations were algebraically calculated in terms of density gradients. In Model 7, where the local values of $\overline{m_a m_b}$, $\overline{T' m_a}$, $\overline{T' m_b}$ and $\overline{T'^2}$ were calculated from the corresponding conservation equations, $\overline{\rho' \phi}$ and $\overline{\rho' u_j}$ were obtained[8,70] from the following equations:

$$\frac{\partial}{\partial x_j}\bar{\rho}\bar{U}_j\,\overline{\rho u} = \frac{\partial}{\partial x_j}\frac{\mu_{\text{eff}}}{\sigma_\rho}\frac{\partial(\overline{\rho' u})}{\partial x_j} + \bar{\rho}\left[\overline{-u^2}\frac{\partial\bar{\rho}}{\partial x} - \overline{uv}\frac{\partial\bar{\rho}}{\partial r} -0.4\,\overline{\rho' u}\left(\frac{\partial\bar{U}}{\partial x}\right) - 0.2\,\overline{\rho' v}\left(\frac{\partial\bar{U}}{\partial r} + \frac{\partial\bar{V}}{\partial x}\right) - 2.5\frac{\varepsilon}{k}\overline{\rho' v}\right] \tag{4.2.67a}$$

$$\frac{\partial}{\partial x_j}\bar{\rho}\bar{U}_j\overline{\rho' v} = \frac{\partial}{\partial x_j}\frac{\mu_{\text{eff}}}{\sigma_\rho}\frac{\partial}{\partial x_j}(\overline{\rho' v}) + \bar{\rho}\left[\overline{-uv}\frac{\partial\bar{\rho}}{\partial x} - \overline{vv}\frac{\partial\bar{\rho}}{\partial r} - 0.4\,\overline{\rho' v}\left(\frac{\partial\bar{V}}{\partial r}\right) - 0.2\,\overline{\rho' u}\left(\frac{\partial\bar{V}}{\partial x} + \frac{\partial\bar{U}}{\partial r}\right) - 2.5\frac{\varepsilon}{k}\overline{\rho' v}\right] \tag{4.2.67b}$$

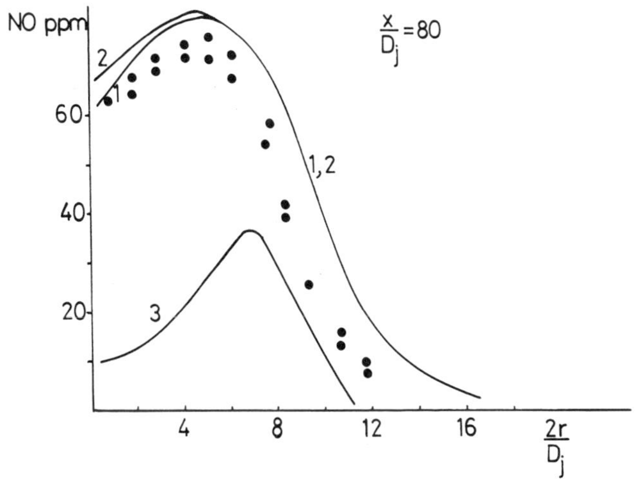

Figure 4.46 Effect of density–velocity terms on NO concentrations[70]

and

$$\frac{\partial}{\partial x_j}\bar{\rho}\bar{U}_j\overline{\rho'\phi} = \frac{\partial}{\partial x_j}\frac{\mu_{\text{eff}}}{\sigma_\rho}\frac{\partial}{\partial x_j}(\overline{\rho'\phi})$$

$$+\bar{\rho}\left[-\overline{u\phi}\frac{\partial\bar{\rho}}{\partial x} - \overline{v\phi}\frac{\partial\bar{\rho}}{\partial r} - \left(\frac{\partial\bar{U}}{\partial x} + \frac{1}{r}\frac{\partial r\bar{V}}{\partial r}\right)\cdot\overline{\rho'\phi} - \overline{\rho' u}\frac{\partial\bar{\Phi}}{\partial x} - \frac{\overline{\rho' v}}{r}\frac{\partial r\bar{\Phi}}{\partial r} + \overline{\rho' S'_\Phi}\right] \quad (4.2.68)$$

where S'_Φ is the fluctuation of the reaction rate. The value of σ_ρ was 0.9. The effects of the entities $\overline{\rho' u_j}$ were also investigated by Kolbe *et al.*[89] for hydrogen diffusion flames and suggested the incorporation of these terms in the conservation equations to obtain physically sound computations. A sample of the obtained comparisons[89] is shown in Figure 4.47 for the flame of Bilger *et al.*[28,29] It is clear that reasonable agreement was achieved.

The influence of density–velocity and density–scalar fluctuation correlations on the flow pattern and flame structure was found to be significant in the vicinity of the reaction zone, and particularly in predicting the NO_x concentrations.

4.3 SUMMARY OF EXISTING MODELS

The previous subsections have been concerned with the modelling of time-averaged conservation equations of scalar entities in turbulent reacting flows. Various modelling assumptions were described; they can mainly be classified according to their appropriateness to diffusion, premixed and arbitrary fired flames. In turbulent diffusion flames, the assumption of infinitely fast chemical reaction was incorporated in the computational scheme. This oversimplified assumption enabled the calculations of the local flame properties through the solution of one conservation equation for the mixture fraction $\bar{f}$. The effect of concentration fluctuation was also considered in conjunction with the mixture fraction and an additional equation for the square of the concentration fluctuation was solved. The time-averaged values and variance of any scalar entity Φ were obtained from $\bar{f}$ and g and the temporal variation of f. Two temporal distributions were considered, namely a square waveform and random distributions. Models incorporating concentration fluctuations were reported to yield better results than Model 1, which ignored those fluctuations.

The modelling assumptions of diffusion flames were found to represent adequately the flow properties and flame contour for engineering design purposes. These models are straightforward and are easy to apply. They lack, however, information regarding flame propagation, flame stretch and blow-out. As this type of flame is diffusion controlled, detailed chemical kinetic information is insignificant. Hence there is no need to pay much heed to these effects in computing the flow fields and main species concentrations, but they will certainly have to be considered for the pollutant formation and burnout computations.

Various other models appropriate to diffusion flames were proposed, but only those which have been applied to furnace flows were considered in subsection 4.2.1. Examples

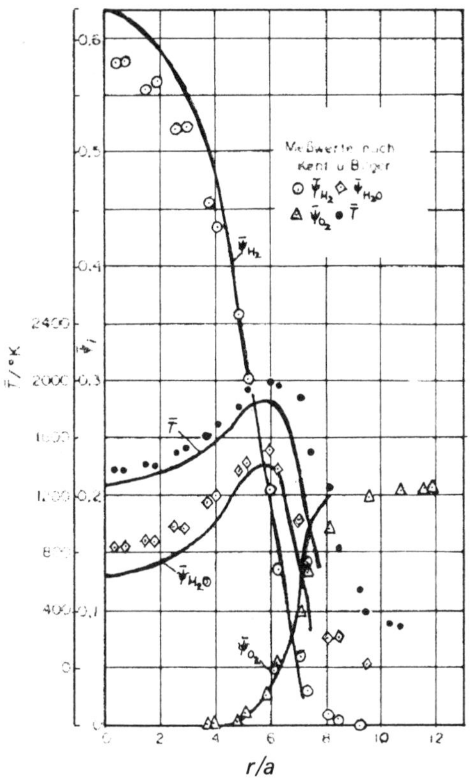

Radial profiles of mean temperature $\overline{T}$ and mean mole fractions $\overline{\psi}_{H_2}$, $\overline{\psi}_{O_2}$, $\overline{\psi}_{H_2O}$ at $X/D = 40$ for $u_J/u_c = 10$.

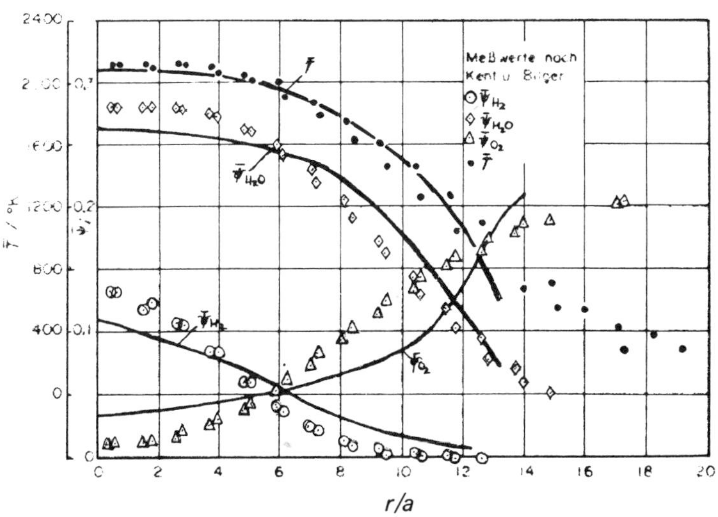

Radial profiles of mean temperature $\overline{T}$ and mean mole fractions $\overline{\psi}_{H_2}$, $\overline{\psi}_{O_2}$, $\overline{\psi}_{H_2O}$ at $X/D = 120$ for $u_J/u_c = 10$.

Figure 4.47 Effect of density–velocity correlations on the flame characteristics[89]

of such models are those of Bilger,[23] Naguib,[21] Bray,[51] Borghi *et al.*[90] and Spalding.[91]

For premixed flames, the above assumptions are not appropriate and finite reaction rate models should be included as a source/sink in the fuel mass fraction transport equations. Models for such flames can be classified, after Spalding,[92] into:

1. *Kinetics-Only Models.* These models neglect the hydrodynamics and molecular transport and turbulence effects. The time-averaged reaction rate is the same function of the time-averaged concentrations and temperature as is the instantaneous rate of the instantaneous concentrations and temperature. Certainly the K–O procedures cannot be applied to real furnace flames and were, therefore, not considered in previous subsections.

2. *Hydrodynamics-Only Models.* These models are applied to furnace and combustor design; they presume that the rates of reaction are influenced only by hydrodynamic properties of the flow, and by the variations of the time average and the fluctuating concentrations which these, in turn, bring into existance. Chemical kinetic data such as the pre-exponential factor and activation energies make no appearance in the formulae used; nor do the diffusion coefficients or other molecular transport properties appear in the reaction rate expression. Such models are the eddy break-up model, and Models 4 and 5. Models of that nature were successfully applied to ducted flames[53] and axisymmetric furnaces.[2,3,4]

3. *Hydrodynamics-Plus-Kinetics Models.* The calculations of reaction rates and the local concentrations in turbulent reacting flows, where the flame is neither diffusion controlled nor chemicaly influenced, can be obtained through Models 6 and 7, which account for the temperature and concentration fluctuations and for their correlations. Models that yield a reaction rate which depends on the chemical–kinetic parameters (such as activation energy and pre-exponential factor) and on the hydrodynamic parameters, but take no explicit account of molecular transport terms, are termed H–K models. The investigations of Donaldson *et al.*,[62,63] Borghi[48,49,50] and Khalil,[5,10,14,15] describe models of that type. Such models are appropriate for flames whose Damkohler numbers are of the order of unity, i.e. where the flame is neither diffusion controlled nor chemically influenced.

These models incorporate presumed probability density functions. They have been criticized for embodying such assumptions. One must necessarily then investigate the fact that the Pdfs do indeed have the assumed shapes; here, more check experimental data are vitally needed. To overcome such shortcomings, an equation for the transport of the probability density function was solved.

4. *Hydrodynamic + Transport + Kinetics Models.* Spalding[92] criticized existing models for neglecting the molecular transport processes and the reactedness variation with reaction rate and for assuming the Pdfs. He initiated research in that line by a series of papers, e.g. references 93, 94 and 95. In these investigations he suggested that a theory on turbulent combustion should combine the Eulerian and Lagrangean

Figure 4.48 The engulfment process in flames

approaches, and named his theory ESCIMO. It is worth while to mention the outline of this theory, which is:

(a) *Engulfment.* It is the process which occurs at the edge of jet or flame. Tongues of fluid from the jet are thrust into free surrounding stream. They then wrap themselves around the free stream fluid element and draw it into the interior of the jet. The process is illustrated in Figure 4.48. The investigation of Roshko[96] has clearly demonstrated this process.

(b) *Stretching.* It is the process of distortion of the interleaved elements of jet fluid and free stream fluid, which causes the thickness of the leaves (also called sandwiches, parcels, layers or blocks) to be continuously reduced. The thickness of the leaves plays a significant role; it diminishes steadily as a consequence of both the shear in the mean motion, and the turbulent fluctuations.

(c) *Coherence.* It appears as a reminder that the parcels of jet and free stream fluids, once interspersed, can never be separated. This concept makes it meaningful to consider a small part of the sandwich consisting of half a layer of jet fluid and half a layer of free stream fluid.[94]

(d) *Interdiffusion.* This term describes the molecular-scale processes of interpenetration of the two cohering fluids, and of their consequent chemical reaction. This process obeys the laminar flow laws.

(e) *Moving Observer.* This term describes the Lagrangean aspects of the ESCIMO theory. Blocks of fluid are considered in one-dimensional transient mode and the information about the age of each block, its birth rate and its death are provided.

This is only a brief note about the ESCIMO theory and the reader is best advised to refer to the original papers for detail. This theory is still in an early stage of development and has not been applied to furnace flow computations to date.

A summary of the models cited in the previous sections and their application in furnace flows is tabulated in Table 4.3; Table 4.4 gives the values of the various constants of the models.

TABLE 4.3

Flame type	Basic assumptions	Model	Equations solved	Remarks
diffusion	infinitely fast chemical reaction	1	$U, V, W, k, \epsilon, P, f, H$	
	infinitely fast chemical reaction with concentration fluctuations	2	$U, V, W, k, \epsilon, P, f, g, H$	square wave species distribution with time
		3	$U, V, W, k, \epsilon, P, f, g, H$	random distribution with time
premixed	finite reaction rate — single step	4	$U, V, W, k, \epsilon, P, \bar{M}_{fu}, g_{fu}, \overline{m_{fu} m_{ox}}, \overline{m_{ox}^2}, H, f$	
	finite reaction rate — multi-step (eddy break-up formulation)	5	$U, V, W, k, \epsilon, P, \bar{M}_a, H, f$	premixed flames; g_a is obtained from algebraic expressions
arbitrary	finite reaction rate with concentration fluctuations	6	$U, V, W, k, \epsilon, P, \bar{M}_{fu}, \overline{m_{fu}^2}, \overline{m_{fu} m_{ox}}, \overline{m_{ox}^2}, H, f$	higher-order correlations were obtained from algebraic expressions
	finite reaction rate with concentration and temperature fluctuations	7	$U, V, W, k, \epsilon, P, \bar{M}_{fu}, \overline{m_{fu}^2}, \overline{m_{ox} m_{fu}}, \overline{m_{fu} T'}, \overline{m_{ox} T'}, \overline{\rho' \phi}, \overline{T'^2}, H, f$	

TABLE 4.4

Combustion Model Constants

Constant	Value
C_{g_1}	2.8
C_{g_2}	2.0
C_R	1.0
C_3	0.127
C_4	0.925
σ_f	0.9
σ_g	0.9
σ_ρ	0.9

References to Chapter 4

1. Spalding, D. B. (1971) Mixing and Chemical reaction in steady confined turbulent flame. *Proc. 13th Symp. (Int.) on Combustion*, p. 649.
2. Khalil, E. E., Spalding, D. B. and Whitelaw, J. H. (1975) The calculation of local flow properties in two dimensional furnaces. *Int. J. Heat and Mass Transfer*, **18**, p. 775.
3. Hutchinson, P., Khalil, E. E., Whitelaw, J. H. and Wigley, G. (1976) The calculation of furnace flow properties and their experimental verification. *J. Heat Transfer*, **98**, p. 276.
4. Pope, S. B. (1976) The calculation of the flows behind bluff bodies with and without combustion. Ph.D. Thesis, London University.
5. Khalil, E. E. (1977) Flow and combustion in axisymmetric furnaces, Ph.D. Thesis, London University.
6. Khalil, E. E. and Whitelaw, J. H. (1976) Aerodynamic and thermodynamic characteristics of kerosene spray flames. *Proc. 16th Symp. (Int) on Combustion*, p. 569.
7. Hutchinson, P., Khalil, E. E. and Whitelaw, J. H. (1976) The calculation of wall heat transfer rate and pollution formation in axisymmetric furnaces. *Proc. 4th Members Conference of the IFRF.*
8. Hutchinson, P., Khalil, E. E. and Whitelaw, J. H. (1977) Measurement and calculation of furnace flow properties. *J. Energy*, **1**, p. 212.
9. Khalil, E. E. and Whitelaw, J. H. (1979) Calculation of turbulent, combusting flows. *Acta Astronautica*, **6**, p. 1011.
10. Khalil, E. E. (1979) On the modelling of turbulent reacting flows in furnaces and combustion chambers. *Acta Astronautica*, **6**, p. 449.
11. Favre, A. (1969) Statistical equations of turbulent gases in *Problems of Hydraulics and Continuum Mechanics*. (Society of Industrial and Applied Mechanics, Philadelphia), p. 231.
12. Bilger, R. (1975) A note on Favre averaging in variable density flows. *Combustion Science and Technology*, **11**, p. 215.
13. Bray, K. N. C. (1974) Kinetic energy of turbulence in flames, *Proc. AGARD Panel Meeting, April 1974*: See also: *AASU Report*, 332, 1974.
14. Khalil, E. E. (1978) Aerodynamic and heat transfer characteristics of axisymmetric confined gaseous flames. *Proc. 6th Int. Heat Transfer Conference*, p. 25.
15. Khalil, E. E. (1981) Numerical computations of turbulent reacting combustor flows in *Numerical Methods in Heat Transfer*, edited by Lewis *et al.* (Pineridge Press), p. 548.
16. Kondratiev, V. N. (1964) Chemical Kinetics of Gas Reactions (Pergamon Press).
17. Arrhenius, S. (1889) Über die Reaktions geschwindigkeit bei der Inversion von rohrzucken durch Sauren. *Z. Phys. Chem.* **4**, p. 226.

18. Baulch, D. L., Drysdale, D. D. and Horne, D. G. (1968–1970) High temperature reaction rate data. University of Leeds, Reports 1–5.
19. Waldman, C. H., Wilson, R. P. and Engleman, V. S. (1974) Analyses of the kinetic mechanism of methane–air combustion with pollutant formation. EPA-68-02-0270-WSS/CI 74–2.
20. Burke, S. and Schumann, T. (1928) *I & E.C.*, **20**, p. 998.
21. Naguib, A. S. (1975) The prediction of axisymmetric free jet, turbulent reacting flows. Ph.D. Thesis, London University.
22. El Ghobashi, S. E. (1974) Characteristics of gaseous turbulent diffusion flames in cylindrical chambers, a theoretical and experimental investigation. Ph.D. Thesis, London University.
23. Bilger, R. (1975) Turbulent jet diffusion flames, *Progress in Energy and Combustion Science*, **1**, p. 87.
24. Shvab, V. A. (1948) Relation between the temperature and velocity fields of the flame of a gas burner, Gos. Engerg. izd. Moscow, Leningrad.
25. Zeldovich, Y. B. (1949) On the theory of combustion of initially unmixed gases. *Zhur. Tekhn. Fiz., 19*. See also: *NACA Tech. Mem. 1296*, 1950.
26. Williams, F. A. (1965) Combustion Theory (Addison-Wesley, Reading, Mass.).
27. Spalding, D. B. (1971) Concentration fluctuation in a round free jet. *Chem. Engineering Sci.*, **26**, p. 95.
28. Bilger, R. and Kent, J. H. (1972) Measurements in turbulent diffusion flames. Sydney University, Dept. of Mechanical Engineering, Report F41.
29. Bilger, R. and Kent, J. H. (1972) Concentration fluctuation in turbulent jet diffusion flame. Sydney University, Dept. of Mechanical Engineering, Report F. 36.
30. Pun, W. M. and Spalding, D. B. (1967) A procedure for predicting the velocity and temperature distribution in confined, steady, turbulent, gaseous, diffusion flames. *Proc. 18th International Astronautical Congress. Belgrade.*
31. Gosman, A. D., Pun, W. M., Runchal, A. K., Spalding, D. B. and Wolfstein, M. (1969) *Heat and Mass Transfer in Recirculating Flows* (Academic Press).
32. Pai, B. R. and Lowes, T. M. (1972) The predictions of flow, mixing and heat transfer in the Ijmuiden furnace. IFRF Doc. G 02/2/22.
33. Hawthorne, W. R., Weddell, D. S. and Hottel, H. C. (1949) Mixing and combustion in turbulent gas jets. *Proc. 3rd Symp. on Combustion, Flame and Explosion Phenomena*, p. 266.
34. Thring, M. W. and Newby, M. P. (1953) Combustion length of enclosed turbulent jet flames. *Proc. 4th Symp. (Int.) on Combustion*, p. 789.
35. Richardson, J. N., Howard, Jr. H. C. and Smith, Jr. R. W. (1953) The relation between sampling-tube measurements and concentration fluctuations in a turbulent gas jet. *Proc. 4th Symp. (Int.) on Combustion*, p. 814.
36. Rhodes, R. P. and Harsha, D. T. (1972) On putting the turbulent in turbulent reacting flow. AIAA Paper 72–68.
37. El Ghobashi, S. E. and Naguib, A. S. (1975) Private communication.
38. Moss, J. B. (1977) Private communication.
39. Lewis, K. J. and Moss, J. B. (1978) Time-resolved scalar measurements in a confined turbulent premixed flame. *Proc. 17th Symp. (Int.) on Combustion*, p. 267.
40. Pope, S. B. (1976) The probability approach to the modelling of turbulent reacting flows. *Combustion and Flame*, **27**, p. 299.
41. El Ghobashi, S. E. and Pun, W. M. (1974) A theoretical and experimental study of turbulent diffusion flames in cylindrical furnaces. *Proc. 15th Symp. (Int.) on Combustion*, p. 1353.
42. Richter, W. (1978) Ph.D Dissertation, University of Stuttgart.
43. Bilger, R. and Beck, R. E. (1974) Further experiments on turbulent jet diffusion flames. *Sydney University, Dept. of Mechanical Engineering*, Report F41.
44. Cernansky, N. P. and Sawyer, R. F. (1974) NO and NO_2 formation in a turbulent hydrocarbon/air diffusion flame. *Proc. 15th Symp. (Int.) on Combustion*, p. 1039.
45. Wu, H. L. and Fricker, N. (1971) An investigation of the behaviour of swirling jet flames in a narrow cylindrical furnace. *Proc. 2nd Members Conference* of the *IFRF*, See also: *J. Inst. Fuel, 1976*, **49**, p. 144.

46. Gunther, R. and Lenze, B. (1972) Exchange coefficients and mathematical models of jet diffusion flames. *Proc. 14th Symp. (Int.) on Combustion*, p. 687.
47. Spalding, D. B. (1967) Development of the eddy break-up model of turbulent combustion. *Proc. 16th Symp. (Int.) on Combustion*, p. 1657.
48. Borghi, R. (1973) Etude théorique de l'evolution residuelle de produits pollutants dans les jets de turboreacteurs AGARD Meeting CP 125.
49. Borghi, R. (1974) Computational studies of turbulent flow with chemical reaction. Project SQUID on Turbulent Mixing.
50. Borghi, R. and Moreau, P. (1977) Turbulent combustion in a premixed flow, *Acta Astronautica*, **4**, p. 321.
51. Bray, K. N. C. The interaction between turbulence and combustion. *Proc. 17th Symp. (Int.) on Combustion*, p. 223.
52. Monin A. S. and Yaglom, A. M. (1971) *Statistical Fluid Mechanics*, edited by J. Lumley. (MIT Press).
53. Mason, H. B. and Spalding, P. B. (1973) Prediction of reaction rates in turbulent premixed boundary layer flows. *European Symp. on Combustion*, p. 601.
54. Beltagui, S. and Maccallum, N. R. L. (1976) Aerodynamics of vane swirled flames in furnaces. *J. Inst. of Fuel*, **49**, p. 183.
55. Steward, F. R., Osuwan, S. and Picot, J. J. (1972) Heat transfer measurements in cylindrical test furnaces. *Proc. 14th Symp. (Int.) on Combustion*, p. 651.
56. Enikolopyan, N. S. (1958) Kinetics and mechanism of methane oxidation. *Proc. 7th Symp. (Int.) on Combustion*, p. 157.
57. Bowman, C. T. (1975) Kinetics of pollutant formation and destruction in combustion. *Progress in Energy and Combustion Science*, **1**, p. 33.
58. Leuchter, O. (1974) Analyse numérique de la phase d'inflammation dans une couche de mélange turbulente. AGARD Meeting, CP 164.
59. Kozlov, G. I. (1958) On high temperature oxidation of Methane. *Proc. 7th Symp. (Int.) on Combustion*, p. 142.
60. Michelfelder, S. and Lowes, T. M. (1974) Report on the M-2 trails. IFRF Doc. F36/2/4.
61. Lenze, B., Milano, M. E. and Gunther, R. (1974) The mutual influence of multiple jet diffusion flames. *Proc. 3rd Members conference of the IFRF.* See also: *Combustion Science and Technology*, 1975, **11**, p. 1.
62. Donaldson, C. dup, Sullivan, R. D. and Rosenbauum, M. (1972) A theoretical study of the generation of atmospheric clean air turbulence. *AIAA Journal*, **10**, p. 162.
63. Donaldson, C. dup, and Varma, A. K. (1976) Remarks on the construction of a second order closure description of turbulent reacting flows. *Combustion Science and Technology*, **13**, p. 55.
64. Bush, W. B., Feldman, P. S. and Fendell, F. E. (1975) On diffusion flames in turbulent shear flows. *Combustion Science and Technology*, **13**, p. 26.
65. Patterson, G. K. (1979) Closure approximation for complex multiple reactions. *Proc. Symp. on Turbulent shear Flows, London.*
66. Dopazo, C. and O'Brien, E. E. (1973) Isochoric turbulent mixing of two rapidly reacting chemical species with chemical heat release. *Physics of Fluids*, **16**, p. 2075.
67. Damkohler, G. Z. (1974) On the effect of turbulence on the flame velocity in gas mixtures. NACA, T.M.1112.
68. Spalding, D. B. (1976) The ESCIMO theory of turbulent combustion. Imperial College, Dept. of Mechanical Engineering, Report HTS/76/13.
69. Spalding, D. B. (1978) A simple model for the rate of turbulent combustion in *Turbulent Combustion*, edited by Kennedy (AIAA), p. 105.
70. Khalil, E. E. (1981) Flow and combustion characteristics of turbulent reacting flames. *Proc. Symp. on Fluid Mechanics of Combustion Systems*, p. 217.
71. Bowman, C. T. and Cohen, L. S. (1975) Influence of aerodynamic phenomena of pollutant formation in combustion, vol. 1 (experimental results). EPA-650/2-75-061-2.
72. Paauw, T. N. T. A. (1974) Some measurements and numerical calculations on turbulent diffusion flames. AGARD meeting, CP 164.

73. Lockwood, F. C., El Mahallawy, F. M. and Spalding, D. B. (1974) An experimental and theoretical investigation on turbulent mixing in cylindrical furnaces. *Combustion and Flame*, **23**, p. 283.
74. Khalil, E. E., Hutchinson, P. and Whitelaw, J. H. (1980) The calculation of the flow and heat transfer characteristics of gas fired furnaces. *Proc. 18th Symp. (Int.) on Combustion*, p. 1927.
75. Spalding, D. B. (1976) Mathematical models of turbulent flames, a review. *Combustion Science and Technology*, **13**, p. 3.
76. Batt, R. G. (1975) Some measurements on the effect of tripping the two dimensional shear layer. *AIAA Journal*, **13**, p. 245.
77. Krambeck, F. J., Katz, S. and Shinnar, R. (1972) The effects of perturbations in flow-rate on a stirred combustor. *Combustion Science and Technology*, **4**, p. 221.
78. Odidi, A. O. O. (1974) The influence of turbulence on the time mean rate of chemical reactions. Ph.D. Thesis, London University.
79. Kennedy, I. M. and Kent, J. H. (1978) Measurements of a conserved scalar in turbulent jet diffusion flames, *Proc. 17th Symp. (Int.) on Combustion*, p. 279.
80. Khalil, E. E. (1978) Measurements of mean and fluctuating temperature in a turbulent spray flame. *J. Engineering for Industry*, **100**, p. 210.
81. Moneib, H. (1980) Fluctuating temperature measurements in a heated round free jet, Ph.D. Thesis, London University.
82. Borghi, R. (1979) Models of turbulent combustion for numerical predictions. ONERA Report T.P. 1979–1.
83. Bonniot, C., and Borghi, R. (1977) Sur la densité de probabilité des fluctuations d'espaces réactives dans la combustion turbulente. ONERA Report T.P. 1977–117.
84. Khalil, E. E. (1980) On the prediction of reaction rates in turbulent premixed confined flames. AIAA Paper 80-0015 (18th Aerospace Sciences Meeting).
85. Khalil, E. E. (1980) Mathematical modelling of Combustion Processes in Combustors. *Proc. 1st Asian Congress of Fluid Mechanics.*
86. Harrison, A. J. (1973) Combustion tests on an 8 inch diameter reheat unit with a V-gutter stabilizer burning propane (premix) or kerosene fuels-velocity traverses in the gutter wake. Lucas Aerospace, Burnley Report B. 49, p. 113.
87. Cushing, B. S., Baucher, J. E., Gandbhir, S. and Shipman, C. W. (1967) Turbulent mass transfer and rates of combustion in confined turbulent flames. *Proc. 11th Symp. (Int.) on Combustion*, p. 817.
88. Howe, N. M., Shipman, C. W. and Vranos, A. (1963) Turbulent mass transfer and rates of combustion in confined turbulent flames, *Proc. 9th Symp. (Int.) on Combustion*, p. 36.
89. Kolbe, W. and Kollmann, W. (1980) Prediction of turbulent diffusion flames with a four equation turbulence model, *Acta Astronautica*, **7**, p. 91.
90. Borghi, R. and Dutoya, D. (1978) On the scales of the fluctuations in turbulent combustion. *Proc. 17th Symp. (Int.) on Combustion*, p. 235.
91. Spalding, D. B. (1977) A general theory of turbulent combustion. AIAA Paper 77-0098 (15th Aerospace Sciences Meeting).
92. Spalding, D. B. (1979) Theories of turbulent combustion. AIAA Conference, New Orleans.
93. Spalding, D. B. (1978) The influence of laminar transport and chemical kinetics on the time mean reaction rate in a turbulent flame, *Proc. 17th Symp. (Int.) on Combustion*, p. 431.
94. Ma, A. S. C., Noseir, M. A. and Spalding, D. B. (1980) An application of the ESCIMO Theory of Turbulent Combustion. AIAA Paper 80-0014, (18th Aerospace Sciences Meeting).
95. Ma. A. S. C., Spalding, D. B. and Sun, R. L. T. (1981) Application of the ESCIMO theory to combustion in a well-stirred reactor. To be published in the HMT series by Pergamon Press.
96. Roshko, A. and Brown, G. (1971) The effect of density difference on the turbulent mixing layer. AGARD Fluid Dynamics Panel Specialist Meeting.

CHAPTER 5

TWO-PHASE TURBULENT COMBUSTION MODELS

5.1 INTRODUCTORY REMARKS

5.1.1 Scope

As is well known, liquid fuels play a big role in fulfilling the different combustion purposes such as in industrial furnaces and gas turbine combustion chambers. It is worthwhile, therefore, to have a clear picture of the behaviour of liquid fuels. This chapter is concerned with determining the interaction of the liquid-air systems with an ultimate purpose of calculating the local values of density, temperature and reaction rate in spray flames. The liquid fuel is injected into the combustion chamber through atomizers, thus producing drops with a wide range of sizes; the mean droplet sizes are generally between 75 and 130 μm,[1] while the maximum droplet size is generally under 250 μm. For high intensity combustors, with highly volatile fuels and fine sprays, liquid droplets may completely vapourize shortly after entering the combustion chamber; subsequent treatment of diffusion and burning of the fuel can be handled in a manner analogous to that of gaseous fuel.[2]

For heavy fuels with coarse sprays, the rates of evaporation and burning are sufficiently slow for the liquid particles to manage to penetrate large distances in the combustion chamber, and the trajectories of droplets become important parameters in assessing combustion performance. The combustion of sprays involves simultaneous heat, mass and momentum transfer and chemical reaction. The main factors affecting the combustion of spray droplets are:

Droplet size
Fuel composition
Surrounding gas composition
Temperature
Pressure
The relative velocity between droplets and surrounding gas and the momentum exchange.

A major difference between liquid and gas fuel combustion is the fact that a delay time must elapse before the liquid droplets are broken down, heated, evaporated and mixed with the surrounding oxidant stream on the microscale level. The mathematical models for turbulent spray flames should provide means to account for the liquid–gas interaction terms in the various conservation equations.

5.1.2 Outline of Chapter

The present chapter describes the spray combustion models that have been developed, and illustrates their validity in furnace configurations. Critical areas where more research is needed in order to improve the model effectiveness are suggested. The available spray combustion models can be classified in two major classes:

1. Locally homogeneous flow models, where the phases are assumed to be in kinetic and thermodynamic equilibrium at each point in the flow.
2. Two phase flow models, where finite rates of transport between phases are considered.

Section 5.2 of this chapter reviews the various models of turbulent spray flames and their validation tests in practical furnace and combustor situations. Section 5.3 describes some modelling features of solid fuel combustion with a view to large-scale power station boilers. The chapter ends with a summary of available models.

5.2 SPRAY COMBUSTION MODELS

5.2.1 General

The prediction of the aerodynamic and thermal characteristics of spray flames requires the consideration of the two phases in the flow field which are coupled through exchange processes of mass, heat and momentum. The main mathematical burden is the modelling of the discontinuous liquid-phase and the coupling between the two phases. Moreover, the modelling of combustion processes requires consideration of the spray characteristics, which act as additional combustion-rate controlling parameters.

5.2.2 Locally Homogeneous Models

5.2.2.1 *Model 1*

Model Assumptions

The transport between phases is fast in comparison to the development of the flow field as a whole, i.e. all phases have the same velocity
Thermodynamic equilibrium is attained
The droplet size distribution is ignored
The exchange coefficients are equal.

The first assumption of negligible transport, implies that at each point in the flow field all phases have the same velocity. Such a requirement can easily exist when the continuous phase is a liquid and the dispersed phase is a gas, due to the relatively low inertia and thermal capacity of the gas bubbles (Collier,[3] Weimer *et al.*[4] and Avery *et al.*[5] In turbulent sprays, the application of this assumption is questionable. Similarity between the structure of gas fuelled flames and those with well atomized sprays was observed by, among others, Onuma *et al.*[6] and Komiayama *et al.*[7] Onuma *et al.*[6]

investigated the transition between the situations where the droplets burn individually or in clouds. The work of Komiayama *et al.*[7,8] was concerned with the effects of droplets on fuel and air mixing in a burner; they suggested an effective rate of turbulent mixing which increases with the increase of droplet size. On the whole the above model assumptions are most appropriate when:

(1) the droplet size is small, and
(2) the densities of phases are nearly equal.

Molecular rates of reaction are infinitely fast so that local thermodynamic equilibrium is maintained. The assumption of equal exchange coefficients for the species is a reasonable approximation for gases, but is more questionable for gas–liquid mixtures. A further simplification is obtained when the assumption of equal exchange coefficients is applied to the energy transport as well, while neglecting the radiation and heat losses effects. The local state of the mixture is determined from the specification of pressure and mixture fraction alone.[6]

Model Formulation

The formulation of such a model, with the associated modelling assumptions, involves a number of empirical parameters which must be specified in order to obtain a solution; beside these assumptions the equation of state is specified for the two-phase mixture. The equations of the model are:[2]

Continuity

$$\frac{\partial}{\partial x_j}\bar{\rho}\bar{U}=0 \tag{5.2.1}$$

Momentum

$$\frac{\partial}{\partial x_j}\bar{\rho}\bar{U}_j\bar{U}_i=\frac{\partial}{\partial x_j}\mu_t\frac{\partial \bar{U}_i}{\partial x_j}+\bar{S}_{ui} \tag{5.2.2}$$

Mixture Fraction

$$\frac{\partial}{\partial x_j}\bar{\rho}\bar{U}_j\bar{f}=\frac{\partial}{\partial x_j}\frac{\mu_t}{\sigma_f}\frac{\partial \bar{f}}{\partial x_j} \tag{5.2.3}$$

The equations for k, ε and g are those described earlier, i.e. Equations 3.3.6 and 3.3.7 and 4.2.25 respectively. The turbulent viscosity is obtained from the $k-\varepsilon$ model formulae, Equation 3.3.2. The effect of the concentration fluctuations is represented by Equation 4.2.25 for the square of the concentration fluctuations.[2] The velocity–density fluctuations correlations were obtained through the Boussinesq approximation.[9] Alternatively this density fluctuation can be taken care of through the Favre-averaging approach (Bilger[10] and Kent *et al.*[11]).

The probability density function of f, $P(f)$ is known in terms of position; examples of such forms are given by Naguib;[12] measurements of the $P(f)$ were reported by, among others, Bilger,[10] Kent *et al.*[11] and Kennedy *et al.*[13] The mean values are given by:

$$\bar{\Phi} = \int_0^1 \Phi(f)P(f)df \tag{5.2.4}$$

The probability distribution can be conveniently characterized by two parameters, the mean value $\bar{f}$ and its fluctuation g.[12–14]

Reaction Model

When an air atomizing injector is used, a portion of the combustion process is premixed. The chemical equilibrium assumption is less effective, particularly near the injector where fuel and oxidant can coexist at relatively low temperatures. It is essential, then, to take into account the finite reaction rate aspects and drop evaporation processes in order to close the set of equations.[15] In these flow situations, it is no longer sufficient to represent the state of the mixture by f only; in addition scalar conservation equations for the stagnation enthalpy, fuel mass fraction, etc, must be solved.

The rate of chemical reaction in turbulent spray flames, where the oxygen and the fuel concentrations have finite values at the same location, involves the application of the eddy break-up model of Spalding.[16, 17] The reaction rate is expressed as

$$\bar{R}_{EBU} = -C_R \bar{\rho} \frac{\varepsilon}{k} (\overline{m^2_{fu}})^{\frac{1}{2}} \tag{5.2.5}$$

where C_R is a constant having a value of 1.07 (Khalil[14]). The entity $\overline{m^2_{fu}}$ is the square of the fluctuations of the fuel mass fraction. The governing equation for mf^2u has been previously described (Equation 4.2.40). Faeth[8] suggested the use of Equation 5.2.5 as the needed source term in the fuel vapour mass fraction conservation equation. This mass fraction can be linked to other species by stoichiometry, when one step chemical reaction is assumed.

Methods of treating rate of reaction in turbulent flows are rapidly developing, yet there are conceptual problems with the eddy break-up model and the treatment of flow properties by these methods (Spalding,[19] Pope,[20] Pope[21] and Spalding[17]). The eddy break-up model has achieved some success in a number of the current spray combustion models.

Assessment of Model

Shearer and his co-workers[15, 22] investigated a variety of turbulent axisymmetric jets and sprays flowing in a stagnant environment. Figures 5.1 and 5.2 are illustrations of mean velocity and mixture fraction for various jets, in the fully developed region of the flow. The agreement between theory and experiment is shown[22] to be reasonably good; the turbulence quantities $\overline{uv}$ and k are also well predicted.

This model does not allow for the slip between phases, and does not allow for the loss of thermodynamic equilibrium. The decay of the velocities of the droplets is slower than that of the gas velocity, hence slip is encountered and phases interaction should be incorporated. It should be noted that droplet temperature responds too slowly to the

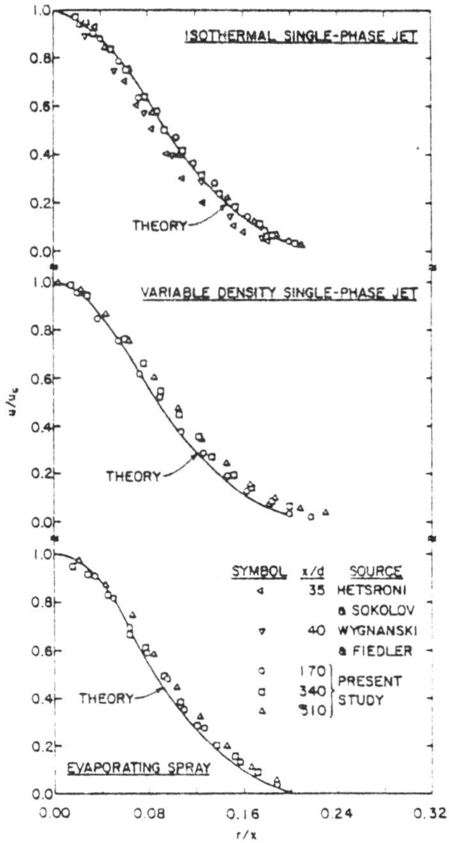

Figure 5.1 Locally homogeneous flow predictions of mean velocity in an evaporating spray[18]

variations in the gas temperature, a phenomena that is completely ignored by the present model.

5.2.2.2 *Model 2*

Model Assumptions

The reaction is single step and irreversible
The exchange coefficients are equal
Thermodynamic equilibrium is attained
The physical properties of the fuel are incorporated as in vapour state
The fuel is completely vapourized at the exit from the burner.

Model Formulation

The present model involves the solution of the conservation equations for mass, three components of momentum, k, ε, fuel mass fraction $\bar{M}_{fu}$, and the square of its fluctuation

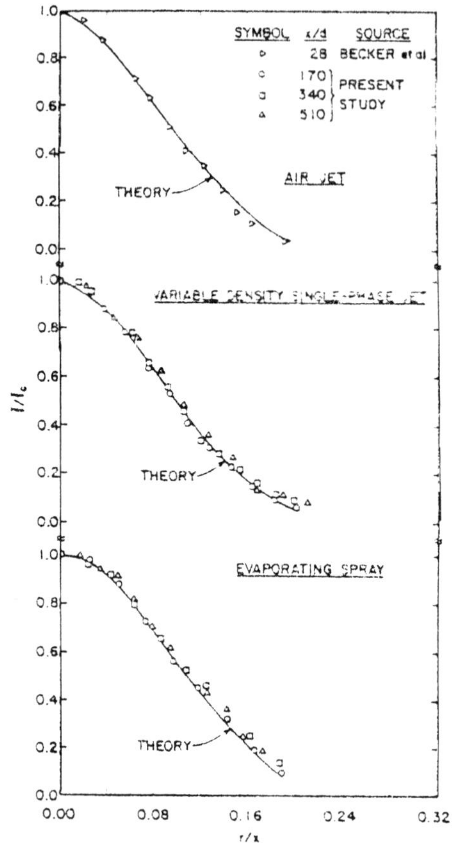

Figure 5.2 Locally homogeneous flow predictions of mean mixture fractions in an evaporating spray[18]

$\overline{m_{fu}^2}$. The general form of these equations is

$$\frac{\partial}{\partial x_j}\bar{\rho}\bar{U}_j\bar{\Phi} = \frac{\partial}{\partial x_j}\Gamma_\Phi\frac{\partial\bar{\Phi}}{\partial x_j} + \bar{S}_\Phi \tag{5.2.6}$$

The rate of chemical reaction is selected to be the minimum of two rates, namely the eddy break-up rate of Equation 5.2.5 and the Arrhenius rate of Equation 4.2.33.

Assessment of Model

To allow comparison with spray flames, initial conditions corresponding to the vapourized kerosene issuing from the jet are incorporated in the solution procedure. The turbulent spray flames of Khalil *et al.*[23] are considered here. Two flame situations were considered, shown in Figures 5.3 and 5.4, and corresponded to Sauter mean diameters of 45 and 100 μm respectively. These free spray flames have cone angles of 52° and

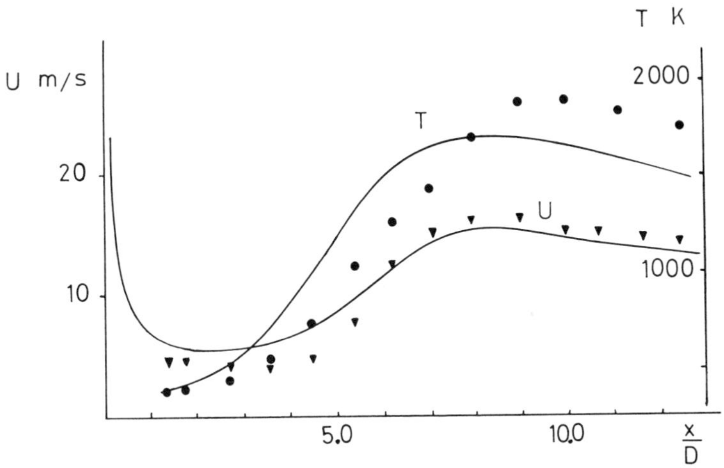

Figure 5.3 Measured and calculated centreline distributions of mean axial velocity and temperature[23]

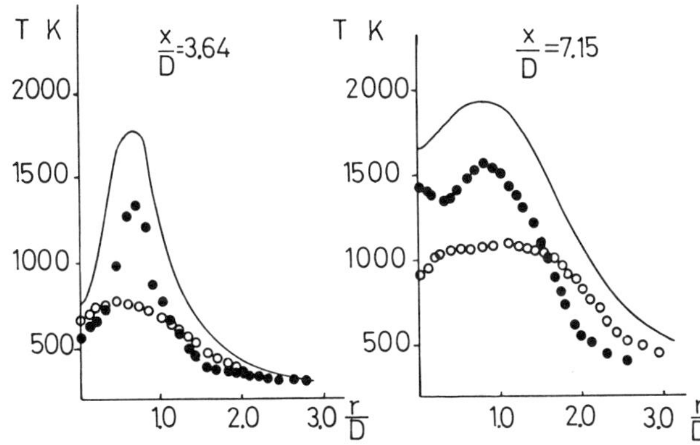

Figure 5.4 Measured and calculated temperature profiles in a free spray flame[23]
● denotes $D_{32} = 45$ μm
○ denotes $D_{32} = 100$ μm
—— represents calculations[23]

68° respectively. The results shown in Figures 5.3 and 5.4 display the radial profiles of mean temperature at two different axial locations from the burner exit, and for the two flame situations. The maximum temperatures in the two experimental flames and in the predicted ones are achieved at some distance from the centreline. This stems from the tendency of a 'hollow cone' flow to exist, which is caused by the angle of spray and the corresponding gaseous simulation. The predicted temperatures are in reasonable agreement with the measured temperature for the fine spray. This observation is consistent with the inadequacy of the assumptions of the models to represent the coarse spray.

Another example of the application of Model 2 to turbulent reacting flames is that reported by Abou Ellail *et al.*[24] Liquid kerosene was supplied to a preheated perforated pipe into the surrounding preheated air. The fuel–air mixture passed through a swirler to the main air cooled axisymmetric chamber, through a diverging conical section. The swirler was fitted with vanes inclined 45° to the flame axis. The diameter of the chamber was 0.15 m and its length was 0.46 m. Three different flames were considered and corresponded to various excess air factors, and main air velocity.

The assumption of fully vapourized fuel at inlet to combustion chamber was made and the radiative heat transfer contribution to the energy balance was represented by the four flux model.[24] The comparisons comprised mean temperature, flame envelope and burned fuel fraction profiles. Figure 5.5 demonstrates the general agreement obtained between measured and calculated gas temperature profiles for the flame situations given in Table 5.1. The temperature profiles exhibited off-centre peaks which identify the reaction zone. For example, flame 2, at x/D_C up to 0.5, spread out and occupied nearly 60% of the combustion chamber cross-sectional area. It should be noted that the maximum discrepancies observed were less than 100 K, except in the burner vicinity. The differences observed in the burner vicinity can be attributed to the inadequacy of the turbulence–chemistry interaction modelling assumptions in regions of very high turbulence and swirl intensities.

This group of investigators also introduced the capability of the model to predict the local velocity profiles. Figure 5.6 shows the predicted mean axial velocity profiles and identifies the regions of high velocity in the wall vicinity, and the presence of a large central reversed flow region. The effect of swirl and divergence section on the flow pattern is characterized by the reversed flow region.

Measured and predicted flame envelopes are shown in Figure 5.7 for the three flame situations; the flame length is always over-predicted while the flame radius is underpredicted. As the excess air factor increases from 0.866 to 1.0, the flame length decreases because more fuel will be burned in a shorter distance.

The measured and calculated burned fuel fraction profiles are shown in Figure 5.8. The local burned fuel rate is characterized by a sharp peak in a layer whose thickness ranges between 0.02 and 0.05 m, surrounding the regions of high heat release rate. The measured and calculated profiles are in qualitative agreement, with discrepancies of up to 20% in the main flow regions. They concluded that there is an urgent need to develop combustion model assumptions as well as to perform more detailed comparisons.

TABLE 5.1

Property	Flame 1	Flame 2	Flame 3
Excess air	0.866	1.00	1.00
Preheat temperature	548	548	548
Main Air Velocity, m/s	14.38	18.38	14.38

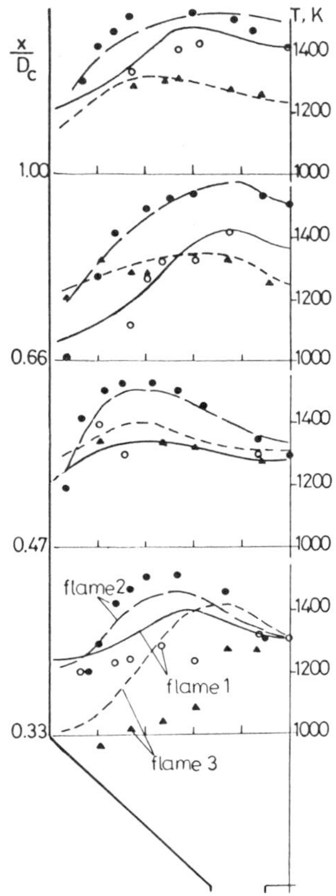

Figure 5.5 Measured and calculated temperature profiles
●, ○, ▲ Measurements
– –, –, --- Calculations of reference 24

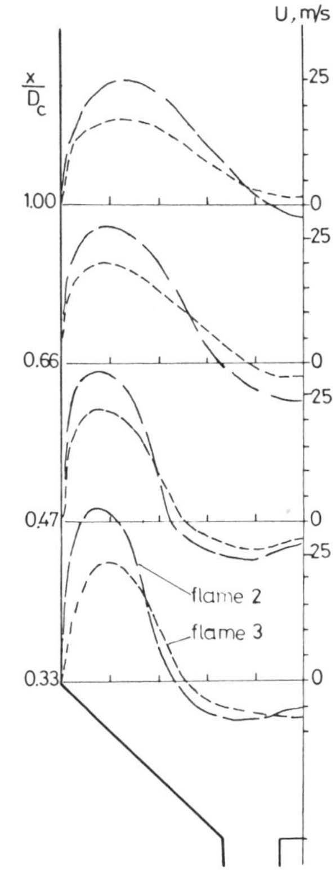

Figure 5.6 Calculated mean axial velocity profiles

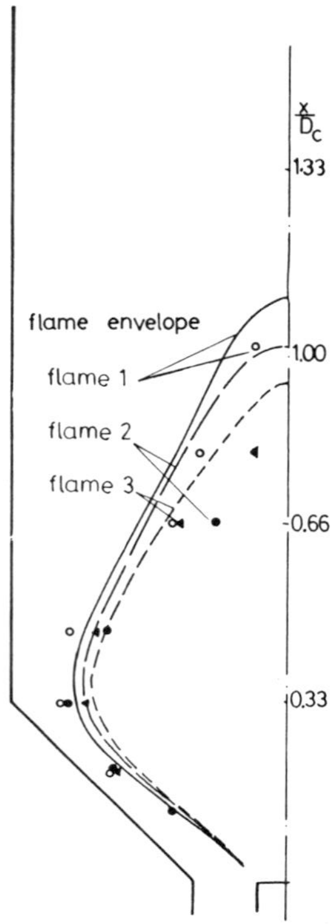

Figure 5.7 Measured and calculated flame envelope
●, ○, ▲ Measurements
– –, –, –– Calculations of reference 24

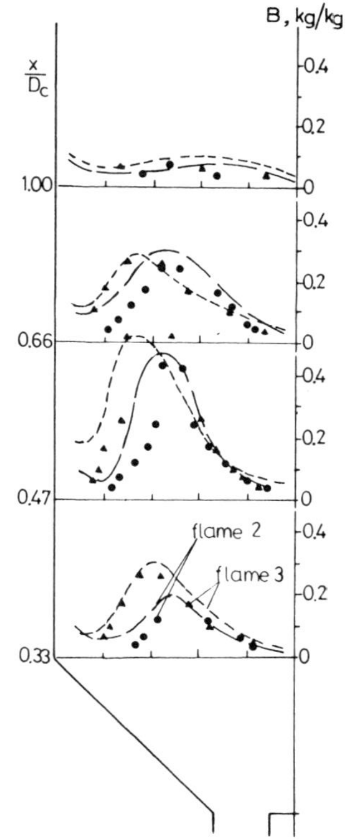

Figure 5.8 Measured and calculated burned fuel fraction profiles
●, ▲ Measurements
––, –– Calculations of reference 24

5.2.2.3 *Model 3*

Model Assumptions

The reaction is single step and irreversible

The transport between phases is fast in comparison to the development of the flow field as a whole, i.e. all phases have the same velocity, at the same location

Thermodynamic equilibrium is attained

The exchange coefficients are equal

An empirical formula for burning speed is used.

Model Formulation

The model[2] solves the equations for the conservation of mass, three components of momentum, $\underline{k \text{ and } \varepsilon}$, stagnation enthalpy, fuel mass fraction $\bar{M}_{fu}$, mixture fraction $\bar{f}$, g and $\overline{m_{fu}^2}$ and $\overline{m_{fu} m_{ox}}$. The model calculates the rate of chemical reaction from a modified eddy break-up model.

The form of the eddy break-up model[16] has been modified[2] to include the effects of the burning velocity, which is controlled in spray flames by the local flame properties. The relation between the burning velocity, the initial air stream turbulence, air to fuel mass flow rate ratio ζ and droplet Sauter mean diameter D_{32} was proposed[25] as

$$\mathrm{FS} = 5380\,(\zeta - 0.012)\, k^{0.575}/D_{32} \tag{5.2.7}$$

This expression is in agreement with the findings of Polymeropoulos *et al.*[26] in the range of droplet Sauter mean diameters of 30 to 100 μm, and accounts for the effects of air to fuel ratio and the turbulence levels.

The parcels of vapour pass through regions of both high and low shear rate, and the rates of mixing and combustion depend on the mixing time, which is defined as

$$\tau_{\mathrm{mix}} = \left(\left| \frac{\varepsilon}{k} + \frac{\partial \bar{U}_i}{\partial x_i} \right| \right)^{-1} \tag{5.2.8}$$

This expression accounts for the shearing rate due to mean flow motion and the random fluctuations. The rate of chemical reaction is expressed[2] in terms of burning velocity and the thickness of the interface between fuel and oxidant parcels, λ_0, as

$$\tau_{\mathrm{chem}} = \lambda_0/FS \tag{5.2.9}$$

The effective rate of fuel consumption was taken as the smallest of the two rates given by:

$$\bar{R}_{\mathrm{fu,mix}} = -\bar{\rho}(\overline{m_{\mathrm{fu}}^2})^{\frac{1}{2}}/\tau_{\mathrm{mix}} \tag{5.2.10}$$

and

$$\bar{R}_{\mathrm{fu,chem}} = -\bar{\rho}(\overline{m_{\mathrm{fu}}^2})^{\frac{1}{2}}/\tau_{\mathrm{chem}} \tag{5.2.11}$$

Local values of fuel mass fraction $\bar{M}_{\mathrm{fu}}$ and the corresponding fluctuations $\overline{m_{\mathrm{fu}}^2}$ are obtained from their respective transport equations. The form of these equations and the equations of $\overline{m_{\mathrm{fu}} m_{\mathrm{ox}}}$ are similar to those discussed earlier in Chapter 4. The probability distribution of fuel mass fraction was assumed to be clipped Gaussian, and the thermodynamic properties of the reacting species were obtained from Spiers.[27]

Assessment of Model

The aerodynamic and thermodynamic characteristics of confined swirling spray flames were predicted[2] for the geometry of Khalil *et al.*[28] The experimental investigations of reference 28 relate to the flow of swirling air, and a spray of kerosene fuel, in an axisymmetric water-cooled furnace. Different swirl intensities were imparted to the annulus combustion air, through the use of guide vane cascade swirlers. At the end of the combustion air supply pipe, where vane swirlers were located, the liquid atomizer was axially aligned to the upstream end of the test section. Swirl intensities of 1.24 and 1.98 only are discussed here. Measurements of axial velocity, tangential velocity and temperature were reported.[28]

Radial profiles of the measured and calculated mean axial velocity are shown in Figure 5.9, for a swirl number of 1.98 and air to fuel ratio of 20.17. These measured and

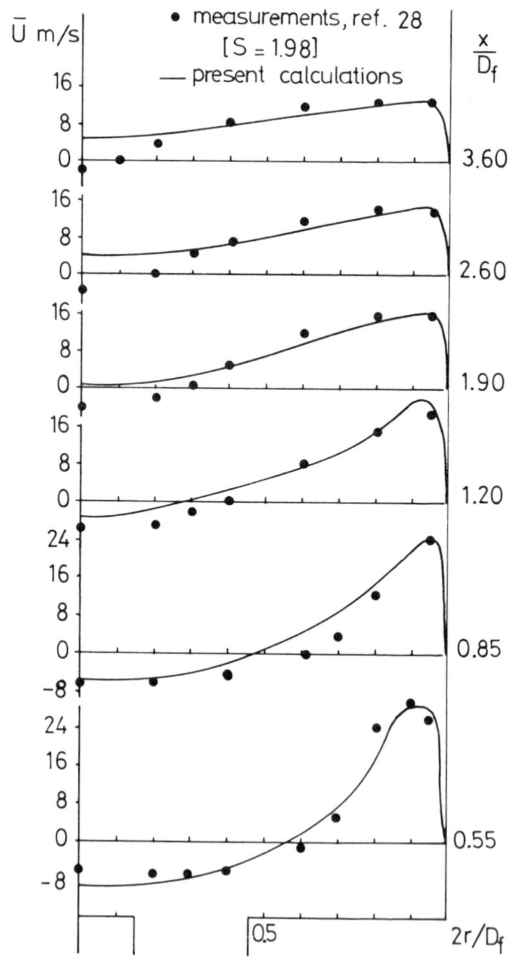

Figure 5.9 Radial profiles of mean axial velocity for the confined flame of reference 28; $S = 1.98$

calculated velocities were in general agreement in the first 1.5 furnace diameters. The size of the central recirculation zone, created due to the effect of swirl, was slightly under-predicted in the vicinity of the burner. At further downstream location the calculated size of the central recirculation zone is nearly 70% of the measured value. These observed discrepancies are attributed mainly to the unknown precision of the measured velocities, particularly near the recirculation zone. Radial profiles of mean temperature at various axial locations are shown in Figure 5.10. The figure shows both measured and calculated temperature profiles. The obtained agreement is satisfactory for engineering design purposes and the maximum discrepancies were less than 150° K. The under-prediction in the gas temperature can be mainly attributed to the radiation model assumptions. However, on the whole, the agreement is reasonable.

Similar comparisons were also shown[2] for the lower swirl number of 1.24 and the same air to fuel mass ratio. Figures 5.11 and 5.12 demonstrated the profiles of mean axial velocity and mean temperature at various axial locations. The comparisons shown

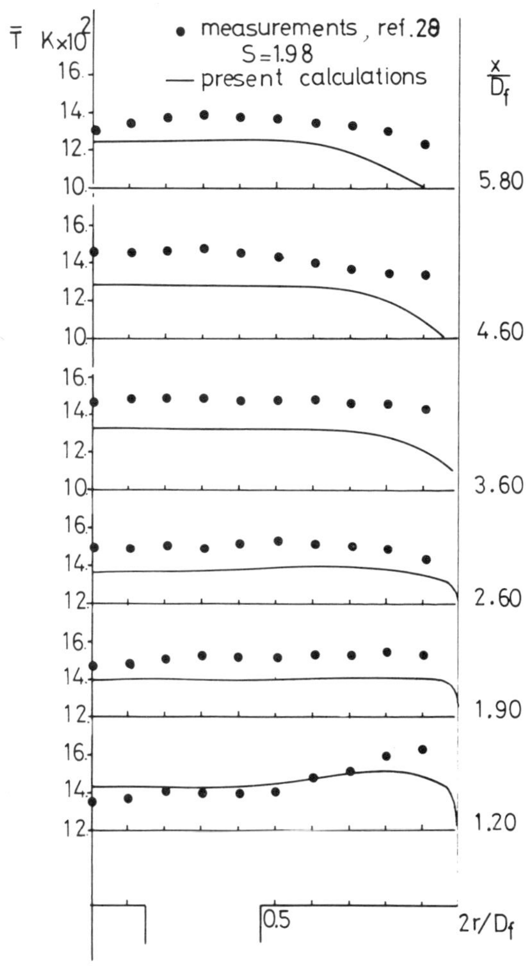

Figure 5.10 Radial profiles of mean temperature for the confined flame of reference 28; $S = 1.98$

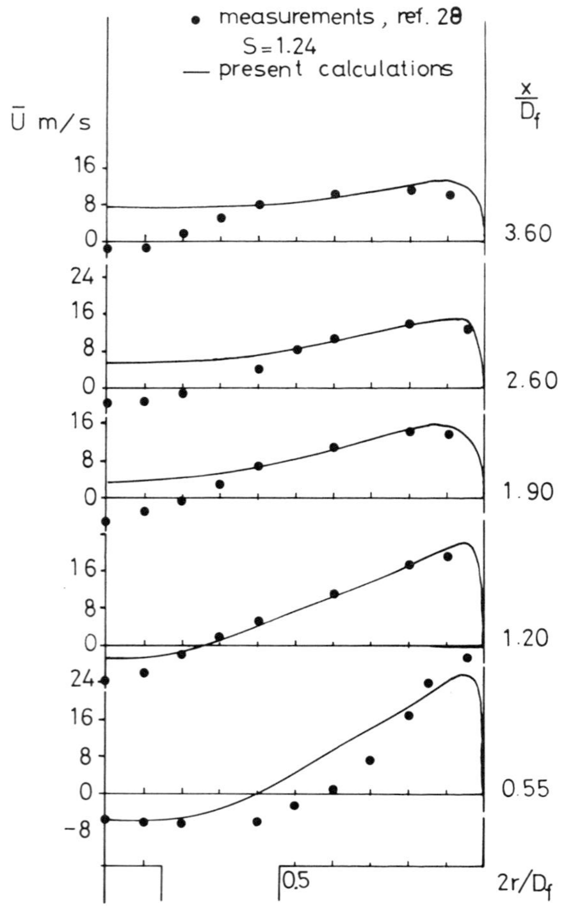

Figure 5.11 Radial profiles of mean axial velocity for the confined flame of reference 28; $S = 1.24$

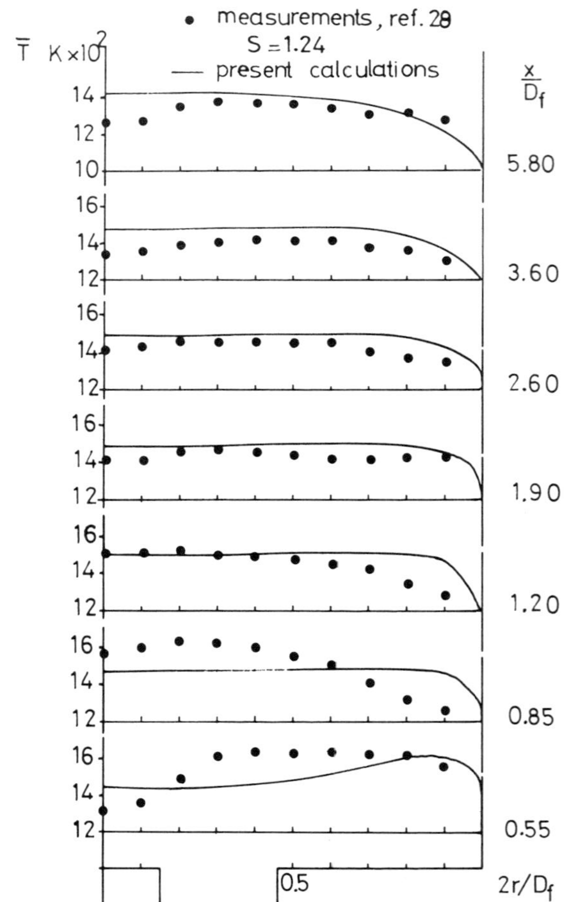

Figure 5.12 Radial profiles of mean temperature for the confined flame of reference 28; $S = 1.24$

TABLE 5.2

Summary of LHF Models of Spray Evaporation and Combustion (Faeth[18])

Reference	Model[1]	Experiment[2]	Assessment
Thring and Newby (1953)	integral, BQ	combustion of air and steam atomized spray, 0.1 MPa	qualitative agreement
Newman and Brzustowski (1971)	integral, BQ	evaporation of pressure atomized liquid near its critical point, 6–9 MPa	good agreement for spray boundary
Shearer and Faeth (1977)	integral, BQ	combustion of a pressure atomized spray, SMD = 30 μm, liquid injection temp. 298 K, 0.1–9 MPa	theory overestimates spray and flame boundary by 30–50%, poor estimation of flow width
Shearer *et al.* (1978)	$k - \epsilon$, BQP	evaporation of an air atomized spray, SME = 29 μm, 0.1 MPa	theory overestimates rates of development of process
Khalil and Whitelaw (1976)	$k - \epsilon$, EQP	combustion of a pressure atomized spray, swirl, SMD = 45, 100 μm, 0.1 MPa	theory overestimates rate of development of process, error decreases as SMD decreases
Khalil (1978)	$k - \epsilon$, MEBU	same as above reference	improved prediction in some cases

1. BQ implies local thermodynamic equilibrium, BQP implies equilibrium with probability distribution function for properties, MEBU implies modified eddy break-up model.
2. Pressures are those of spray environment; all flows are steady.

in Figure 5.11 for the axial velocity demonstrated qualitative agreement and the size of the central recirculation zone was still under-predicted. The agreement is not as good as that of Figure 5.9. This is because of the increase in the swirl intensity which results in a higher burning velocity, and a greater tendency of the flame towards homogeneous phase combustion.

Measured and calculated mean temperature profiles are shown in Figure 5.12; fair agreement is observed away from the vicinity of the atomizer where the effect of the droplets' combustion is still persistent.

A summary of the available models of the locally homogeneous flow is shown in Table 5.2, which indicates previous investigations and their assessment.

It can be concluded that the present models provide a useful qualitative description of the turbulent spray flames. The models are relatively easy to use and they provide reasonable description of the flow and flame when high turbulence intensity and fine droplets are involved.[2] The ability of the models to provide details of the flame structure and burning velocities is limited. It is essential, therefore, to investigate models that account for the slip and for the finite rate transport processes between phases.

5.2.3 Two-Phase Flow Models

5.2.3.1 *General Assumptions.* The general features of the models that account for droplet characteristics and transport properties[29, 30] are:

1. The droplets are assumed to be spherical and well mixed. Ballastic effects are ignored.
2. The gas phase is assumed to be quasi-steady; i.e. it adjusts immediately to the mean boundary conditions at each instant of time. In other words, the turbulent velocity and mixture fraction fluctuations are neglected.
3. The heat and mass transfer exchange by convection is empirically determined.
4. The chemical reaction in the liquid phase is neglected. The liquid and gas phase compositions, and the surface temperature are related by phase equilibrium criterion.
5. The pressure field is equal.
6. An effective binary diffusion coefficient is used.
7. Radiation is neglected.

Several investigations incorporated the above assumptions; see, for example, references 29–36.

The equations governing the liquid phase are as follows[29, 30, 35]

$$\frac{\partial M_i}{\partial t} = -m_{\text{evap}} e_i, \quad i = 1, N \tag{5.2.12}$$

where m_{evap} is the total evaporation rate and e_i is the mass flux fraction. The momentum equations are represented as:

$$\frac{d^2X}{dt^2} = C_l \left| U - \frac{dX}{dt} \right| \cdot \left(U - \frac{dX}{dt} \right)$$

$$\frac{d^2r}{dt^2} = C_1 \left| V - \frac{dr}{dt} \right| \cdot \left(V - \frac{dr}{dt} \right) + \frac{1}{r} \left(\frac{dz}{dt} \right)^2 \qquad (5.2.13)$$

and

$$\frac{d^2z}{dt^2} = C_1 \left| W - \frac{dz}{dt} \right| \left(W - \frac{dz}{dt} \right) - \frac{1}{r} \frac{1}{dt} \frac{dz}{dt}$$

and $C_1 = 3/4 C_D \bar{\rho}/(\bar{\rho}_l D)$, x, r and z are coordinate directions.

The energy equation for a droplet is expressed[36, 37] as

$$mC_p \frac{dT_l}{dt} = \dot{q} + L \frac{dM}{dt} \qquad (5.2.14)$$

where L is the latent heat of vapourization of liquid droplets, and $\dot{q}$ is the rate of heat transfer to the droplet:

$$\dot{q} = \pi D^2 h_0 (T_F - T_l) \qquad (5.2.15)$$

The evaporation rate $\dot{m}_{\text{evap}}$ is expressed as

$$\dot{m}_{\text{evap}} = \frac{\pi D^2 h_0}{C_{p,v}} U(t - t_h) \ln(1 + B) \qquad (5.2.16)$$

where h_0 is the local heat transfer coefficient and $U(t - t_h)$ is a unit step function and equals zero when the value of the time is less than the heating time t_h. The constant B is a transfer number given[38] as:

$$B = C_{p,v} (T_F - T_s)/L \qquad (5.2.17)$$

Several correlations for the drag coefficient C_D can be found in the literature, for example in references 29, 32 and 34. The correlation of Ingebo has been used by several investigators, and was found to be reasonably accurate; its formula is:

$$\begin{aligned} C_D &= 27\ \text{Re}^{-0.84} && 0 < \text{Re} < 80 \\ C_D &= 0.271\ \text{Re}^{0.217} && 80 < \text{Re} < 10^4 \\ C_D &= 2.0 && \text{Re} > 10^4 \end{aligned} \qquad (5.2.18)$$

The effects of evaporation and droplet distortion on C_D are included implicitly in Equations 5.2.18. For Re less than 2000, a general expression was given by Canada *et al.*[32] as

$$C_D = \frac{24}{\text{Re}} \left(1 + \frac{\text{Re}^{2/3}}{6} \right) \qquad (5.2.19)$$

In the present model Equation 5.2.18 is used as it covers a wider range of droplets' Reynolds number.

The heat transfer coefficient is expressed in terms of the Nusselt number as

$$\mathrm{Nu} = 2\ \ln[(1 + B_Y)(1 + B_0)]/B_Y \tag{5.2.20}$$

where $B_Y = (1 - \bar{M}_{a,s})/\bar{M}_{0,s}$; $B_0 = \bar{M}_{0,\infty}/\Sigma_N e_i/i$; i is the stoichiometric coefficient; and $\bar{M}_{0,\infty}$ is the mass fraction of the surrounding stream.

The above expressions for m_{evap} and Nu were obtained for flow of droplets with Re = 0. They should, therefore, be modified to account for the relative velocity between phases as:

$$\dot{m}_{\text{evap}} = \dot{m}_{\text{evap,Re=0}}\left[1 + \frac{0.278\mathrm{Re}^{1/2}P_r^{1/3}}{(1 + 1.232/(\mathrm{Re}P_r^{4/3}))^{1/2}}\right] \tag{5.2.21}$$

and

$$\mathrm{Nu} = \mathrm{Nu}_{\mathrm{Re}=0}\left[1 + \frac{0.278\ \mathrm{Re}^{1/2}\ P_r^{1/3}}{(1 + 1.232/(\mathrm{Re}P_r^{4/3}))^{1/2}}\right] \tag{5.2.22}$$

The droplet temperature during preheating is determined from the heat balance Equation 5.2.14. The decrease of the droplet diameter due to evaporation is obtained from equation 5.2.12 and:

$$\frac{d\dot{M}}{dt} = \frac{\pi}{2}\rho_l D^2 \frac{dD}{dt}$$

and

$$\frac{dD}{dt} = -\frac{2N_u K}{D\rho_l C_{p,F}} U(t - t_h)\ln(1 + B) \tag{5.2.23}$$

where K is the thermal conductivity.

The flame temperature T_f can be obtained from the following equation, provided that the dissociation is neglected:

$$\frac{T_F - T_s}{T_{\text{amb}} - T_s} = \left[\frac{Q_R}{C_P(T - T_s)} + 1\right]\left[1 + \frac{B_0}{B_Y}(B_Y + B_0)\right] \tag{5.2.24}$$

where Q_R is the heat release rate per unit mass.

5.2.3.2 *Model 4*

Model Assumptions

The assumptions of Section 5.2.3.1 are applied to the liquid phase. The droplets evaporate during their motion in the chamber as a result of convective heat transfer from products of combustion. The vapour is convected and diffused by the turbulent motion of the gases, and may burn partially if the local conditions of the hot gases are favourable. The degree of reaction is obtained from the solution of the equation of transport of the mixture fraction f and fuel mass fraction M_{fu}.[36]

The gas phase equations are those described earlier in Sections 3.2 and 4.2; the interaction between the phases is being taken into account as described later in the chapter.

Model Formulation

The liquid phase conservation equations, discussed in Section 5.2.3.1, are appropriate to a single droplet in a hot stream. The droplets from the atomizer are heated due to the convective heat transfer from the surrounding products of combustion. The time taken by the heating process is denoted by t_h. This time depends on the temperature difference and the droplet diameter. It also depends on the fuel properties, the local turbulence and heat transfer coefficient. This dependency takes the form of

$$t_h = 0.166 D \bar{\rho}_F C_{P,F} \ln \frac{T_a - T_0}{T_a - T_S} / h_0 \tag{5.2.25}$$

As the droplet temperature reaches the saturation temperature T_s, evaporation starts, with still more heat addition to the droplet. The droplet diameter decreases at a constant temperature according to Equation 5.2.23 and the evaporation rate is obtained from Equation 5.2.16. The time elapsed during the evaporation of droplet is t_{evap}. The variation of droplet diameter with time is shown in Figure 5.13 and is expressed in terms of D^2 (Spalding[38]).

The spray is approximated by a finite number of droplet size groups up to 300 μm, with an increment of 10 μm.[36] The incremental number dn of each size group is calculated from the droplet size distribution of the spray which can be expressed as:

$$\mathrm{d}n = aD^{\alpha} \exp(-bD^{B}) dD \tag{5.2.26}$$

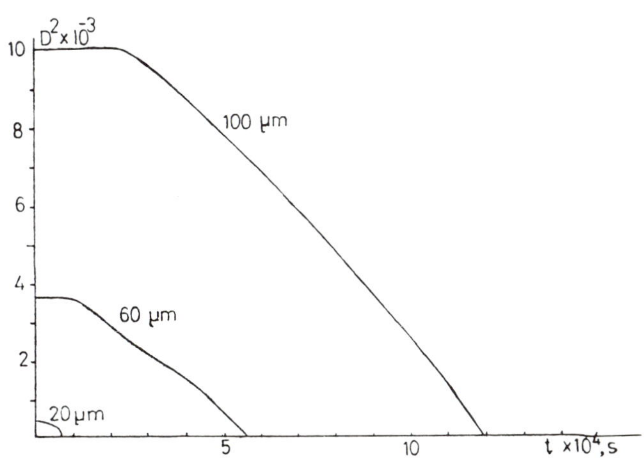

Figure 5.13 Droplet size variation with time[36]

Where α, B, a and b are constant, dependent on the type of the atomizer and operating conditions. An example of the droplet size distribution of a pressure atomizer is shown in Figure 5.14, after Abou Ellail.[30]

Equations for each size group have to be solved in the flow domain. Each of these size groups is tackled through the flow field by solving their Lagrangian equations of motion, and the droplet position is given by

$$\vec{X}_j = \int_0^t U_{j,D}\,\mathrm{dt}$$

The effect of turbulent droplet diffusion has been ignored because the response of the droplets to the mean motion and, at a lesser extent, to the turbulent fluctuations, is relatively slow. This lasts until the end of their lifetime. Since the bulk of their mass has been, by that time, transferred to vapour, the turbulent motion of the droplet does appear not to be the primary means of transporting the fuel within a spray.[18]

Gas Phase

The model solves the conservation equations of mass, momentum and energy for the gas phase. The form of the conservation equation for the gas phase is as follows[36]

$$\frac{\partial}{\partial x_j}\bar{\rho}\bar{U}_j\bar{\Phi} = \frac{\partial}{\partial x_j}\Gamma_\Phi\frac{\partial\bar{\Phi}}{\partial x_j} + S_\Phi + S_{L\Phi} \tag{5.2.28}$$

where the general dependent variable $\bar{\Phi}$ represents the gas phase velocity components, $\bar{U}$, $\bar{V}$, $\bar{W}$, k, ε, $\bar{f}$, $\bar{M}_{fu}$ and $\bar{H}$. S_Φ is the source/sink term of the entity Φ and Γ_Φ is the turbulent exchange coefficient.[14] The source term $S_{L\Phi}$ denotes the interaction of the liquid droplet phase on the gaseous phase.

In the continuity equation, the source term S_{Lm}, which accounts for dropl evaporation, is expressed[36] as

$$S_{Lm} = \sum_l\left(\sum_n m_{\mathrm{evap}n,l}\cdot\dot{g}_{n,l}\cdot\mathrm{dt}_n\right)\mathrm{dn}_l \tag{5.2.29}$$

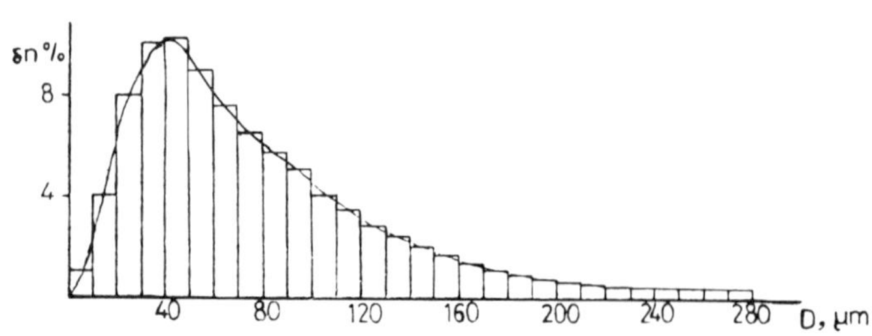

Figure 5.14 Droplet size distribution[30]

where Σ_l and Σ_n denote the summation over size groups and time steps respectively. The parameter $\dot{g}_{n,l}$ varies from 1 to 0, which correspond to full and no interaction respectively. $\dot{m}_{\text{evap}n,l}$ is calculated from Equation 5.2.16 and dn from Equation 5.2.26.

In the momentum equations, the source terms, S_{LU}, S_{LV} and S_{LW}, are obtained as the sum of two terms. The first term is due to the convection of the fuel vapour momentum and the second is due to drag forces, namely

$$S_{LU} = \sum_l \left[\sum_n \left[\dot{m}_{\text{evap}n,l}(U_{Dn,l} - U) - F_{D,Xn,l} \right] g_{n,l}\,\mathrm{dt} \right] \mathrm{dn}_l \tag{5.2.30}$$

where $F_{D,Xn,l}$ is the drag force and is expressed[3] as

$$F_{D,Xn,l} = \frac{C_D}{2} \pi \rho D^2 (U - U_{Dn,l}) \cdot |U - U_{Dn,l}| \tag{5.2.31}$$

The first term of Equation 5.2.30 results from the difference between the momentum of the fuel vapour as it leaves the surface of the droplet, and the corresponding momentum when the fuel vapour mixes and flows with a velocity equal to the local velocity of the gas phase. The second term is due to drag action. Similar expressions for S_{LV} and S_{LW} are incorporated. The effect of droplet motion on the equations characterizing the turbulence is generally ignored.[18] The mixture fraction, fuel mass fraction and enthalpy interaction source terms (S_{Lf}, S_{Lfu} and S_{LH}) are due to convection only, and are simply given by:

$$\begin{aligned} S_{Lf} &= S_{Lm}(1 - f) \\ S_{Lfu} &= S_{Lm}(1 - M_{fu}) \\ S_{LH} &= S_{Lm}(H_{fu} - H) \end{aligned} \tag{5.2.32}$$

The treatment of radiation heat transfer follows the previous modelling assumptions of the four flux model,[14] but no account is taken of the droplet radiation.

Assessment of Model

The assumptions of this model were applied to predict the local flow properties in axisymmetric furnaces by Abou Ellail *et al.*[36, 39] and El Banhawy *et al.*,[35] Anasoulis *et al.*[40] and Gibeling *et al.*[41] The calculations of Abou Ellail *et al.*[36, 39] are the ones discussed here. The experimental set-up is that of Khalil *et al*;[28] it is simply a kerosene spray issuing into an axisymmetric, water-cooled furnace. The combustion air, surrounding the fuel atomizer, is subject to a swirl number of 0.721. The furnace is 2 m long and has a diameter D_f = 0.205 m.

Radial profiles of mean axial and tangential velocity components are shown in Figure 5.15, and are compared with the corresponding measurements. The measured and calculated axial velocity profiles are in general agreement, but the size of the central recirculation zone is under-predicted. The scatter of the experimental measurements of the tangential velocity component may be held as partly responsible for the discrepancies observed in the comparisons of the W-profiles; the remaining responsibility for these lies in the turbulence model assumptions. Measured and

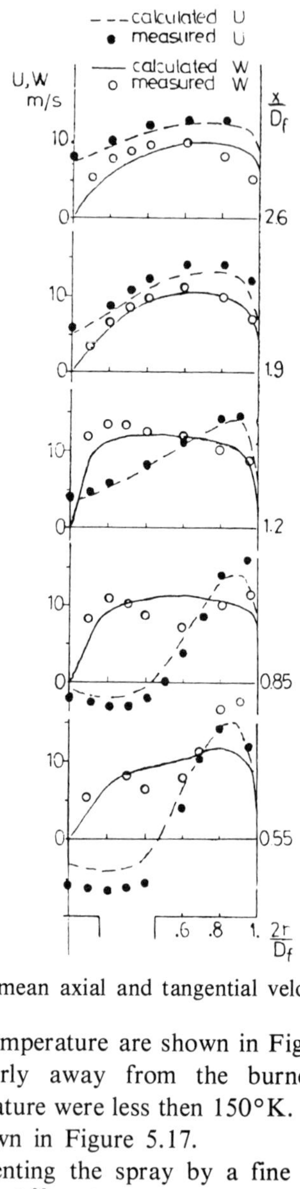

Figure 5.15 Radial profiles of mean axial and tangential velocities in an oil-fired furnace[39]

calculated profiles of mean temperature are shown in Figure 5.16. Reasonable agreement is observed, particularly away from the burner vicinity. The maximum discrepancies in mean temperature were less then 150°K. The droplet trajectories were also predicted[39] and are shown in Figure 5.17.

The assumption of representing the spray by a fine number of size ranges was examined by El Banhawy *et al.*[35] and concluded that increasing the number of size ranges from 5 to 20 caused an averaged reduction of 5% in the observed differences between the experimental and calculated results. This insignificant influence of number of size ranges was attributed[35] to the fact that the flame investigated was of the hollow cone spray, where most droplets follow the same path.

The current state in the development of the two phase flow combustion models is listed in Table 5.3.

In more recent work, Boysan *et al.*[42] extended and applied the assumptions of the two phase flow models to three dimensional flow situations. The combustor considered was of the gas turbine type and is sketched in Figure 5.18. The air stream through the

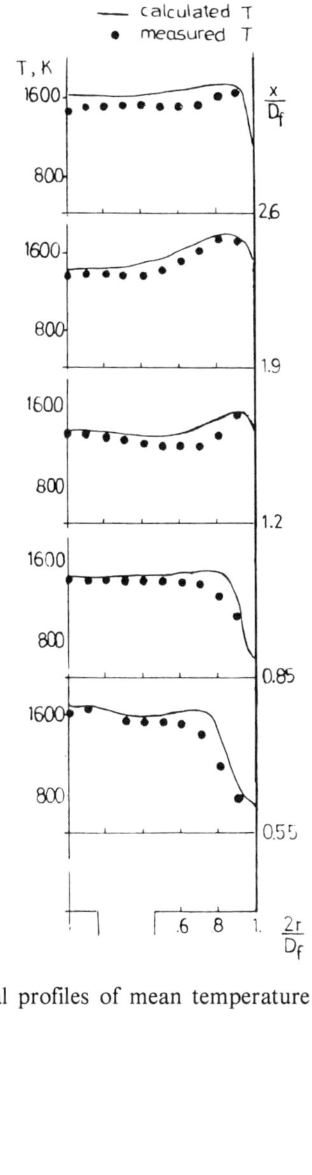

Figure 5.16 Radial profiles of mean temperature in an oil-fired furnace[39]

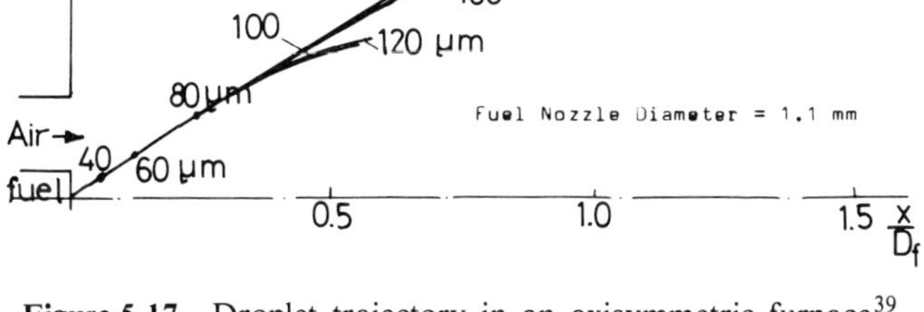

Figure 5.17 Droplet trajectory in an axisymmetric furnace[39]

TABLE 5.3

Summary of TPF Models of Spray Evaporation and Combustion[1], (Faeth[18])

Reference and date	Dimensions	Turbulence model[2]	Combustion model[3]	Pollutant prediction	Radiation[4]	Drop model
No-Slip Models						
Elghobashi *et al.* (1976)	$2D + t$	$k - \epsilon$	EBU	none	none	steady drop evaporation or combustion
Swithenbank and coworkers (1978)	$2D$, $3D$	$k - \epsilon$	EBU	none	MF	steady drop evaporation or combustion
Jones and Pridden (1978)	$3D$	$k - \epsilon$, Favre averaging	EQGP	CO, HC, NO	none	steady drop evaporation
Slip Models						
Anasoulis *et al.* (1974)	$2D$	algebraic	EQG	NO	MF	steady evaporation corrected for convection
Gibeling *et al.* (1975–76)	$3D$	$k - \epsilon$	EQG	NO	MF	steady evaporation corrected for convection
Crowe and coworkers (1974–78)	$2D$	$k - \epsilon$	evaporation	none	none	transient heat-up and evaporation with convection; treats drop diffusion
Abou Ellail and Khalil (1978)	$2D$	$\frac{k - \epsilon}{\rho' u_i}$	EB	none	MF	transient heat-up (no evap.), evaporation with convection
Mongia and coworkers (1977–78)	$3D$	$k - \epsilon$	MEBU	CO, HC, NO_x	MF	transient heat-up and evaporation with convection

1. Finite difference calculations assuming a dilute spray.
2. Reynolds averaging unless noted otherwise; no direct effect of drops on turbulence.
3. EBU denotes the eddy break-up model of Spalding; EQGP denotes equilibrium gas phase and use of probability distribution function; EQG denotes equilibrium gas phase with drop evaporation; EB denotes drop evaporation of Barghi; MEBU denotes eddy break-up model but drops evaporate or combust depending on whether mean mixture ratio is greater or less than stoichiometric.
4. MF denotes multiflux radiation model.

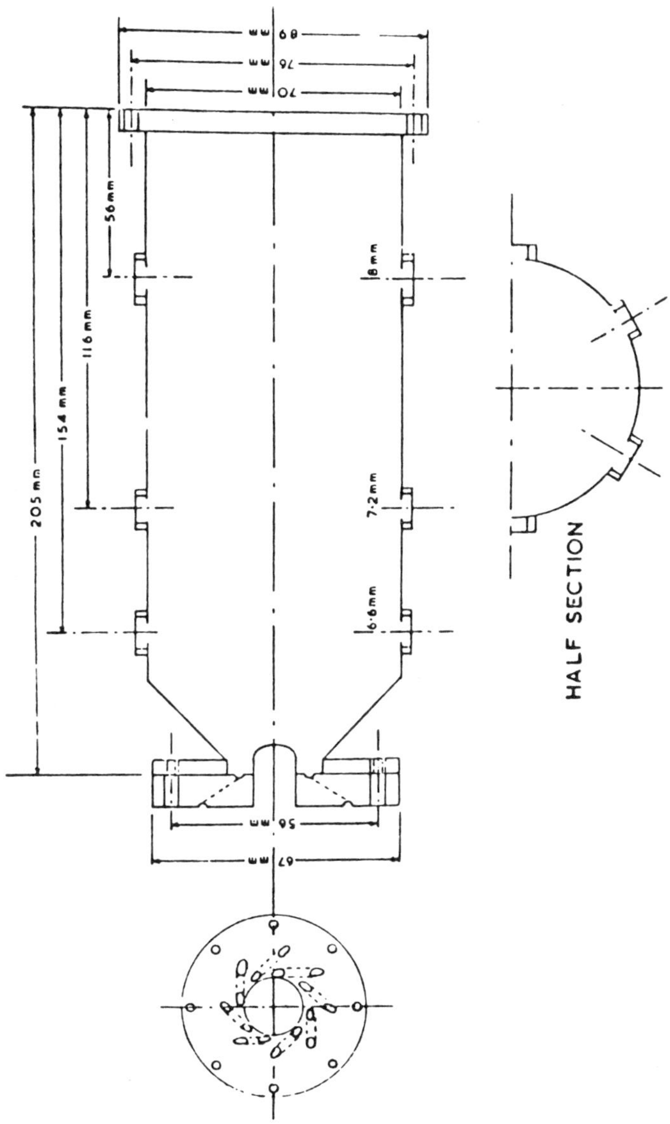

Figure 5.18 Gas turbine combustor can[42]

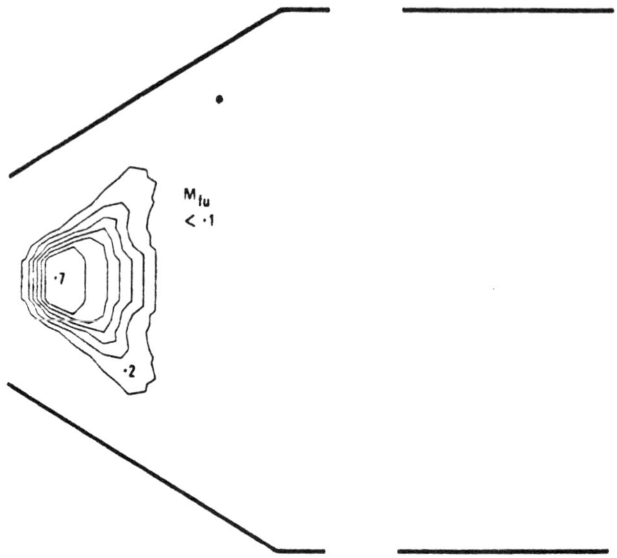

Figure 5.19 Computed distribution of mass fraction of unburnt fuel, Boysan *et al.*[42]

swirler enters the chamber with a finite swirl velocity. Primary, secondary and dilution air streams were introduced via three sets of injection holes, each consisting of six equally spaced circular orifices, thus giving rise to the three-dimensional nature of the problem. Fuel spray, which was of the hollow cone type with an included angle of 60 degrees, was introduced from an axially located fuel nozzle. A total of 3402 grid nodes was used and the sizes of droplets were classified in 20 ranges.

Predicted distribution of evaporated but unburned fuel mass fraction is shown in Figure 5.19 and demonstrates that the majority is found in the vicinity of the fuel spray. Fuel rich regions, near the axis, are formed due to the high-intensity recirculation zone along the axis. This distribution is of great significance in investigations of the quenching phenomenon leading to minimization of unburnt hydrocarbons in idle conditions.

5.3 COMBUSTION MODELS FOR SOLID FUELS

5.3.1 General

Coal is the most used form of the available solid fuels and may be described as black, heterogeneous rock that is frail to hard, of variable vegetable origin with extraneous mineral additions of equally variable composition and quantity. Elemental analysis by mass of different coals shows anything from 65% to 95% carbon, 2% to 7% hydrogen, up to 25% oxygen, up to 10% sulphur, and typically 1% to 2% nitrogen. Mineral matter, that is converted to ash on burning, can be up to 30% by mass. Even this is considered high in commercial coals.

The search for laws and equations governing the combustion characteristics of solid fuels, and from which the performance of practical systems can be predicted, is complex enough, even when the fuel is reasonably well characterized. It is very difficult with coal to argue with certainty from one reaction mechanism to another. By comparison, elucidation of the methane reaction, for example, or behaviour in a rocket, is simplicity itself.

5.3.2 Mechanism of Coal Combustion

The overall behaviour of coal during combustion may be summarized by a brief consideration of the burning times, which in turn further points to the problem of knowing the rate mechanism for small particles. For particles whose size is above 100 μm, the burning time is proportional to the square of the diameter. Below 100 μm, the burning times in furnaces are generally stated to be in the range from 0.1 to 1 second.[43] In small flames, the burning times reported are 1 to 10 milliseconds; it is generally believed that only the volatiles were burnt in these experiments.[44]

The char particles formed burn out by an internal reaction mechanism. The reaction of these particles can be at a constant diameter or at a constant density, or in a combination of both modes. On the one hand there is evidence of some particles burning at constant diameter. On the other hand there is also evidence of other particles burning in the combination mode, with partial penetration of the oxygen into the particles. This causes the subsequent burning by the surface retreat, with constant penetration depth of the reaction.[45]

Generally, the combustion of coal particles can occur in three regimes,[46] as shown in Figure 5.20. The first regime is the wakes trailing the particles; the fuel and oxidant are premixed, though imperfectly, before combustion starts. Burning is then accomplished by turbulent flame fronts sweeping through the gas. Such a regime is important for excessively high rates of heating, causing particles to explode and release volatiles.

In the second regime, the rate of vapourization is slow enough to allow a diffusion flame to attach itself either around or behind the particle. Combustion occurs in the flame sheet surrounding the particle and its wake. The reaction zone is dependent on the rate of oxygen diffusion into this zone.

The third regime is the char burnout region, where flame front approaches the particle until oxygen molecules can attack the char. Carbon is then converted to carbon monoxide.

5.3.2.1 *Volatilization Model*

Model Assumptions

The particles are spherical
The combustion takes place in an unsteady state
Gas is evolved throughout the particle at a rate which depends on temperature and concentration of volatiles in the char
The pressure inside the particles builds up
The volatiles are driven out of the particles because of the pressure driving force.

A. *Totally Detached Flame*. Rapid mixing, rapid volatilization

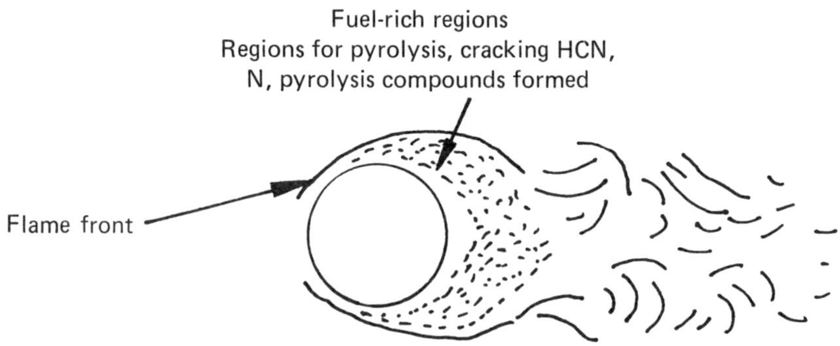

B. *Attached, Diffusion Flame*. Pyrolysis occurs in fuel-rich region char is being formed

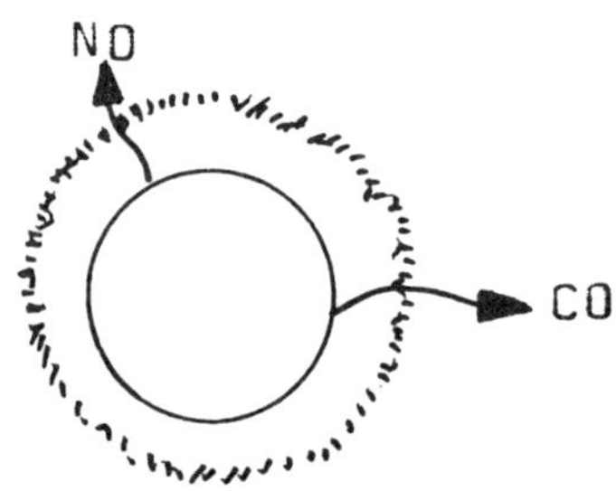

C. *Char Burn-out*. Surface combustion occurs when volatiles have been driven out. CO formed as result of surface reaction of carbon and O_2. This CO burns externally to the solid particle in a diffusion flame.

Figure 5.20 Regimes of coal particle combustion

During this time the volume of the particle (solids and space), and hence the porosity, are changing.

Model Formulation

The governing equations are as follows:

$$\eta \frac{\partial \rho}{\partial t} + \frac{1}{r^2} \frac{\partial}{\partial r}(r^2 \rho V) = R_v(T) \tag{5.2.33}$$

and

$$\frac{\partial P}{\partial r} + \frac{\mu}{\kappa} V = 0 \qquad (5.2.34)$$

$$P = \rho RT/M_0 \qquad (5.2.35)$$

The porosity η is expressed as

$$\eta = 1 - \left(\frac{R_0}{r_0}\right)^3 \qquad (5.2.36)$$

where R_0, r_0 are the initial and current particle radii and κ is the Darcy permeability defined as

$$\kappa = \beta \cdot \eta^3/(1 - \eta)^2,\ \beta \text{ is a constant} \qquad (5.2.37)$$

The rate of generation of volatile $R_v(T)$ is expressed as

$$R_v(T) = A\ \exp(-E/RT)\ C_{vol}(1 - \eta) \qquad (5.2.38)$$

where A, E are constants of the volatile generation rate expression and C_{vol} is the concentration of volatiles in solid phase and M_0 is the mean mixture molecular weight of volatiles. The rate at which the volatile concentration changes is given as

$$\frac{d}{dt} C_{vol} = -A\ \exp(E/RT)\ C_{vol} \qquad (5.2.39)$$

These equations are solved simultaneously throughout the time radius domain. The main equation to be solved is expressed as

$$\eta \frac{\partial \rho}{\partial t} - \frac{\kappa}{\mu} \left(\frac{RT}{M_0}\right) \frac{1}{r^2} \frac{\partial}{\partial r} \left(r^2 \rho \frac{\partial \rho}{\partial_r}\right) = R_v \qquad (5.2.40)$$

This equation is solved by a fully implicit forward-marching solution procedure.[46] The model has been applied only to demonstrate the effect of reducing η for two different values of the maximum temperature. It may be deduced that in some cases these physical phenomena strongly affect the volatile evolution rate.

4.3.2.2 *Char Burnout Model.* In the char burnout model, the strong heating causes the residue of the particle, containing the heavy ends, to be pyrolysed very strongly, and then converted into char. As the volatiles are depleted, the flame front approaches the particle until oxygen molecules can attack the char directly. This is the char burnout regime where carbon is converted into carbon monoxide at the surface, and the fuel nitrogen is converted to nitric oxides. The gas/solid reactions are heterogeneous. Heterogeneous surface reactions occur on the surfaces of the pores inside the coal particles, and on the external particle surface; while homogeneous reactions occur in the space of the pores and in the film surrounding the particle.[47]

The dominant modelling features of internal burning of the char are that carbon and fuel nitrogen are converted into carbon monoxide and to nitric oxide, respectively, at the particle surface. Further reactions to carbon dioxide and molecular nitrogen occur homogeneously inside the particle pore.[48] The reactions can be classified into:

1. Surface reactions consuming carbon and producing CO.

$$C + \tfrac{1}{2}O_2 \rightarrow CO \qquad (5.2.41)$$

2. Surface reactions consuming fuel nitrogen and producing NO.

$$N_2 + O_2 \rightarrow 2\,NO \qquad (5.2.42)$$

3. Surface reactions catalytically reacting NO, CO and carbon to reduce NO to N_2.
4. Homogeneous reactions oxidizing CO to CO_2.
5. Homogeneous reactions reducing NO to N_2 in local fuel rich environment. The reaction mechanism is expressed in ten reactions[49] as:

$$\begin{array}{rlcl}
\text{(i)} & C + OH & \rightleftharpoons & CO + H \\
\text{(ii)} & H + O_2 & \rightleftharpoons & OH + O \\
\text{(iii)} & O + H_2 & \rightleftharpoons & OH + H \\
\text{(iv)} & O + H_2O & \rightleftharpoons & OH + OH \\
\text{(v)} & H_2 + OH & \rightleftharpoons & H_2O + H \\
\text{(vi)} & O_2 + M & \rightleftharpoons & O + O + M \\
\text{(vii)} & H_2O + M & \rightleftharpoons & H + OH + M \\
\text{(viii)} & C + \tfrac{1}{2}O_2 & \rightleftharpoons & CO \\
\text{(ix)} & C + CO_2 & \rightleftharpoons & 2CO \\
\text{(x)} & C + H_2O & \rightleftharpoons & H_2 + CO
\end{array} \qquad (5.2.43)$$

Forward and backward reaction rates are given by:

$$K_f = A_f T^{n_f} \exp(-E_f/RT) \qquad (5.2.44)$$

and

$$K_b = A_0 T^{n_b} \exp(-E_b/RT) \qquad (5.2.45)$$

The values of A_f, A_b, n_f, n_b, E_f and E_b are given in Table 5.4 for the reaction mechanism 5.2.43.

5.3.2.3 *Simple Coal Combustion Model*

Model Assumptions

Fuel is composed of fixed spherical carbon particles
The reaction is composed of one mechanism

TABLE 5.4

Values of Reaction Rate Parameters for the Reactions of 5.2.43

Reaction	A_f	n_f	E_f	A_b	n_b	E_b
R1	5.60*E*11	0	1.080*E*3	7.29*E*13	0	23.41*E*3
R2	2.24*E*14	0	16.8*E*3	1.3*E*13	0	0
R3	1.70*E*13	0	9.45*E*3	7.3*E*12	0	7.3*E*3
R4	5.75*E*13	0	18.0*E*3	5.75*E*12	0	0.76*E*3
R5	2.19*E*13	0	5.15*E*3	8.41*E*73	0	20.1*E*3
R6	51.19*E*18	−1	118.7*E*3	25.63*E*17	−1	0.34*E*3
R7	3.00*E*15	0	105.0*E*3	15.75613	0	−14.628*E*3
R8	376.1	1	40.0*E*3	0	0	0
R9	3.6*E*9	0	85.3*E*3	0	0	0
R10	1.2*E*9	0	80.0*E*3	0	0	0

Note: In this table 5.60*E*11 = 5.60×10^{11} etc.

The reaction occurs in two steps
The reactions are of the first-order type.

Model Formulation

The present model was proposed by Richter.[50] The only reactions considered are the heterogeneous reaction between carbon and oxygen to yield carbon monoxide, and the homogeneous reaction of CO to CO_2 outside the boundary layer around the particles. The latter reaction is very fast, and in excess air combustion, no special equation for CO concentration is needed. The overall reaction rate R_s of the heterogeneous reaction between carbon and oxygen is assumed to be of first order. The rate is expressed as:

$$R_s = P_{ox}\left(\frac{1}{K_D} + \frac{1}{K_s}\right)^{-1} \tag{5.2.46}$$

and

$$K_s = A_s \exp(-E/RT_s) \tag{5.2.47}$$

where P_{ox} is the oxygen pressure and K_D is the diffusional rate coefficient, T_s is the particle temperature.

The rate of reaction refers to the external surface of the particles, but because of their porous structure, eventual internal combustion can occur. All effects of internal combustion on the reaction rate are included in the factor A_s and the apparent activation energy '*E*'.

The continuous size distribution of the pulverized fuel is approximated by *n* mono-sized fractions. For each size group, a transport equation is derived for the carbon concentration M_c. The source term S_c in each of these equations is divided into three parts:[50]

$$S_c = -S_{n\to n-1} + S_{n+1\to n} - S_{n,f} \tag{5.2.48}$$

where S_c, $S_{n\to n-1}$, $S_{n+1\to n}$ are associated with the transfer of mass between adjacent size groups because the particles burn with decreasing diameter. The term $S_{n,f}$ is the loss of mass of the nth size group per unit volume and unit time due to consumption of fixed carbon. The rate $S_{n,f}$ is obtained as

$$S_{n,f} = R_s \times A_n \tag{5.2.49}$$

where A_n is the external surface area of the nth fraction locally available per unit volume; it is given as

$$A_n = 6\rho_m m_c/(\rho_c D_{n,av}) \tag{5.2.50}$$

where ρ_m is the density of the suspension, ρ_c is the density of carbon and $D_{n,av}$ is the average particle diameter in the nth mass fraction.

Assessment of Model

Richter[50] investigated the effect of the activation energy on the calculated burnout on the axis of an anthracite flame. Figure 5.21 illustrates this effect. He also reported tests on the furnace of the International Flame Research Foundation; the furnace had a height of 1.6 m, a width of 1.5 m and a length of 10.5 m. The fuel was pulverized anthracite burnt with 5% excess air. The total heat input was equal to 2.5 MW. Figure 5.22 shows the measured and calculated temperatures; O_2 and CO_2 concentrations; and the degree of mixing and burnout. It can be seen from the figure that reasonable agreement was obtained.

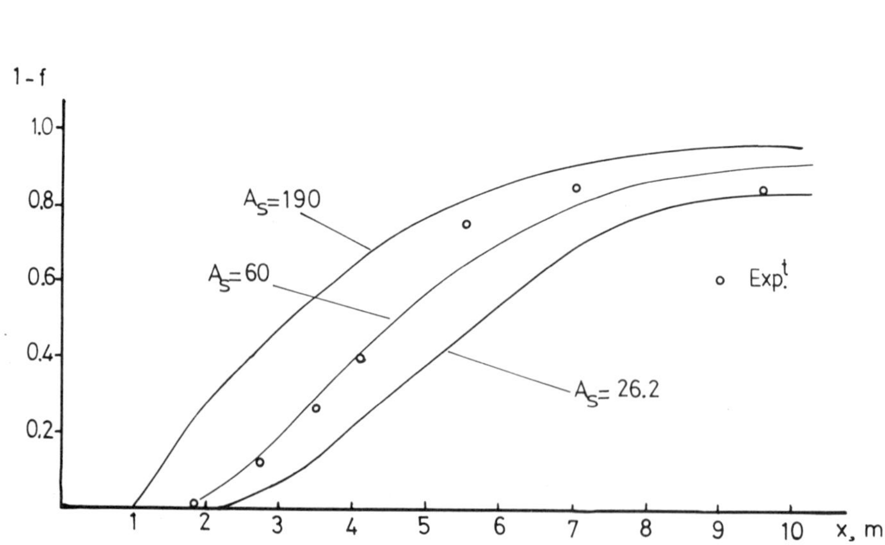

Figure 5.21 Effect of activation energy on burnout in an anthracite flame[50]

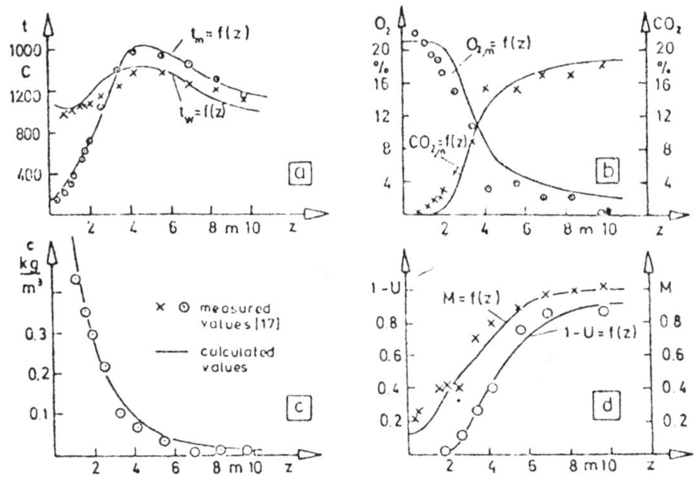

Figure 5.22 Measured and calculated distributions of temperature, oxygen and carbon dioxide concentrations in a coal-fired furnace[50]

(a) Temperature
(b) O_2 and CO_2 concentration
(c) Concentration of solids
(d) Aerodynamic degree of mixing and burnout

References to Chapter 5

1. Chigier, N. A., McCreath, C. G. and Roett, M. F. (1974) The structure of sprays in liquid fuel flames. *Acta Astronautica*, **1**, p. 687.
2. Khalil, E. E. (1978) A simplified approach for the calculation of free and confined spray flames. AIAA Paper 78-0029 (16th Aerospace Sciences Meeting).
3. Collier, J. G. (1973) *Convective Boiling and Condensation* (Mc-Graw-Hill), p. 7.
4. Weimer, J. C., Faeth, G. M. and Olson, D. R. (1973) Penetration of vapour jets submerged in subcooled liquids. *A.I.Ch.E. J.*, **19**, p. 552.
5. Avery, J. F. and Feath, G. M. (1974) Combustion of a submerged gaseous oxidizer jet in a liquid metal. *Proc. 15th Symp. (Int.) on Combustion*, p. 419.
6. Onuma, Y. and Ogaswara, M. (1974) Studies on the structure of a spray combustion flame. *Proc. 15th Symp. (Int.) on Combustion*, p. 453.
7. Komiyama, K., Flagan, R. C., and Heywood, J. B. (1976) The influence of droplet evaporation on fuel air mixing rate in a burner. *Proc. 16th Symp. (Int.) on Combustion*, p. 549.
8. Komiyama, K., Heywood, J. B. and Flagan, R. C., (1975) *Proc. 2nd European Synposium on Combustion*, p. 641.
9. Bousinesq, V. J. (1877) Théorie de l'écoulement tourbillant. *Mem. Pres. Acad. Sci.*, **23**, p. 46.
10. Bilger, R. W. (1976) Turbulent jet diffusion flames. *Progress Energy and Combustion Science*, **1**, p. 87.
11. Kent, J. H. and Bilger, R. W. (1976) The prediction of turbulent diffusion flame fields and nitric oxide formation. *Proc. 16th Symp. (Int.) on Combustion*, p. 1643.
12. Naguib, A. S. (1975) The prediction of axisymmetric free, jet, turbulent, reacting flows. Ph.D. Thesis, London University.
13. Kennedy, I. M. and Kent, J. H. (1978) Measurements of a conserved scalar in turbulent jet diffusion flames. *Proc. 17th Symp. (Int.) on Combustion*, p. 279.
14. Khalil, E. E. (1977) Flow and Combustion in axisymmetric furnaces. Ph.D. Thesis, London University.

15. Shearer, A. J., Tamura, H. and Faeth, G. M. (1978) Evaluation of a locally homogeneous model of spray evaporation. AIAA Paper 78-1042 (16th Aerospace Sciences Meeting).
16. Spalding, D. B. (1971) Mixing and chemical reaction in steady confined turbulent flame. *Proc. 13th Symp. (Int.) on Combustion*, p. 649.
17. Spalding, D. B. (1976) Development of the eddy break-up model of turbulent combustion. *Proc. 16th Symp. (Int.) on Combustion*, p. 1657.
18. Faeth, G. M. (1979) Spray combustion models – a review. AIAA paper 79-0293 (17th Aerospace Sciences Meeting).
19. Spalding, D. B. (1976) Mathematical models of turbulent flames, a review. *Combustion Science and Technology*, **13**, p. 3.
20. Pope, S. B. (1977) The implications of the probability equations for turbulent combustion models. *Combustion and Flame*, **29**, p. 235.
21. Pope, S. B. (1976) The probability approach to modelling of turbulent reacting flows. *Combustion and Flame*, **27**, p. 299.
22. Shearer, A. J. (1979) Evaluation of a locally homogeneous flow model of spray evaporation. Ph.D. Thesis, Pennsylvania State University.
23. Khalil, E. E. and Whitelaw, J. H. (1976) Aerodynamic and thermodynamic characteristics of kerosene spray flames. *Proc. 16th Symp. (Int.) on Combustion*, p. 569.
24. Abou Ellail, M. M. M., EL Kotb, M. and Khalil, E. E. (1979) On the structure of turbulent swirling premixed flames. *Proc. 2nd Turbulent Shear flow Symposium*, p. 3.24.
25. Mizutani, Y. and Nishimoto, T. (1972) Turbulent flame velocities in premixed sprays, Part I. Experimental study. *Combustion Science and Technology*, **6**, p. 1.
26. Polymeropoulos, C. E. and Das, S. (1975) *Combustion and Flame*, **25**, p. 247.
27. Spiers, H. M. (1962) *Technical Data on Fuel*, 6th Ed. (British National Committee, World Power conference).
28. Khalil, K. H., EL Mahallawy, F. M. and Moneib, H. S. (1976) Effect of combustion air swirl on the flow pattern in a cylindrical oil fired furnace. *Proc. 16th Symp. (Int.) on Combustion*, p. 135.
29. Faeth, G. M. (1977) Current status of droplet and liquid combustion. *Progress on Energy and Combustion Science*, **3**, p. 191.
30. Abou Ellail, M. M. M. (1974) Theoretical and experimental investigation of the precombustion period events of fuel droplets in gas turbine combustion chamber. Ph.D. Thesis, Cairo University.
31. Faeth, G. M. and Lazar, R. S. (1971) Fuel droplet burning rates in a combustion gas environment. *AIAA Journal*, **9**, p. 2165.
32. Canada, G. S. and Faeth, G. M. (1974) Combustion of liquid fuels in a flowing combustion gas environment at high pressures. *Proc. 15th Symp. (Int.) on Combustion*, p. 419.
33. Putnam, A. (1961) Integrable form of droplet drag coefficient, *ARS Journal*, **31**, p. 1467.
34. Ingebo, R. D. (1962) Heat transfer and drag coefficients for ethanol drops in a rocket chamber burning ethnol and liquid oxygen, *Proc. 8th Symp. (Int.) on combustion*, p. 1104 (Williams & Wilkins).
35. El Banhawy, Y. and Whitelaw, J. H. (1978). The calculation of flow properties of a confined kerosene-spray flame, *Proc. 2nd Int. Mech. Power Eng. Conference*, **2**, paper 18, Cairo.
36. Abou Ellail, M. M. M. and Khalil, E. E. (1980) A mathematical model for gas-liquid interaction in the turbulent reacting flows. *Applied Mathematical Modelling*, **4**, p. 136.
37. Crowe, C. T. (1978) A numerical model for the gas-droplet flow field near an atomizer, *Proc. 1st Int. Conference on Liquid Atomization and Spray Systems*.
38. Spalding, D. B. (1953) The combustion of liquid fuels. *Proc. 4th Symp. (Int.) on Combustion*, p. 847.
39. Abou Ellail, M. M and Khalil, E. E. (1978) Mathematical modelling of turbulent spray flames. *Proc. 2nd Int. Mech. Power Eng. Conference,* **2**, paper 19, Cairo.
40. Anasoulis, R. F., McDonald, H. and Buggelin, R. C. (1974) Development of a combustor flow analysis, Part I: theoretical studies. AFAPL-TR-73-98.
41. Gibeling, H. J., McDonald, H. and Briley, W. R. (1975) Development of a three dimensional combustor flow analysis. AFAPL-TR-75-59.

42. Boysan, F., Ayers, W. H., Swithenbank, J. and Pan, Z. (1980) Three dimensional model of spray combustion in gas turbine combustors. University of Sheffield Report HIC 354.
43. Beer, J. M. (1960) Investigation into the combustion mechanism of pulverised fuel flames with special respect to the mixing of flames with secondary air. Ph.D. Thesis, University of Sheffield.
44. Smoot, L. S., Horton, M. D., and Williams, G. A. (1976) Western States Section of Combustion Institute, Paper 76-3.
45. Essenhigh, R. H. (1970) Dominant mechanisms in the combustion of coal. ASME Paper 70-WA/Fu-2.
46. Lilley, D. G. and Wendt, J. O. L. (1976) Modelling pollutant formation in coal combustion. *Proc. 25th Heat Transfer and Fluid Mechanics Institute*, p. 196.
47. Wendt, J. O. L. and Schulze, O. E. (1976) On the fate of fuel nitrogen during coal char combustion. *A.I.Ch.E., J.*, **22**, p. 102.
48. Krazinski, J. L., Buckins, R. O., and Krier, H. (1979) Coal dust flames: a review and development of a model for flame propagation. *Progress in Energy and Combustion Science*, **5**, p. 31.
49. Lilley, D. G. (1978) Methods for rapid solution of stiff systems of diffusion/reaction equations; analysis of internal burning during coal char combustion. Presented at Combustion Institute (Canadian Section) Meeting.
50. Richter, W. (1974) Prediction of heat and mass transfer in a pulverised fuel furnace. *Letters in Heat and Mass Transfer*, **1**, p. 83.
51. Gibson, M. M. and Morgan, B. B. (1970) Mathematical model of combustion of solid particles in a turbulent stream with recirculation. *J. Inst. Fuel*, **43**, p. 517.

CHAPTER 6

HEAT TRANSFER AND POLLUTANT FORMATION MODELS

6.1 INTRODUCTION

6.1.1 Scope

In furnaces and combustors, it is most essential to represent adequately the characteristics of heat transfer and energy balance in a mathematical model of the flow and reaction processes. This is done to be able to determine the actual heat flux distribution to the furnace walls, and to predict the local gas temperature distribution. Since chemical kinetics involved in the combustion process are strongly temperature dependent, accurate calculation of the local gas temperature is needed if the evaluation of the NO_x pollutant formation rate and concentration is to be meaningful. In this chapter, the phenomena of heat transfer and pollution formation are discussed.

In real furnaces and combustors, two modes of heat transfer exist, namely radiation and convection. A fundamental calculation of heat transfer requires the simultaneous solution of fluid flow, chemical reaction and energy transfer as explained earlier in Chapter 3. The inter-dependence and interaction of these processes make the problem extremely complex; the general approach adopted is to develop simplified models that deal with the turbulence and the reaction characteristics, as described earlier in Chapters 3, 4 and 5. The third and last characteristic of the flow, i.e. radiative and convective heat transfer, is discussed here in terms of available models, their assumptions, formation and validation in furnaces and combustion chambers.

The emphasis is on the emission control, and in particular on the urgent desire for simultaneous elimination of both reducing species (CO and hydrocarbons) and oxidizing species (nitrogen oxides). High efficiency, reasonable size and low emission are the goals to be achieved. The formation of nitric oxides, nitrogen dioxides and carbon monoxides in turbulent reacting furnace flows are considered; the model assumptions, formulation and assessment are discussed.

6.1.2 Outline of the Chapter

The first part of the chapter deals with radiation models with emphasis on those applied in practical furnaces, such as the zone method, flux models, spherical harmonics and the discrete ordinate models. Critical review of each of the models is given with various examples shown to assess the validity of the model. For the sake of completeness, the process of heat transfer by convection is decribed briefly.

The second part of the chapter deals with the pollution formation and describes the formation reactions and modelling assumptions. Samples of calculated NO_x profiles are shown in comparison to the corresponding measurements.

6.2 HEAT TRANSFER MODELS

6.2.1 Equation of Radiant Energy Transfer

The basis on which all the methods of solving the radiation problems stand is the equation of radiant energy transfer. It is derived by writing a balance equation for monochromatic radiant energy, passing in a specified direction, through a small volume, in an emitting–absorbing–scattering medium. For steady-state conditions, and for coherent isotropic scattering this equation is expressed[1] as

$$(\boldsymbol{\Omega}\cdot\nabla)I_\lambda(\mathbf{r},\boldsymbol{\Omega}) = -(K_{a,\lambda}+K_{s,\lambda})I_\lambda(\mathbf{r},\boldsymbol{\Omega}) + K_{a,\lambda}I_{b,\lambda}(\mathbf{r}) + \frac{K_{s,\lambda}}{4\pi}\int_{\Omega=4\pi} I_\lambda(\mathbf{r},\Omega')\,d\Omega' \tag{6.2.1}$$

where $I_\lambda(\mathbf{r},\boldsymbol{\Omega})$ is the radiation intensity; it is a function of the wavelength, position and direction. $I_{b,\lambda}$ is the monochromatic intensity of black body radiation at the temperature of the medium. $K_{a,\lambda}$ and $K_{s,\lambda}$ are the spectral absorption and scattering coefficients of the medium respectively, for a wavelength λ.

The left-hand side of the above equation represents the gradient of the radiation intensity in the specified direction $\boldsymbol{\Omega}$. The right-hand side terms represent, in turn, the change of radiative intensity due to absorption and out-scattering, emission and in-scattering. The intensity gradient in the direction $\boldsymbol{\Omega}$ can be represented as:

$$(\boldsymbol{\Omega}\cdot\nabla)I_\lambda(\mathbf{r},\boldsymbol{\Omega}) = \mu_0\frac{\partial I}{\partial r} - \frac{\eta_0\partial I}{r\partial\zeta} + \xi_0\frac{\partial I}{\partial x} + \frac{\eta_0}{r}\frac{\partial I}{\partial\theta} \tag{6.2.2}$$

and

$$I = I(\mathbf{r},\theta,x)$$

μ_0, η_0 and ξ_0 are the directional cosines, and are defined as shown in Figure 6.1 by:

$$\mu_0 = \sin\theta\cos\zeta$$
$$\eta_0 = \sin\theta\sin\zeta$$
$$\xi_0 = \cos\theta$$

6.2.2 Representation of Real Furnace Gas

In natural gas and oil fired combustors, only three species contribute significantly, in the infra-red region, to radiation. These species are the carbon dioxide and the water vapour, which are products of combustion, and hot soot particles within the flame, produced by the thermal decomposition of fuel under oxygen deficient situations.[2] The gases radiate in several bands, while the soot emits continuously over a wide range of

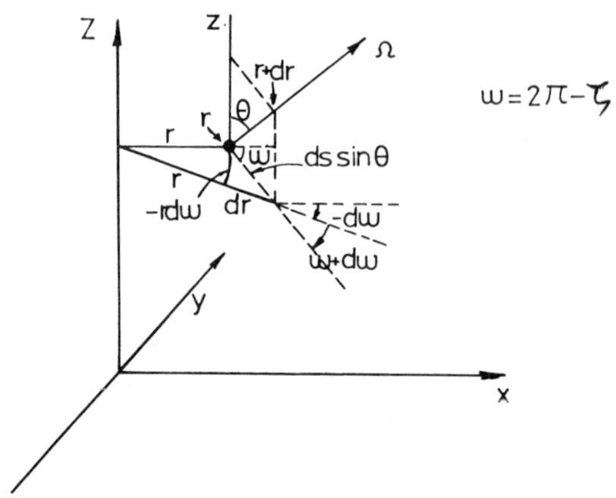

Figure 6.1 Radiation intensity distribution

wavelengths. The total emittance of a gas is expressed as

$$\varepsilon(T, L) = \frac{1}{\sigma T^4} \int_0^\infty E_\lambda(T)[1 - \exp(-K_{a,\lambda}L)]\, d\lambda \tag{6.2.3}$$

where L is the radiation path length, $E_\lambda(T)$ is the Planck spectral distribution of emissive power, and σ is the Stefan–Boltzmann constant.

For real gas situations, the following grey gas emittance-path length relation does not hold:

$$\varepsilon_{\text{grey}} = 1 - \exp(-KL) \tag{6.2.4}$$

Instead, a modified expression is deduced by representing the real gases by a mixture of grey gases; this expression is given by Truelove[2] as

$$\varepsilon(T, L) = \sum_n a_n(T)[1 - \exp(-K_n L)] \tag{6.2.5}$$

and

$$\sum_n a_n(T) = 1$$

a_n is a weighting coefficient and is the fraction of black body energy in the wavelength region where the absorption coefficient is K_n. The coefficient a_n is given as

$$a_n = \frac{1}{\sigma T^4} \int_{K=K_n} E_{b,\lambda}(T)\, d\lambda \tag{6.2.6}$$

For most engineering calculations adequate accuracy may be obtained by using only three or four terms of Equations 6.2.5 and 6.2.6. The partial pressures of carbon dioxide and water vapour are introduced into the expression for the emittance of gas mixture in the form of:

$$\varepsilon_g = \sum_n a_{g,n}(T)[1 - \exp(-K_{g,n}(P_W + P_C)L)] \tag{6.2.7}$$

where P_W and P_C are the partial pressures for water vapour and carbon dioxide respectively.

As many industrial furnaces and combustors use heavy fuels, luminous radiation from suspended soot particles contributes to the energy balance. The absorption coefficient is given as

$$K_{\lambda,s} = \frac{C}{\rho_s \lambda} F \tag{6.2.8}$$

where C is the soot concentration, F is a function depending on soot optical properties, and ρ_s is soot density.

For a combination of carbon dioxide, water vapour and soot, a more complex expression emerges.[3]

It should be emphasized that the success or failure of any of the radiation models depends on how easy the solution of the model can be coupled with the aerodynamic solution procedure.

6.2.3 Radiation Models

6.2.3.1 *Zone Method.* Several methods have been developed for the sake of solving the radiative heat transfer Equation 6.2.1. Of these, the zone method[3,4,5] has been applied both extensively and successfully to the prediction of radiant heat transfer in furnaces and combustion chambers.[6,7,8] The combustion chamber is subdivided into a volume (in this case the gas) and surface zones; these are sufficiently small for the local temperature and other physical properties to be considered uniform within each of the zones. The zone size and geometry are dependent on the furnace geometry. Cubical or square zones are both appropriate to rectangular furnaces, while for furnaces with cylindrical symmetry, the zones are chosen to be coaxial cylindrical gas annuli and coaxial cylinder wall and end rings.

The incident flux at a surface zone j is obtained from the integral of the transfer equation; the integration is performed over the zone surface A_j. The resultant expression is

$$A_j q_j^+ = \sum_i \overline{s_i s_j}\, \overline{q_i} + \sum_K \overline{g_K s_j}\, E_K \tag{6.2.9}$$

where

$$\overline{s_i s_j} = \int_{A_i}\int_{A_j} \frac{\cos\theta \cos\theta' \, \tau(\mathbf{r}, \mathbf{r}')}{\pi |\mathbf{r} - \mathbf{r}'|^2} dA\, dA' \tag{6.2.10}$$

and

$$\overline{g_k s_j} = \int_{V_k}\int_{A_j} \frac{K(\mathbf{r}'') \cos\theta \, \tau(\mathbf{r}, \mathbf{r}'')}{\pi |\mathbf{r} - \mathbf{r}''|^2} dA\, dV'' \tag{6.2.11}$$

where $\mathbf{r}'$ is the position at the boundary point; τ is the transmission factor; $\mathbf{r}''$ is the position inside the medium; dV'' is the volume increment; i and j refer to surface zones; and E_k is the black body emissive power.

The geometrical factors $\overline{ss}$ and $\overline{gs}$ refer to the surface-to-surface and gas-to-surface direct exchange areas of Hottel *et al.*[9] The net flux into surface zone j is given as

$$A_j q_j = \sum_i \overline{S_i S_j} E_i + \sum_K \overline{G_k S_j} E_k - A_j \varepsilon_j E_j \tag{6.2.12}$$

and the geometrical factors $\overline{SS}$ and $\overline{GS}$ are the total exchange areas.[9]

The direct radiative exchange between two finite zones, with an absorbing grey gas between them, is proportional to the difference in emissive powers (σT^4) of the two zones. The direct exchange areas are the constants of proportionality relating the net radiative flux to the difference in the emissive power. These exchange areas are functions of the zone dimension, and gas absorption coefficient.[10] When the furnace walls are not black, radiation from one zone can reach the second zone, both directly and indirectly, after one or more surface reflection; the total radiant flux between zones depends on those total exchange areas $\overline{S_i S_j}$.

The temperature of each zone is obtained from the solution of total energy balances, one for each of the zones with the unknown temperature. The directed flux areas enable a radiation energy balance to be drawn up. For a total energy balance, other forms of energy transport are considered, i.e. convection, diffusion and gas mass flow due to combustion heat release.

Assessment of Model

Extensive tests of the capabilities of the zone method and its application to furnace flows were carried out. The prediction of the radiative heat flux distribution along the furnace walls, for this model, requires the specification of local gas and wall temperatures, densities and mass velocities. These were obtained from the available measurements.

Validation tests vary in their complexity and completeness; such tests were reported for axisymmetric and three-dimensional enclosures. Three-dimensional furnace flows calculations, were reported by, among others, Johnson *et al.*,[5] Zuber *et al.*,[11] Hottel *et al.*,[4] Lowes *et al.*[12] and Richter *et al.*[13] These investigations indicated the good agreement obtained between the measured and the calculated wall radiative heat flux and radiation intensity at various locations in the flame. The effect of non-grey gases approximation and scattering heat transfer were emphasized.[5] A second and equally important analysis is that of Hyde *et al.*,[14] which discussed the predictions in large furnaces.

Calculations of the incident radiation were carried out for the furnace of the International Flame Research Foundation,[11,15] whose dimensions were 2 x 2 x 6 m, firing natural gas at a rate of 2.96 MW with excess air of 4%. The particular case shown here is for flame 29 with zero swirl. Figure 6.2 shows the furnace geometry divided into zones. Figure 6.3 shows the distribution of the incident radiant flux along the furnace. Measured and calculated fluxes are in reasonable agreement, which is rather satisfactory considering that the experimental error and averaging procedure can have a cumulative effect on the computations.

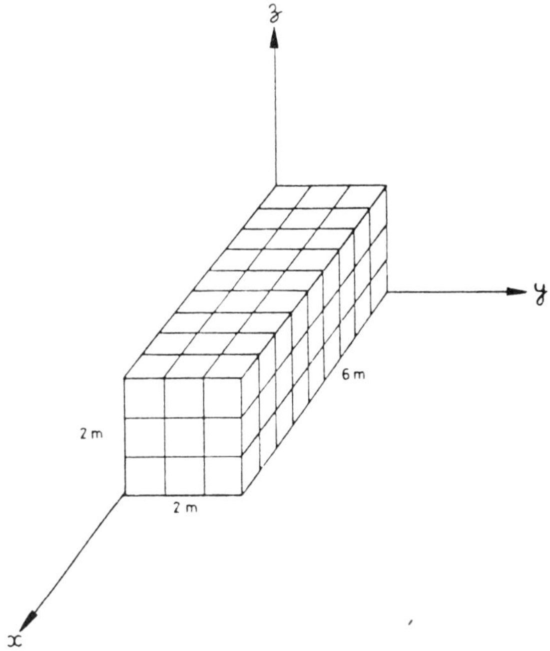

Figure 6.2 Furnace geometry divided into zones

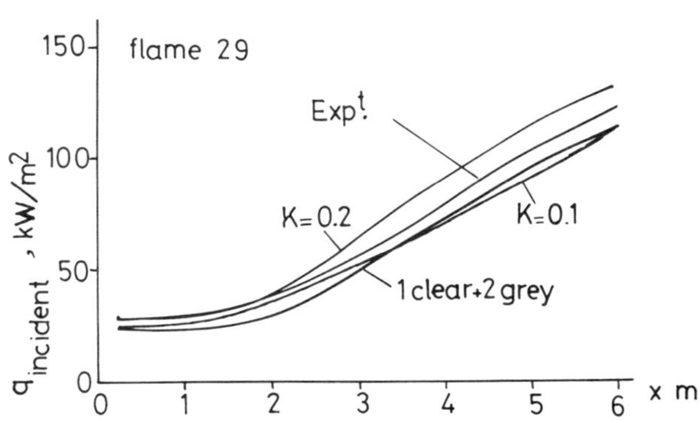

Figure 6.3 Incident radiant flux distribution along the furnace of reference 12. Zone method calculations; $S = 0$

A second example of the capability of the zone method is shown in Figure 6.4 for the same furnace, but for a highly swirling flame (flame 11). The flame is shorter due to the action of swirl ($S = 1.95$). This is consistent with the location of the peak wall heat flux which is nearer to the burner than the peak observed in Figure 6.3. Hyde *et al.*[14] calculated the incident radiation in an experimental furnace of reference (16), M3 experiments (flame 10). The furnace was fired with natural gas (3 MW) using an inclined tunnel burner mounted in the centre of the end wall ($x = 0$). The measured and calculated incident radiation fluxes are shown in Figure 6.5 for the furnace roof and

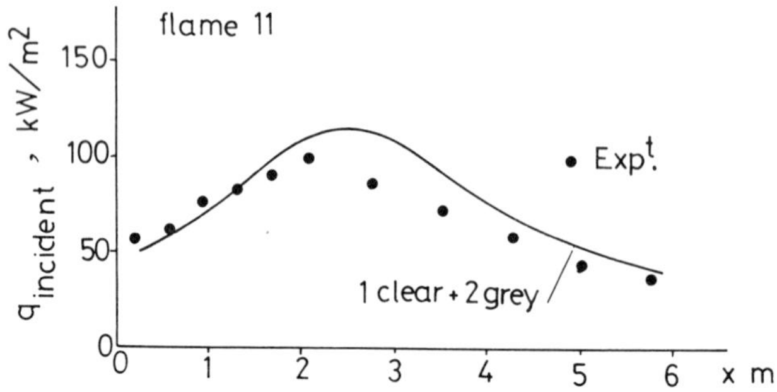

Figure 6.4 Incident radiant flux distribution along the furnace of reference 12. Zone method calculations; $S = 1.95$

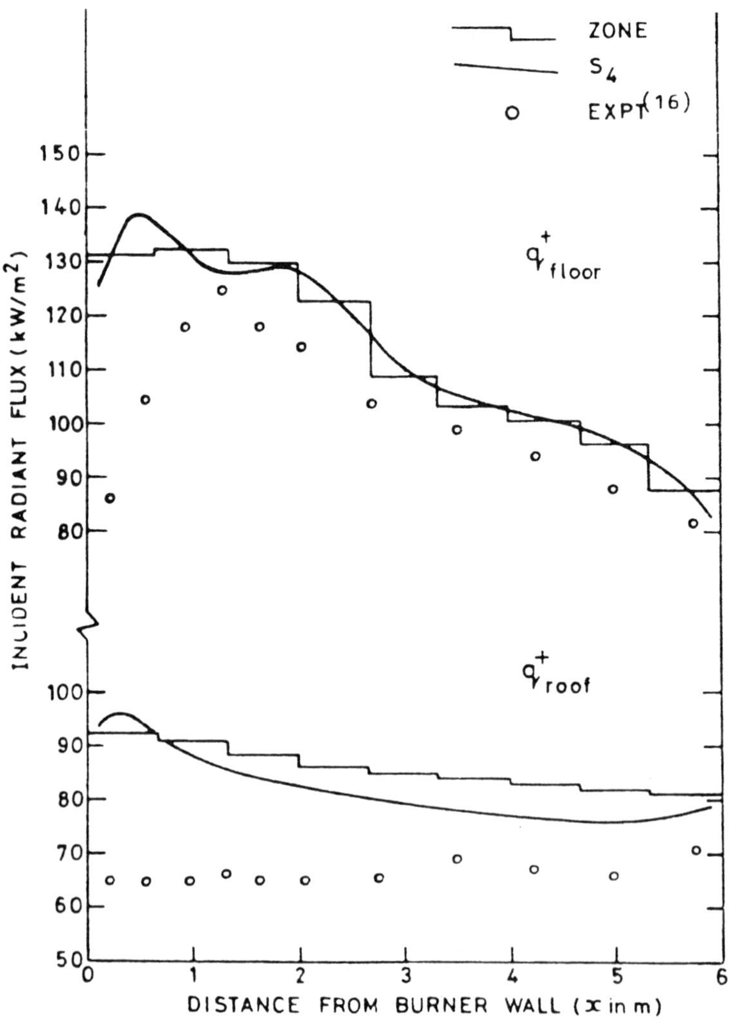

Figure 6.5 Comparison between zone method predictions[14] and the experiments for the incident radiation flux in the furnace of reference 16

floor. The discrepancies observed can be attributed to the inaccuracy of the measurements and the approximate representation of the temperature distributions.

In axisymmetric flows, radiative heat transfer calculations were reported by Steward *et al.*,[6] Heap *et al.*,[17] Wu *et al.*[18] and Selcuk *et al.*[19] The calculations of the cross-sectional average gas temperature in the furnace of Wu *et al.*[18] are shown in Figure 6.6. The calculated average cross-sectional temperatures[19] are in good agreement with the corresponding measurements,[18] for $S = 0.84$ and a furnace load of 1.06 MW.

The measured and the calculated wall heat flux for the same furnace are shown in Figure 6.7. The measured wall heat flux distribution was obtained from measuring the thermal load of the ten water-cooled sections which constitute the furnace. The zone method predicts correctly the wall heat flux distribution; such agreement is attributed to the way the model depends on measured temperatures, densities and velocities. These parameters are not predicted via a numerical procedure but are specified at each zone from available measurements.

Investigators who carried out calculations of radiative heat flux using the zone method described the merits of this approach and illustrated its capability in calculating the radiative heat transfer characteristics. It is, however, very important to discuss the disadvantages and limitations of this method when applied to combustion chambers of steep property gradients. These limitations can be sumamrized as:

1. Due to the integro-differential nature of the transfer equation, the zone method cannot be easily coupled to the aerodynamic and combustion calculation procedures, such as the ones described in Chapter 7.

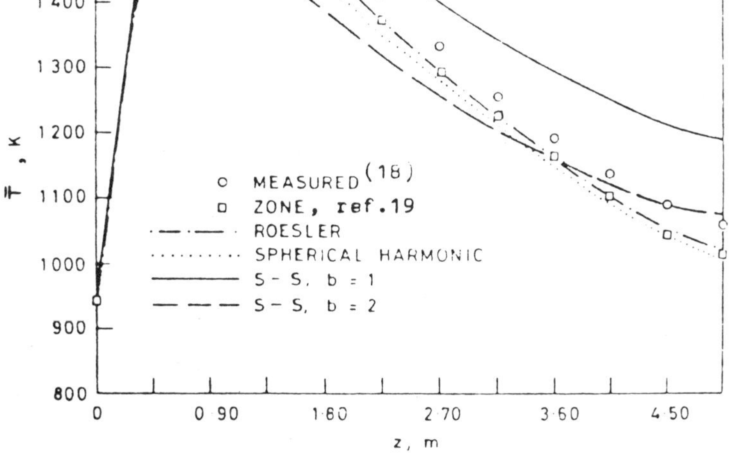

Figure 6.6 Comparison between predicted and measured cross-sectional average gas temperatures

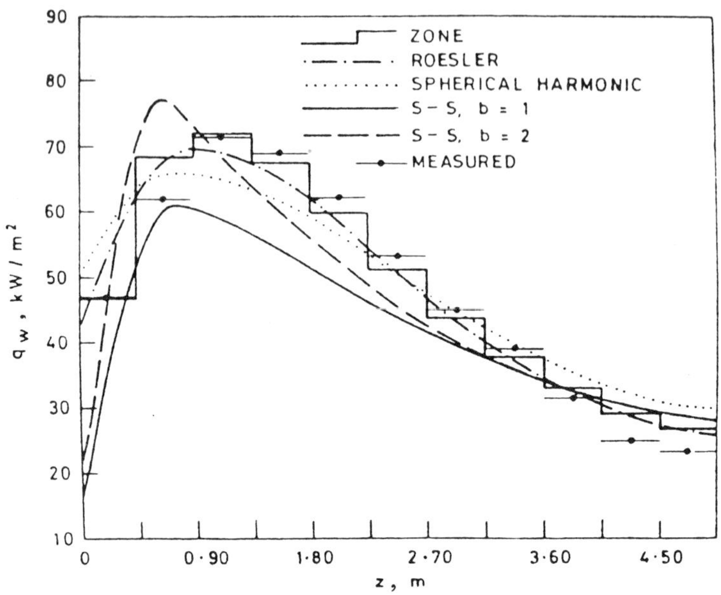

Figure 6.7 Comparison between predicted and measured total net flux densities to the water-cooled side walls (furnace of reference 18)

2. Limitation is also imposed on the model because of the sizes of the zones, and the necessity of matching them with the fluid control volumes.

In the discussions of Khalil *et al.*[20] and Khalil,[21] it was shown that in many practical furnace applications, the cross-stream properties vary sharply, particularly near the reaction zones. This observation obviates the support of uniform property assumption commonly used in zone method calculations. The zone method, however, uses measured gas temperature, and hence eliminates a source of error that stems from using predicted temperatures, which are, in turn, obtained using combustion models and simultaneous solution of conservation equations.

6.2.3.2 *Spherical Harmonics Model.* The spherical harmonics method was first suggested by Jeans[22] in connection with certain problems in astrophysics. A general description of this method is given in the books by Kouganoff[23] and Davidson.[24] Here, the intensity can be represented by a series of spherical harmonics as:

$$I(\mathbf{r}, \Omega) = \sum_{n=0}^{\infty} \sum_{m=0} [A_n^m(r, \xi, x) \cos m\zeta' + P_n^m(r, \zeta, x) \sin m\zeta'] P_n^m(\cos \zeta') \quad (6.2.13)$$

where P_n^m are associated Legendre polynomials. As the value of n, which is the number of terms retained, increases to infinity, the solution approaches the exact solution of Equation 6.2.1, after the intensity distribution has been substituted into it. The lowest order approximation P_1, is equivalent to the Eddington approximation.[25] When the distribution of the intensity, given by Expression 6.2.13, is substituted into the transfer equation, a relation which is equivalent to an infinite set of differential equations involving the A_n^m, is obtained. Two different shcemes of approximation have been proposed to represent the boundary conditions for the spherical harmonics method. For a diffuse reflecting wall, the exact boundary condition is difficult to apply to truncated series, and an alternative approach was given.[26] A common approximation is the P_1, in which the series is truncated after the first four terms. The equation of radiative transfer, for an axisymmetric flow, in a grey non-scattering medium is given by:

$$\frac{1}{r}\frac{\partial}{\partial r} rq_r + \frac{\partial q_x}{\partial x} = K_a(4E_G - G) \tag{6.2.14}$$

$$\frac{\partial}{\partial r} G = -3K_a q_r \tag{6.2.15}$$

and

$$\frac{\partial}{\partial x} G = -3K_a q_x \tag{6.2.16}$$

where G is the total incident flux density at a point, and q is the net radiant density in the positive coordinate direction denoted by the subscript r, x.

The two flux spherical harmonics model of Selcuk *et al.*[26] represent the radiant flux densities as:

$$q_\delta^+ = \frac{G}{4} + \frac{q_\delta}{2} \tag{6.2.17}$$

and

$$q_\delta^- = \frac{G}{4} - \frac{q_\delta}{2} \quad \text{for} \quad \delta = r, x \tag{6.2.18}$$

Assessment of Model

Several investigators, for example Whitacre *et al.*[8] and Selcuk *et al.*[19,26] incorporated the spherical harmonics models in their computations of furnace flows. Comparisons were reported between the results obtained using the spherical harmonics model and those obtained using the zone method, by Selcuk *et al.*[26] for an axisymmetric heater. Figure 6.8 shows the comparisons of gas temperature distributions and the net heat flux density to the tube surfaces.

Further comparisons of the dimensionless tube surface temperature and the sidewall refractory temperature are shown in Figure 6.9. Both the zone method and the spherical harmonics model are in good agreement. Selcuk *et al.*[26] recommended the use of this method for furnace design purposes. However this model still suffers from the same drawback encountered in the zone method; it cannot be coupled readily to the flow and reaction solution procedure.

6.2.3.3 *Discrete Ordinate Model.* This method was first suggested by Chandrasekhar[27] for one-dimensional astrophysics problems. More recently, Carlson *et al.*[28] developed the discrete ordinate approximation system to solve multi-dimensional

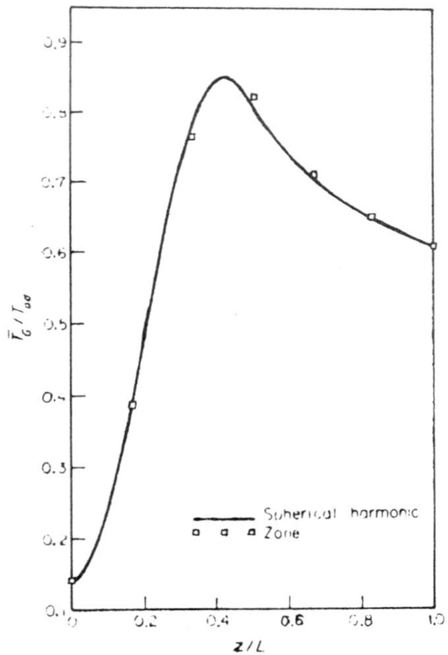

Comparison of dimensionless furnace gas temperature distributions for $L_f = 0.48L$.

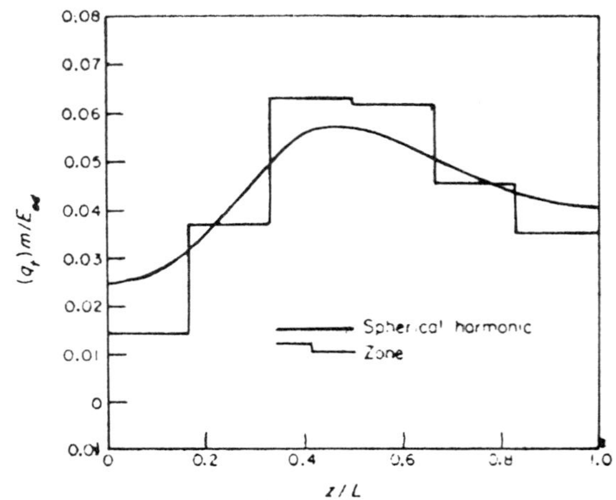

Comparison of dimensionless average net flux densities to the tube surfaces for $L_f = 0.48L$.

Figure 6.8 Gas temperature distributions and net heat flux in a furnace (calculations of reference 26 using the spherical harmonics model)

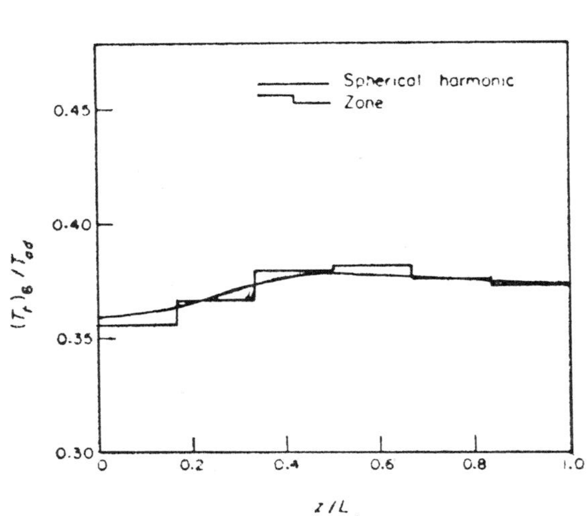

Comparison of dimensionless tube surface temperatures for $i = 6$ and $L_f = 0.48L$.

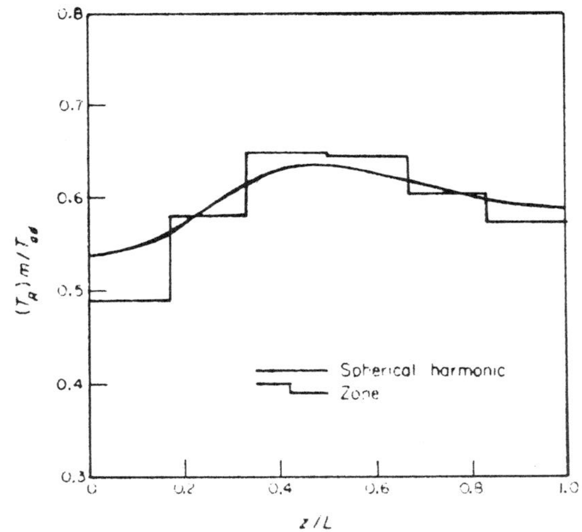

Comparison of dimensionless sidewall refractory temperatures for $L_f = 0.48L$.

Figure 6.9 Comparisons of dimensionless tube surface temperature and heat flux (calculations of reference 26 using spherical harmonics model)

neutron transport problems. The basic concept of this method is very simple. The angular integrals in the radiant energy transfer equation are approximated using a numerical quadrature scheme. The incident radiation integral of

$$\int_{4\pi} I(\mathbf{r}, \Omega)\, d\Omega = \sum_{i}^{N} a_i I(\mathbf{r}, \Omega_i) \tag{6.2.19}$$

is evaluated by splitting the integral into Ω_i directions. The so called S_N equations are obtained by evaluating Equation 6.2.1 for a set of discrete directions, *i.e.* Ω_i, $i = 1, \ldots, N$. Associated with each direction $\mathbf{\Omega}_i$, is a weight factor a_i; these factors serve as the angular quadrature weights and are used to evaluate the angular integrals of the intensity distribution. Each direction $\mathbf{\Omega}_i$ is viewed as a point on the surface of a unit sphere with which a surface area a_i is associated; Figure 6.10 illustrates this configuration. The sum of the weights to the surface area of the unit sphere is given by

$$\sum_{i}^{N} a_i = 4\pi$$

The discrete ordinate equations for the intensities $I(\mathbf{r}, \mathbf{\Omega}_i)$ are obtained by evaluating the exact transfer Equation 6.2.1 at each of the discrete directions $\mathbf{\Omega}_i$. Similarly, the boundary conditions are obtained by evaluating the exact boundary conditions[20] at the quadrature points Ω_i.

The radiative heat flux distribution is obtained from an overall energy balance. The net radiative heat flux is calculated by evaluating the exact boundary conditions, thus yielding

$$\mathbf{q}(\mathbf{r}) = \int_{4\pi} \Omega I(\mathbf{r}, \Omega)\, dr = \sum_{i}^{N} a_i \Omega_i I(\mathbf{r}, \Omega_i) \tag{6.2.20}$$

The choice of quadrature scheme is arbitrary; nevertheless, restrictions are imposed on the $\mathbf{\Omega}_i$ and a_i in order to preserve symmetries and invariance properties of the physical system. The accuracy of the model depends on the accuracy of the quadrature formula chosen to evaluate the integrals in Equations 6.2.19 and 6.2.20. The quadrature scheme suggested by Carlson *et al.*[28] is applied.

The quadrature scheme used by Khalil *et al.*[20] and Hyde *et al.*[14] is a fully symmetric S_4 scheme. The order N relates the number of discrete direction cosines on each coordinate axis; with $N = 4$ there are four values on each axis. The values of the direction Ω_i and associated weight factors a_i are given by Hyde *et al.*[14] for axisymmetric and three-dimensional geometries.

Model Formulation

The discrete ordinates equations are given as:

$$\xi_i \frac{\partial I_i}{\partial x} + \frac{\mu_i}{r} \frac{\partial r}{\partial r} I_i - \frac{1}{r} \frac{\partial}{\partial \zeta} (\eta_i I_i) = -K I_i + K \frac{\sigma \bar{T}^4}{\pi} \tag{6.2.21}$$

and $i = 1, \ldots, N$, $I_i = I(r, \zeta, x, \Omega_i)$.

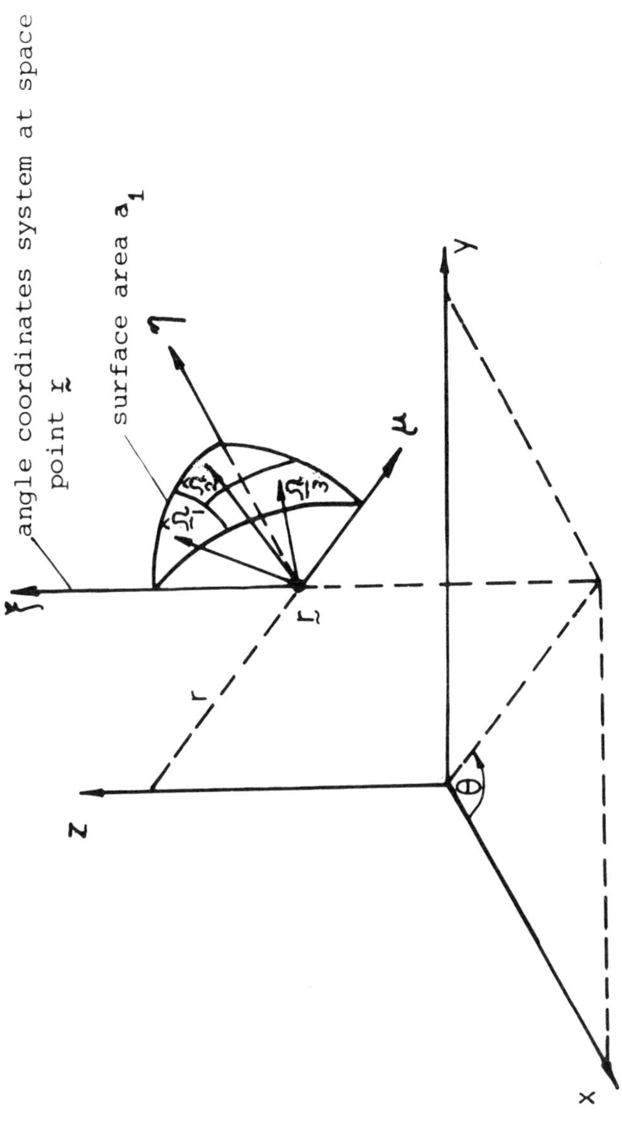

Figure 6.10 Pictorial representation of the angular quadrature weights a_i as elements of area on the surface of the unit sphere in angle space. To retain clarity only one octant of the sphere is shown

Equation 6.2.21 is a system of N coupled partial differential equations for the N intensities I_i. When Equation 6.2.21 is solved, and the intensities determined, the radiant heat flux is calculated as

$$\mathbf{q} = \int_{4\pi} \Omega I \, d\Omega = \sum_{i=1}^{N} a_i \Omega_i I_i \tag{6.2.22}$$

The components of the vector $\mathbf{q}$ are the net fluxes in the three coordinate directions. Hemispherical fluxes may be determined using the numerical quadrature to evaluate the half-range angle integrals.

The radiation source term in the enthalpy equation is expressed as

$$\bar{S}_H = K \sum_{i=1}^{N} a_i I_i - 4K\sigma \bar{T}^4 \tag{6.2.23}$$

The total number of independent discrete directions, for S_4, is 24 and is reduced to twelve for axisymmetric cylindrical geometries.[20]

Assessment of Model

The above model was used to calculate the radiative heat transfer characteristics in large gas-fired furnaces, by Hyde *et al.*,[14] Khalil *et al.*[20] and Khalil.[21] The computations carried out in reference 14 are for two different furnaces. The first furnace is that of reference 16. The flame situation was described before in Section 6.2.3.1; Figure 6.11 shows the incident radiant flux at the roof and floor of the furnace. The measured and calculated values are in reasonable agreement; discrepancies are observed in the upstream region, in the burner vicinity, and are attributed to the assumptions of the model, together with the uncoupling of the aerodynamic and reaction governing equations with those of the radiative transfer. On the same figure, the zone model calculations previously discussed in Figure 6.5 are also shown.

The second furnace is that of Wu *et al.*;[18] this furnace consists of axisymmetric vertical enclosure, segmented into 10 water-cooled sections. The burner was centred at the bottom of the furnace; the measured temperature distribution was fed into the present model to obtain the calculations of the wall heat flux. Figure 6.12 shows comparisons between the S_4 predictions[14] of radiant head flux distribution and the corresponding measurements.[18] The S_4 model predicts the wall heat flux satisfactorily. This agreement is helped by the insertion of the measured temperature profiles, rather than predicting them through solving the flow and energy differential equations. Also shown on the figure are the calculations obtained[14] using the zone method. The S_4 model has a number of advantages over the zone method. First it is computationally simpler and more economical (although, of course, somewhat less accurate), requiring very little computer storage and run time. The second important feature is that the S_4 model can be easily coupled to the equations for fluid flow and chemical reaction in a complete mathematical model, used for furnace flow predictions.

An example of the application of the S_4 model to furnace flow predictions was reported by Khalil *et al.*[20,29] The model was applied to predict the radiative heat flux distribution in the furnace of reference 30. The furnace dimensions were 4 x 4 x 11 m,

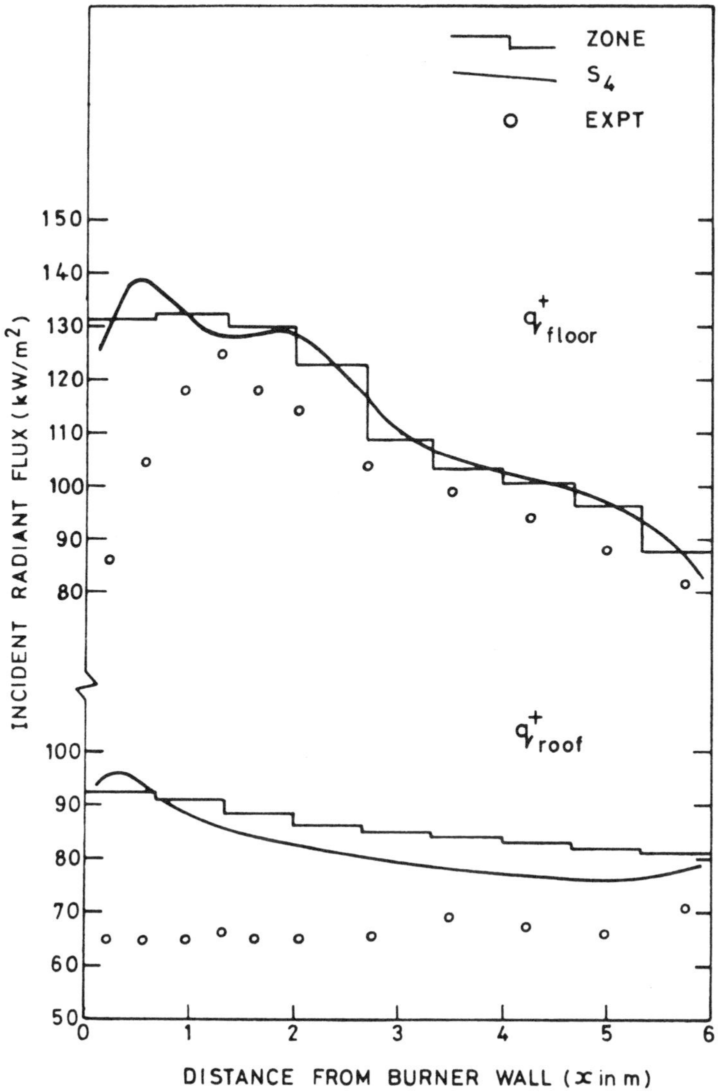

Figure 6.11 Comparison between discrete ordinate predictions[14] and the experimental distributions of incident radiant flux in the furnace of reference 16

and the thermal load was 12.88 MW. The flow pattern and reaction characteristics were obtained with the aid of a numerical procedure solving the conservation equations governing the transport of mass, momentum and energy. The solution involving the radiation model was coupled with that of the flow governing equations. The source term $\bar{S}_H$ obtained from Equation 6.2.23 is substituted in the enthalpy Equation 3.2.8. The radiative characteristic of the furnace gas was represented[20] by a single grey with an absorption coefficient K, $K = 0.3\ \mathrm{m}^{-1}$, obtained by matching a one grey gas representation to the multi-grey gas representation described earlier in this chapter. The effect of

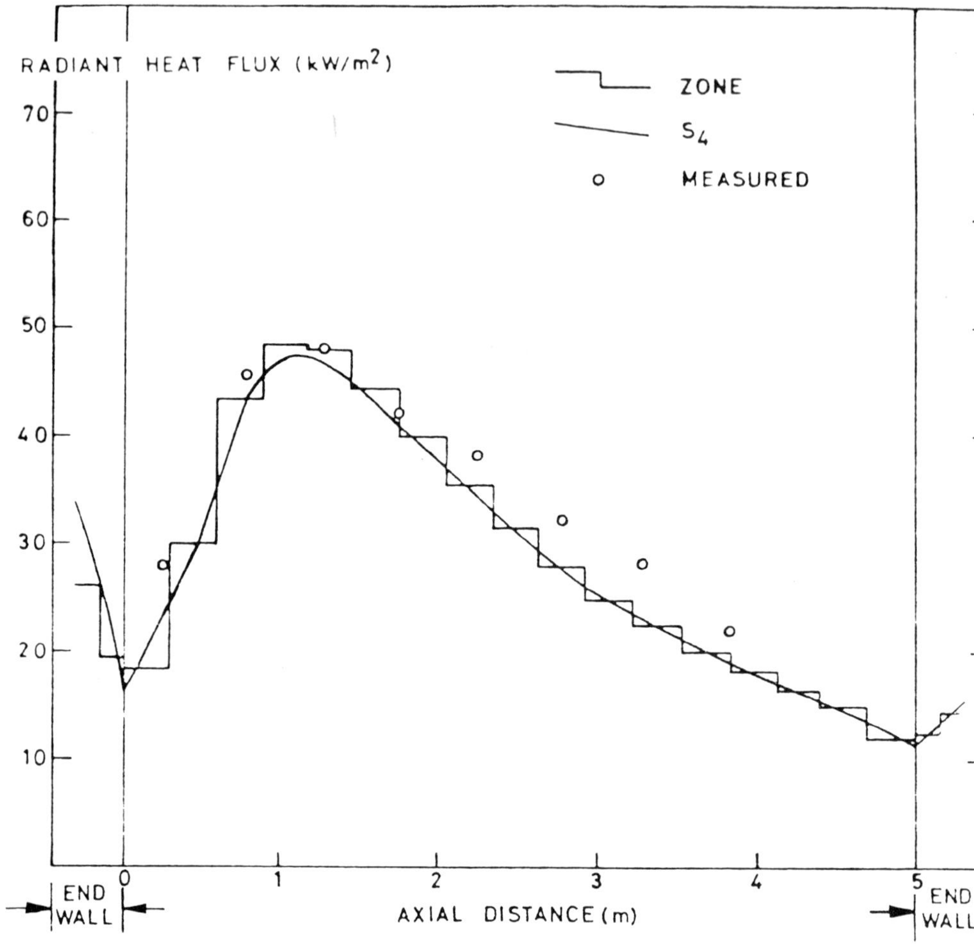

Figure 6.12 Comparison between calculated wall heat flux, obtained with the aid of discrete ordinate S_4 model of reference 14 and the experimental distributions of reference 18

luminous radiation was included[29] by assuming a mean soot concentration of 5×10^{-5} kg/m^3 at the flame temperature.

The measured and calculated profiles of mean gas temperature are compared in Figures 6.13 and 6.14 at two axial locations and for a swirl number of 1.7. The agreement is satisfactory with maximum discrepancy of less than 150 K. The local gas temperature distribution[29] indicated a long flame, and is consistent with the wall heat flux distribution shown in Figure 6.15. Khalil *et al.*[29] demonstrated that for this furnace 92% of the total heat transfer to the wall was by radiation. The radiative wall heat flux distributions shown in Figure 6.15 correspond to the measurements of Hasanack[30] and the calculations of Khalil *et al.*[29] At a swirl number of 1.7, the radiation model predicts a gradual increase in wall heat flux as the distance from the burner increases. This can be interpreted as the existence of long flame. An approach to the ultimate requirement of a uniform wall heat flux is observed.

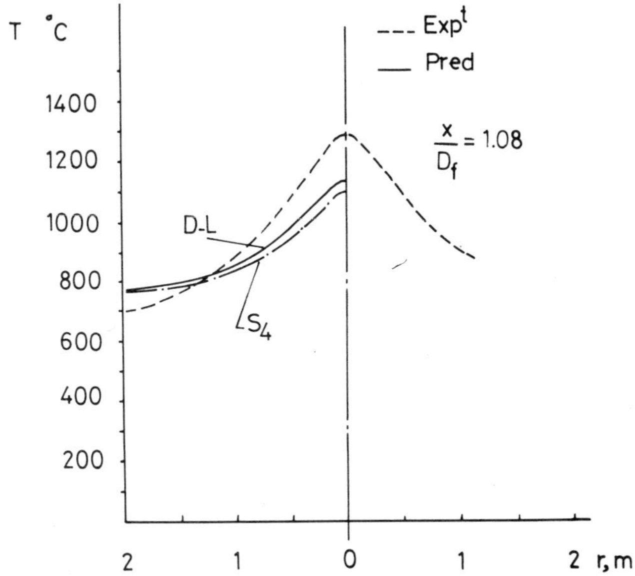

Figure 6.13 Comparison between measured and calculated[29] gas temperature profiles in the furnace of reference 30; $S = 1.7$

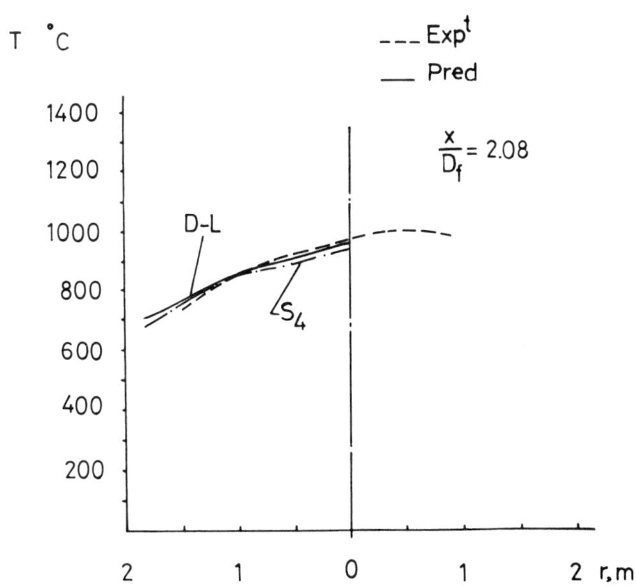

Figure 6.14 Comparison between measured and calculated[29] gas temperature profiles in the furnace of reference 30; $S = 1.7$

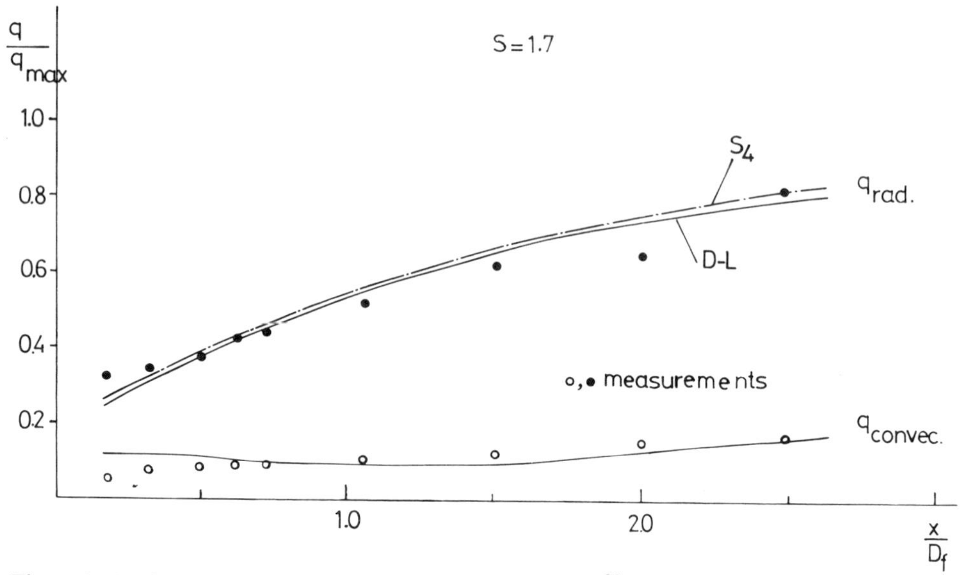

Figure 6.15 Comparisons between calculated wall heat flux[29] and the corresponding measurements obtained in the furnace of reference 30

The present model has not been widely applied to furnace and combustor flows, but a few further validation tests were reported in reference 31. The discrete ordinate method shows promising signs of being able to meet the set criteria of an economical computational procedure of the differential equation type which is capable of giving reasonably accurate predictions of radiant heat transfer in practical systems.

6.2.3.4 *Flux Models.* Another trend of radiation modelling is the direction towards the so-called *flux models*, which are based on the use of some simplifying assumptions of the angular variation of the radiant intensity in space. These assumptions allow the exact integro-differential radiation transfer equations to be reduced to a system of approximate partial differential equations. These are, in turn, amenable to the solution simultaneously with the equations for the flow and reaction.

The simplest type of these flux models is the one-dimensional model, e.g. the Schuster model,[32] and the Hamaker[33] and Eddington[25] models, which have been used in astrophysics. In the approximation of Schuster,[32] the following flux distribution equations are used:

$$\frac{dI}{dx} = -K_a I + K_a E \tag{6.2.24}$$

and

$$\frac{dJ}{dx} = +K_a J - K_a E \tag{6.2.25}$$

where I and J are positive fluxes in the positive and negative x directions. This one-dimensional model has been extended to axisymmetric geometries by Lockwood and Spalding,[34] thus accounting for the radiation fluxes in both axial and radial directions. Hence this model has been known as the 'L-S four flux model'.

The Flux Model of Lockwood and Spalding

The assumptions of this model allow the energy transfer in four principal coordinate directions. The modelled equations are obtained by formulating the radiant energy balances on a small volume of the medium. The resultant flux equations are:

$$\frac{d}{dx} F_x^+ = -K_a F_x^+ + K_a E \tag{6.2.26}$$

$$\frac{d}{dr} F_r^+ = -K_a F_r^+ - \frac{(F_r^+ - F_r^-)}{r} + K_a E \tag{6.2.27}$$

$$\frac{d}{dx} F_x^- = K_a F_x^- - K_a E \tag{6.2.28}$$

and

$$\frac{d}{dr} F_r^- = K_a F_r^- - K_a E \tag{6.2.29}$$

where F^+ and F^- are the radiation fluxes in the positive and negative coordinate directions, i.e.

$$F_x^+ = \int_{\xi>0} \xi I \, d\Omega \tag{6.2.30}$$

The above flux equations are rearranged to yield the fluxes in the x and r directions:

$$\frac{d}{dx} \frac{1}{K_a} \frac{d}{dx} R_x = K_a [R_x - E] \tag{6.2.31}$$

and

$$\frac{1}{r} \frac{d}{dr} \frac{r^2}{rK_a + 1} \frac{d}{dr} R_r = K_a [R_r - E] \tag{6.2.32}$$

where

$$R_x = (F_x^+ - F_x^-)/2$$

and

$$R_r = (F_r^+ - F_r^-)/2$$

The radiant energy boundary conditions take the form: at the wall, $F_r^= = F_r^-$ and

$$F_0 = \varepsilon_w E_w + (1 - \varepsilon_w)F_i \tag{6.2.33}$$

F_i and F_0 are the incident and leaving fluxes. At the wall, the resulting boundary conditions are given[35] as:

$$\frac{d}{dx}\frac{1}{K_a}\frac{d}{dx}R_x + \frac{\varepsilon_w}{2-\varepsilon_w}\frac{E_w}{\Delta x} - \frac{\varepsilon_w}{2-\varepsilon_w}\frac{R_x}{\Delta x} = 0 \tag{6.2.34}$$

and

$$\frac{1}{r}\frac{d}{dr}\left(\frac{r^2}{rK_a + 1}\right)\frac{d}{dr}R_r + \frac{\varepsilon_w}{2-\varepsilon_w}\frac{E_w}{\Delta r} - \frac{\varepsilon_w}{2-\varepsilon_w}\frac{R_r}{\Delta r} = 0 \tag{6.2.35}$$

The existence of the unusual second term of the right-hand side of Equation 6.2.27 for F_r^+ is due to curvature. The reason for the absence of a similar term in Equation 6.2.29 is not clear and it has been shown[36] that this equation leads to erroneous results.

The radiative contribution to the energy balance is represented by the expression

$$\bar{S}_H = 2K_a[R_x + R_r - 2E] \tag{6.2.36}$$

Assessment of Model

This model has been widely applied in combustion chambers and furnaces: Lockwood *et al.*,[34] Gosman *et al.*,[37] Khalil *et al.*,[38] Whitacre[8] and Hutchinson *et al.*[39,40,41] are some of those who have reported the application of this model.

Samples of the predicted heat flux obtained with the aid of this model are shown in Figures 6.16 to 6.20. Figure 6.16 shows the measured and calculated wall heat flux distribution for the furnace flow of reference 15; the calculations are those of Khalil *et al.*[38] for the two swirl intensities of $S = 0$ and 0.50. Both the measured and calculated heat flux distributions are in reasonable agreement, except near the inlet and exit sections. Figure 6.17 shows the wall heat flux distribution for the furnace of reference 39;

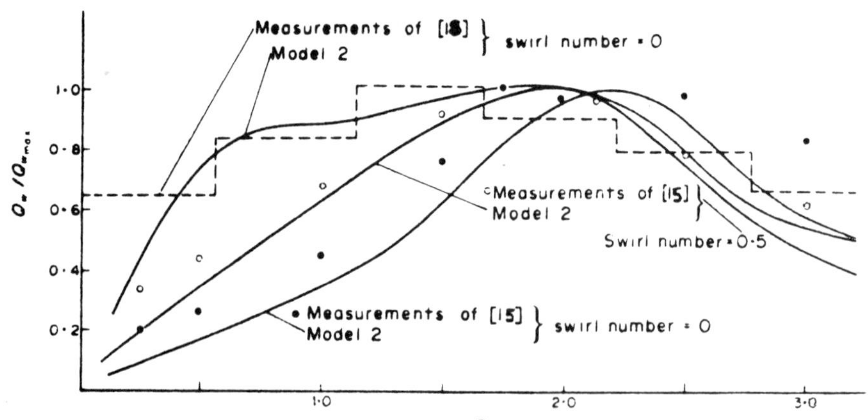

Figure 6.16 Measured and calculated wall heat flux distribution in the furnace of reference 15

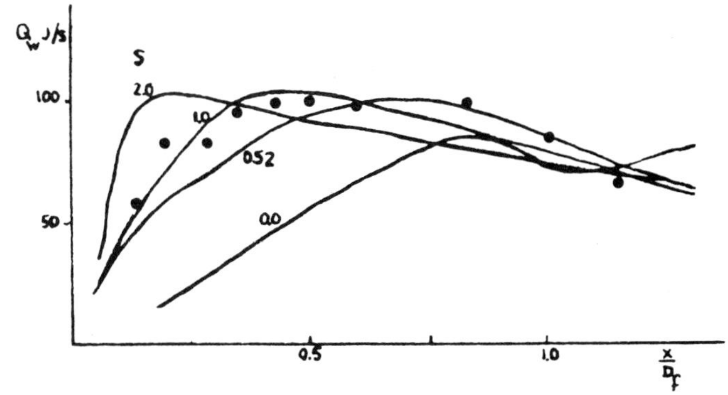

Figure 6.17 Measured and calculated wall heat flux distribution in the furnace of reference 39

the calculated distribution was in reasonable agreement with the corresponding measurements, which were obtained through the use of water-cooled sections. In this case the flame uses natural gas as fuel and the swirl number is 0.52. The obtained wall heat flux distribution is consistent with the flame length and the location of wall recirculation zone.

Another situation for which the model was used is the flame situation in an axisymmetric furnace with a diverging burner quarl. The wall heat flux distribution is shown in Figure 6.18. Discrepancies were observed[41] at the inlet section, and can be attributed to the uncoupled flux equations of the model. The corresponding centreline temperature distribution was also predicted, and is shown in Figure 6.19. The wall heat flux and the flame temperature distributions are consistent with each other, and indicate a short

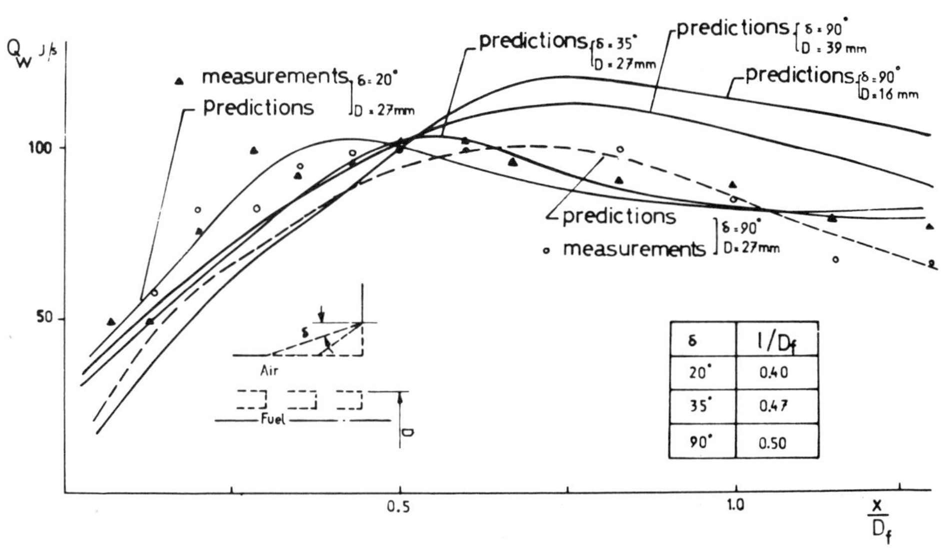

Figure 6.18 Measured and calculated wall heat flux distributions in the furnace of reference 40, for various burner quarl angles and swirl number of 0.52

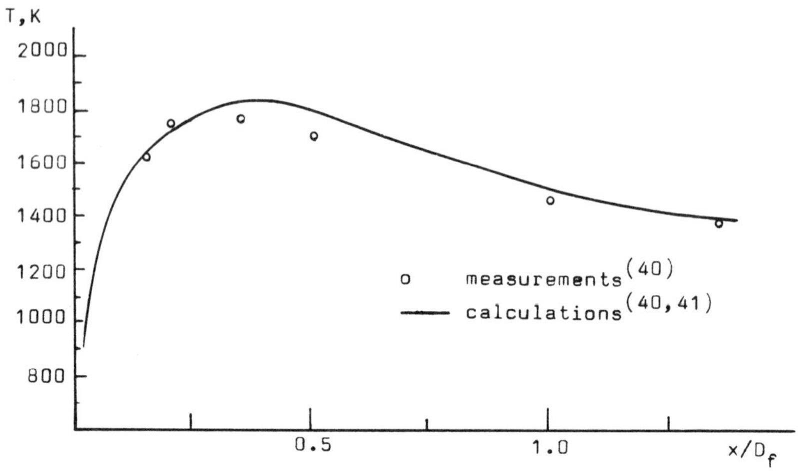

Figure 6.19 Measured and calculated centreline temperature distribution in the furnace of reference 40 with a burner quarl and swirl number of 0.52

flame as shown by the location of the peaks in the heat flux and temperature distributions.

The application of the model to large scale furnaces, was reported by Lowes *et al.*[12] for a natural gas non-swirling flame, in the furnace of the International Flame Research Foundation. The model was applied for different absorption coefficients K_a as shown in Figure 6.20; the various calculations were in qualitative agreement with the measurements. The agreement is less satisfactory near the furnace exit.

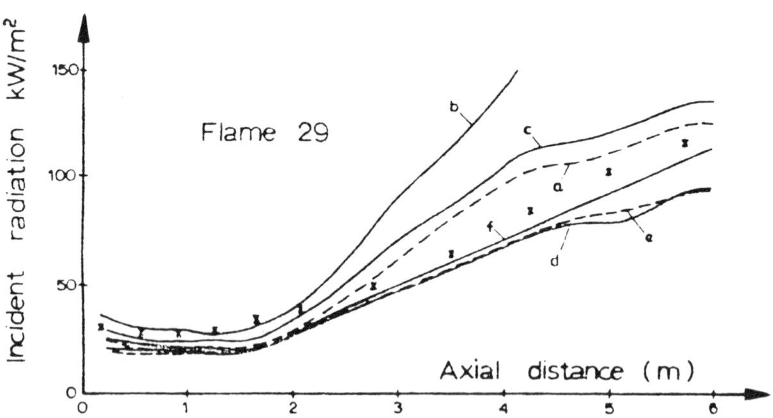

x	represents the measurements.
a, b, c, d	represent calculations using various values of K in the two-flux-Milne–Eddington model
e	represents a two-flux model of Schuster
f	represents the zone model calculations

Figure 6.20 Comparison between measured and calculated incident radiation flux, reference 12

The model was applied to three dimensional furnace flows by Patankar *et al.*,[42] Serag Eldin,[43] Serag Eldin *et al.*[44] and Turan.[45]

There are two objections to this model approximation. The first is that there is no interlinkage (coupling) between the axial and radial fluxes, a feature which is physically unrealistic. The second objection lies in the form of the intensity distribution, from which the flux equations are obtained. It is not obvious how the model equations follow systematically by approximating Equation 6.2.1. Further development to this model approximations was reported by Gibb *et al*;[46] their modification couples the flux in the coordinate directions but assumes that the intensity is unidirectional.

The DeMarco Flux Model

To cure the above mentioned defects, modifications and development to the flux models have been reported.[47] The intensity distribution was assumed to have the form:

$$I = A_x(\mathbf{i} \cdot \mathbf{\Omega}) + A_r(\mathbf{j} \cdot \mathbf{\Omega}) + A_z(\mathbf{k} \cdot \mathbf{\Omega}) + B_x(\mathbf{i} \cdot \mathbf{\Omega})^2 + B_r(\mathbf{j} \cdot \mathbf{\Omega})^2 + B_z(\mathbf{k} \cdot \mathbf{\Omega})^2 \quad (6.2.37)$$

where **i**, **j**, and **k** are unit vectors in the x, r and z directions respectively. Six terms are considered in the Taylor series expansion. The intensity distribution of Equation 6.2.37 is substituted in the transfer Equation 6.2.1, which is integrated over solid angles of 2π. After some numerical manipulation and rearrangements, the integrated transfer equations can be represented by the form:

$$\frac{d}{dx}\frac{1}{K_a}\frac{d}{dx}R_x = K_a[3R_x - R_y - R_z - E] \quad (6.2.38)$$

$$\frac{d}{dy}\frac{1}{K_a}\frac{d}{dy}R_y = K_a[3R_y - R_x - R_z - E] \quad (6.2.39)$$

$$\frac{d}{dz}\frac{1}{K_a}\frac{d}{dz}R_z = K_a[3R_z - R_x - R_y - E] \quad (6.2.40)$$

These are the three final equations relating the radiative flux in the coordinate directions for a non-scattering medium. The above equations are of the second order differential type and possess the detailed 'sparse matrix' property; they are amenable to solution by the tridiagonal matrix algorithm.

The boundary conditions at the walls were given[47] by

$$\int_{4\pi} I\mathbf{\Omega} \cdot \mathbf{n} d\Omega = \alpha_w \int_{2\pi} I\mathbf{\Omega} \cdot \mathbf{n} d\Omega - \varepsilon_w \sigma \bar{T}_w^4 \quad (6.2.41)$$

The left-hand side of this equation denotes the net flux towards the wall. On the right-hand side, the first term denotes the absorbed flux, while the second term represents the emitted flux. This analysis was applied to two-dimensional situations and the governing equations are expressed[20,29] as:

$$\frac{d}{dx}\frac{1}{K_a}\frac{d}{dx}R_x = \frac{4}{3}K_a(2R_x - R_r - \sigma\bar{T}^4) \qquad (6.2.42)$$

and

$$\frac{1}{r}\frac{d}{dr}\left(\frac{r}{K_a}\frac{dR_r}{dr}\right) = \frac{4}{3}K_a(2R_r - R_x - \sigma\bar{T}^4) \qquad (6.2.43)$$

where R_x and R_r are the axial and radial total radiation fluxes respectively, defined in terms of the B's.[29] The radiant energy contribution to the source term in the enthalpy equation is expressed as[20]

$$\bar{S}_H = \frac{16}{9}K_a(R_x + R_r - 2\sigma\bar{T}^4) \qquad (6.2.44)$$

Assessment of Model

The present model was applied to furnace and combustor flows in order to predict the radiant heat transfer characteristics. The validity of the model for simple configuration was assessed by DeMarco *et al.*,[48] Khalil,[35] Khalil *et al.*[20] and Hutchinson *et al.*[49] A few sample calculations are shown and discussed in turn. The wall heat flux distribution along the furnace of reference 49 is compared with corresponding measured distribution in Figure 6.21. In this axisymmetric furnace configuration, the fuel used was natural gas; a moderate swirl was imparted to the flow, the swirl number being 0.3. The calculated distribution represents satisfactorily the measured values, and predicts correctly the location and magnitude of peak heat flux.

The calculated wall heat flux distributions and the corresponding measurements are also shown in Figure 6.21. In this instance excellent agreement was obtained. Yet discrepancies of the order of 15% were reported in the vicinity of the burner. For the high swirl number of 0.52, the predicted location of the peak wall heat flux was shifted upstream of the measured location. This is attributed to the under prediction of the wall recirculation zone.

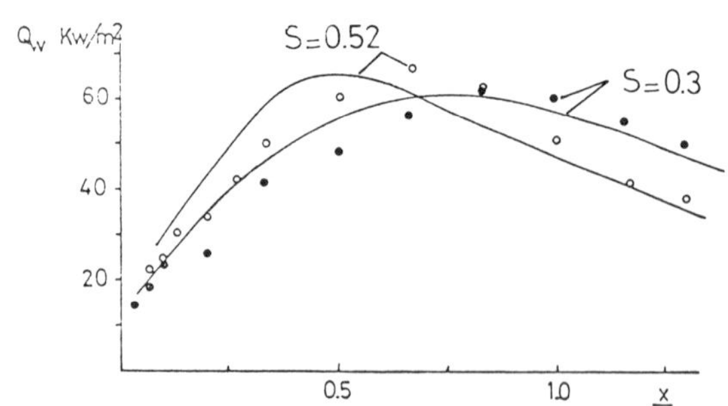

Figure 6.21 Measured and predicted wall heat flux distribution in the furnace of reference 49
—— Predictions using the four flux model[35]
O Experiments of reference 49

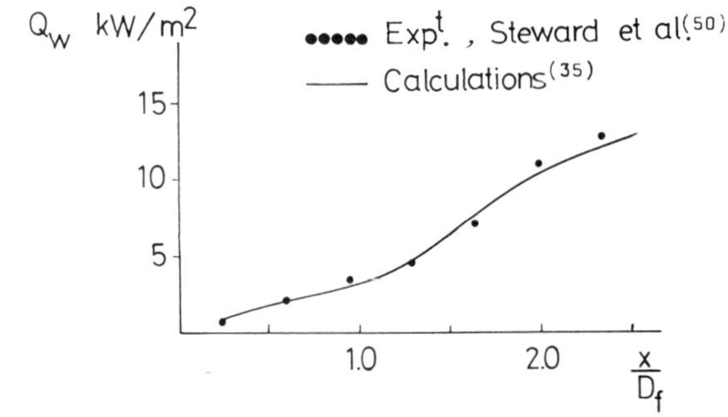

Figure 6.22 Measured and calculated distribution of wall heat flux in the furnace of reference 50

The wall heat flux distributions for the flow of Steward *et al.*[50] were predicted[35] using this model, and are shown in Figure 6.22. Reasonable agreement was observed with this model; it is worthwhile to note that the L-S model results in 40% underprediction of the radiative heat flux. Further comparisons were shown in references 20, 21, and 51 for large scale furnaces; comments on these comparisons are given later in the book.

The above DeMarco's analysis has been modified by Lockwood *et al.*[52,53] The proposed modifications combine features from both the spherical harmonics and the discrete ordinate models. The main modification is that the angles of integration may vary through the introduction of what was termed 'view factors'. These make the optical representation of the true intensity feasible. The relative sizes of the surfaces of integration were readily varied,[53] and were obtained by numerical optimization. For uniform intensity, these factors are unity and the integration surfaces are of equal areas.

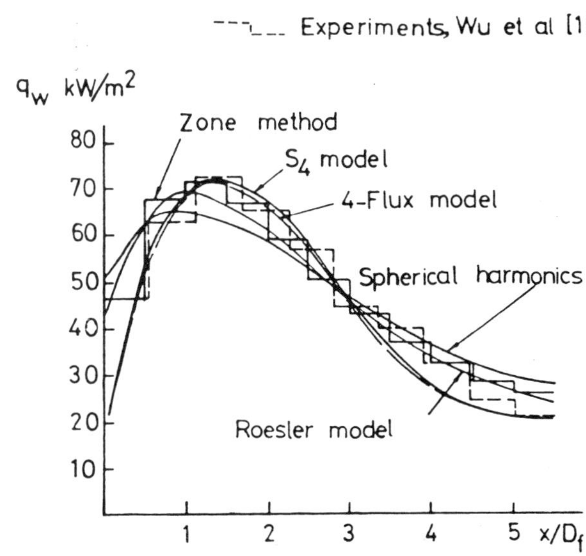

Figure 6.23 Comparison of wall heat flux distribution obtained with the aid of various radiation models in the furnace of Wu *et al.*[18]

Comparisons between the various radiation models are shown in Figure 6.23 for the furnace geometry of reference 18. It is obvious that the four flux, discrete ordinate and zone methods give nearly the same results, but the zone method is hindered by the need of more computer storage and run time than the other two.

The convective heat transfer to the wall is obtained from a gradient type expression:

$$q_{\text{conv}} = -\Gamma_H \frac{\partial \bar{H}}{\partial r} \tag{6.2.45}$$

The value of Γ_H is obtained from the wall function of Chapter 3. Samples of calculated convective wall heat flux are shown in Figures 6.24 and 6.25 with the corresponding measurements. The assumptions of Equation 6.2.45 should be further examined in recirculating flows, but the equation yields general satisfactory agreements.

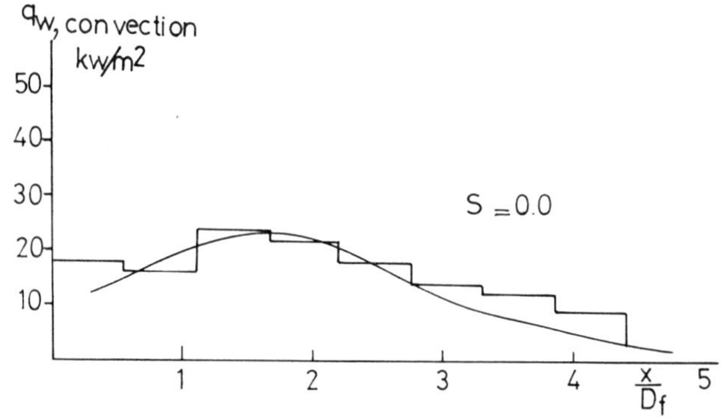

Figure 6.24 Comparisons between measured and calculated wall convective heat flux

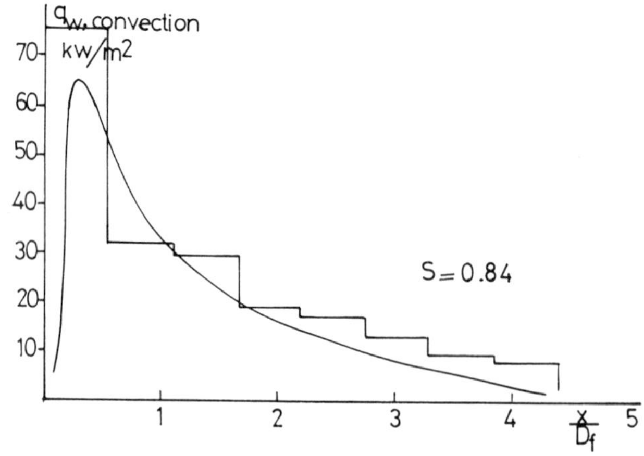

Figure 6.25 Comparisons between measured and calculated wall convective heat flux

6.3 POLLUTANT FORMATION MODELS

6.3.1 General

Having discussed the means of calculating the heat transfer by radiation, it is now time to focus attention on the important phenomenon of pollutant formation. In this section, the formation of nitric oxides and carbon monoxide is discussed. The pollutant emission levels from a particular combustor depend on the interaction between physical and chemical processes occurring within. The concentrations of the pollutant are quite often different from the calculated equilibrium values; this undoubtly indicates the importance of the role of chemical kinetics in pollutant formation process. The various modelling approximations which were found useful by previous investigations,[35,41,54–61] are discussed and extended to furnace flow situations.

The two major pollutants, NO and CO, are tested separately. The formation of NO in flames occurs at very small concentrations relative to the reactants, and hence does not influence the fuel kinetics described earlier in Chapter 4. In contrast, the carbon monoxide formation and destruction reactions are an intimate part of the combustion process.

The present section reviews recent numerical investigations of the kinetics of formation and destruction of nitrogen oxides, and carbon monoxide. The formation and oxidation of soot in combustion have been the subject of several recent reviews, and are therefore not repeated here. Examples of predicted pollutant concentrations are given and the validity of the model assumptions is assessed.

6.3.2 Models for Calculating Nitrogen Oxides

Nitric oxide and nitrogen dioxide are the two major oxides of nitrogen emitted from combustion processes; the sum of these two species is denoted as NO_x. In combustion, nitric oxide is the bigger of the two components; and NO_2 is formed from NO. Consequently, models of NO_x kinetics stress the formation of NO; the nitric oxide can be formed through the following passages:

(a) From reactoins of N_2 with oxygen, (thermal NO)
(b) From nitrogen-containing fuel compounds (fuel NO)
(c) From reactions of fuel-derived radicals with N_2 which ultimately lead to NO (prompt* NO).

Previous investigations of NO emission have been reported for both premixed and diffusion flames and support the proposed Zeldovich mechanism for thermal NO; for example see Zeldovich *et al.*,[62] Iverach *et al.*[63] The mechanism follows the steps:

$$O + N_2 \rightleftharpoons NO + N \tag{6.3.1}$$

$$N + O_2 \rightleftharpoons NO + O \tag{6.3.2}$$

$$N + OH \rightleftharpoons NO + H \tag{6.3.3}$$

*See page 182.

Models for thermal NO depend on an accurante computation of the O-atom concentration. The O-atom was calculated, under equilibrium, from the equation:

$$O_2 + M \rightleftharpoons O + O + M \tag{6.3.4}$$

Equation 6.3.3 has been shown[64] to have a significant influence in the situations where the fuel is rich. Fenimore[65] has critically reviewed the published rate data and has recommended the rate constant expressions for reactions 6.3.1 to 6.3.3, as shown in Table 6.1.

The simplified NO formation mechanism is shown in Figure 6.26.

The rate of the reaction 6.3.1, in the simple Zeldovich mechanism, is much slower than that of the reaction 6.3.2, which indicates that the formation of an NO molecule from Equation 6.3.1 is accompanied by a release of a N-atom. This atom forms rapidly NO according to reaction 6.3.2. Consequently, Equation 6.3.1 controls the formation rate, since Equation 6.3.3 plays a small role in stoichiometric flames. The NO formation rate is approximated by

$$\frac{d}{dt}\,\mathrm{NO} = 2K_1[\mathrm{O}]_{eq} \cdot [\mathrm{N}_2]_{eq} \tag{6.3.5}$$

However, the equilibrium [O] atom concentration is obtained from Equation 6.3.4 as

$$[\mathrm{O}]_{eq} = \frac{K_4}{(RT)^{\frac{1}{2}}}[\mathrm{O}_2]_{eq}^{\frac{1}{2}} \tag{6.3.6}$$

where K_1 and K_4 are rate expressions, and $K_4 = 3.6 \times 10^3 \times \exp(-31090/T)$ atm$^{\frac{1}{2}}$.

The rate of NO formation was found to depend strongly on the mean and the fluctuating temperatures.[41,58,61] Jones[61] integrated numerically the temporal rate of NO formation for various temperature fluctuation intensities. At a temperature of around 2000°K, and for $\sqrt{\bar{T}'^2}/\bar{T}$ of 0.1, the rate of NO formation, when the temperature fluctuations were considered, was nearly five times the rate when the fluctuations were neglected. This ratio increased as the mean temperature decreased, and the fluctuation intensity increased. Similar investigations were reported by Thomson *et al.*[57] These illustrated the influence of temperature fluctuations on NO formation rate and indicated the need to incorporate it in the pollutant formation rate used in the numerical calculations.

TABLE 6.1

Reaction	Rate constant, cm^3/mol sec	Temperature range, K
$O + N_2$	$7.6 \times 10^{13} \exp(-38\,000/T)$	2000–5000
$N + O_2$	$6.4 \times 10^9 T \exp(-3150/T)$	300–3000
$N + OH$	10^{14}	300–2500

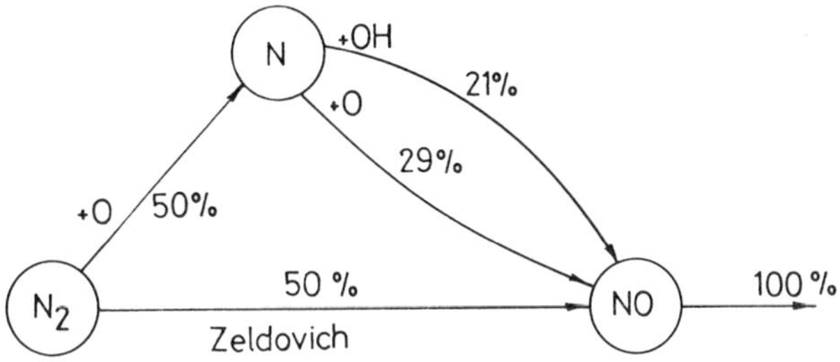

Figure 6.26 Nitric oxide formation mechanism

NO Model Formulation

The calculated NO concentrations are obtained from the corresponding conservation equation which can be expressed in the form:

$$\frac{\partial}{\partial x_j}\bar{\rho}\bar{U}_j\bar{M}_{NO} = \frac{\partial}{\partial x_j}\Gamma_{NO}\frac{\partial \bar{M}_{NO}}{\partial x_j} + \bar{R}_{NO} \qquad (6.3.7)$$

and

$$\Gamma_{NO} = \mu_{eff}/\sigma_f$$

$$\bar{R}_{NO} = 8.39 \times 10^{16}(\bar{T})^{-\frac{1}{2}}\bar{M}_{N_2}\bar{M}_{ox}^{\frac{1}{2}}\exp(-134900/R\bar{T})[1 + F] \qquad (6.3.8)$$

where

$$F = \frac{\overline{m_{N_2}m_{ox}}}{\bar{M}_{N_2}\bar{M}_{ox}} + \frac{E}{R\bar{T}}\left[\left(\frac{1}{2}\frac{E}{R\bar{T}} - 1\right)\left(\frac{\sqrt{\overline{T'^2}}}{\bar{T}}\right)^2 + \frac{\overline{T'm_{N_2}}}{\bar{T}\bar{M}_{N_2}} + \frac{\overline{T'm_{ox}}}{\bar{T}\bar{M}_{ox}}\right]$$

and

$$E = +134900$$

In situations where [O] concentrations are higher than the equilibrium values, the concept of super-equilibrium was suggested[65] and was quantified by Ghezzi *et al.*[58] Iverach *et al.*[63] observed that, for hydrocarbon fuels, the simple Zeldovich mechanism was applicable for equivalence ratios less than 1.5, with the O-atom determined for equilibrated conditions, Equation 6.3.4. For highly rich flames, the rate calculated for Equation 6.3.1 and 6.3.4 required an improbably high O-atom concentration.

Models of the nitric oxide formation from fuel-nitrogen are limited due to the complexity of the chemical steps through which the chemically-bound nitrogen passes. De Soete[66,67] determined the kinetic coefficients in a global mechanism, which describes his experimental observations on NO yields from hydrocarbon flames with fuel nitrogen additives.

The model assumes an instantaneous reaction of fuel-nitrogen to N atoms. These N atoms would then form NO and N_2 by the two steps in the Zeldovich mechanism, i.e. Reaction 6.3.2 and reverse of Reaction 6.3.1.

Fenimore[65] obtained experimental data that showed an appreciable formation of NO in the flame; he called this *prompt NO* and suggested that it might be due to reactions of fuel-nitrogen fragments with N_2, thus producing CN and HCN products. De Soete[67] suggested that, in the flame front, molecular nitrogen acts as a fuel–nitrogen compound, which reacts to form NO which in turn decomposes to form N_2 via an HCN intermediate. The rate of prompt NO formation was expressed in terms of oxygen, nitrogen and fuel concentrations, and was exponentially temperature dependent.

Assessment of Model

The modelled Equations 6.3.7 and 6.3.8 were used to calculate the local NO concentration in furnaces and combustion chambers, by Scheffer *et al.*,[54] Quan *et al.*,[59] Pompei *et al.*,[68] Mizutani *et al.*,[69] Hutchinson *et al.*,[41,49] Khalil[70] and Richter.[71] Figure 6.27 displays the comparison between the measured and calculated NO concentration in an opposed jet flow in a cylindrical duct.[54] Premixed air--propane jet flows into an opposing stream of air–propane mixture. The cylindrical combustor was 0.058 m in diameter and 0.356 m long; the jet had an inner diameter of 3.97 mm and an outer diameter of 6.35 mm. The agreement is qualitative in nature, near the combustor exit, the predicted NO concentrations were under-predicted by 50%. This discrepancy can be attributed to the neglect of the temperature and concentration fluctuations effect, i.e. $F=0$.

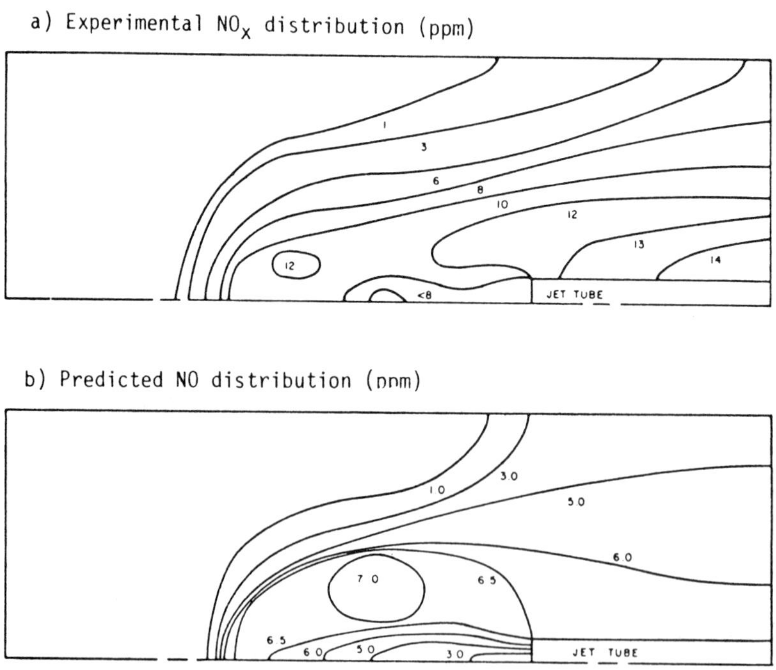

Figure 6.27 Comparison between measured and predicted NO concentration in opposed jet flow in a cylindrical duct[52]

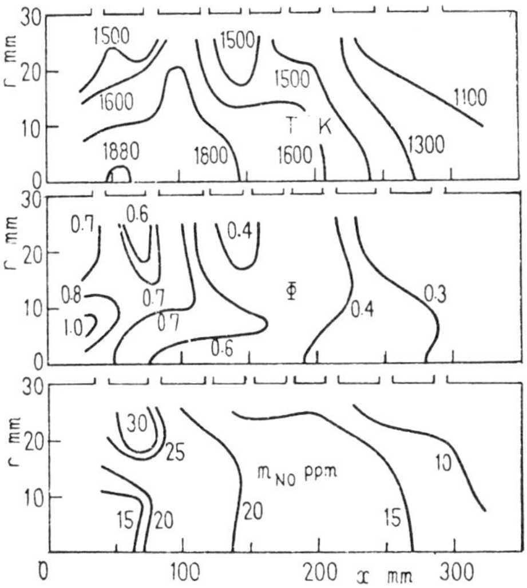

Experimental distribution maps
$(A/F)_o = 66$, $(A/F)_p = 10.6$, $T_i = 320$ K,
$G_f = 1.2$ g/s.

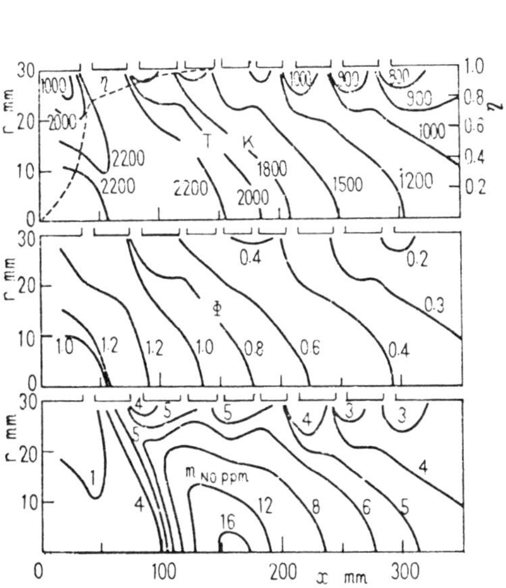

Predicted distribution maps
$(A/F)_o = 65$, $(A/F)_p = 11$, $T_i = 300$ K,
$G_f = 1.2$ g/s.

Figure 6.28 Calculations and measurements of local NO_x concentration contours in the combustor of reference 69

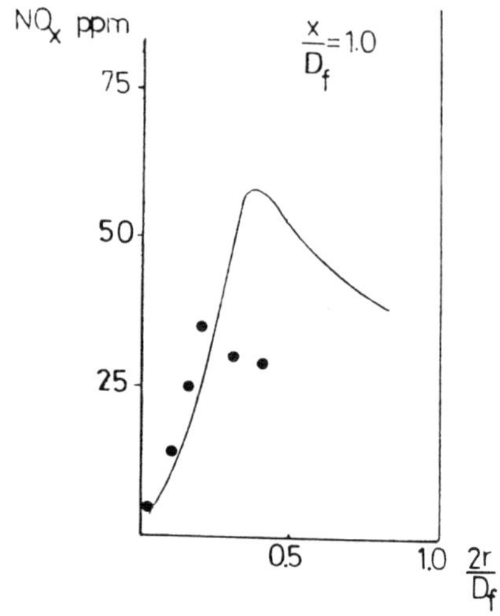

Figure 6.29 Measured and calculated NO_x concentrations in the Ijmuiden furnace[41]

The calculations of the NO concentration in a gas turbine combustor of the can type were reported by Mizutani *et al.*[69] and are shown in Figure 6.28. The calculated NO concentrations are compared with the measurements, and the obtained agreement was reasonably good, except in the primary zone. The effect of turbulence fluctuations on the NO formation was again ignored.

A step forward was achieved by the inclusion of the fluctuation effects as reported by Hutchinson *et al.*[41] for various furnace geometries; they employed Equations 6.3.7 and 6.3.8. Figures 6.29 and 6.30 illustrate the extent to which the NO concentration can be

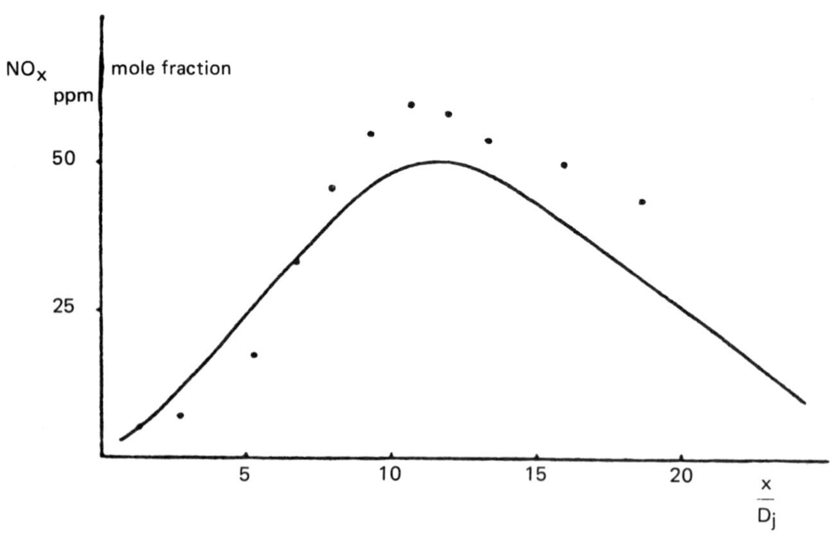

Figure 6.30 Measured and calculated centreline distribution of NO_x concentration in a co-flowing ducted flame[35,41]

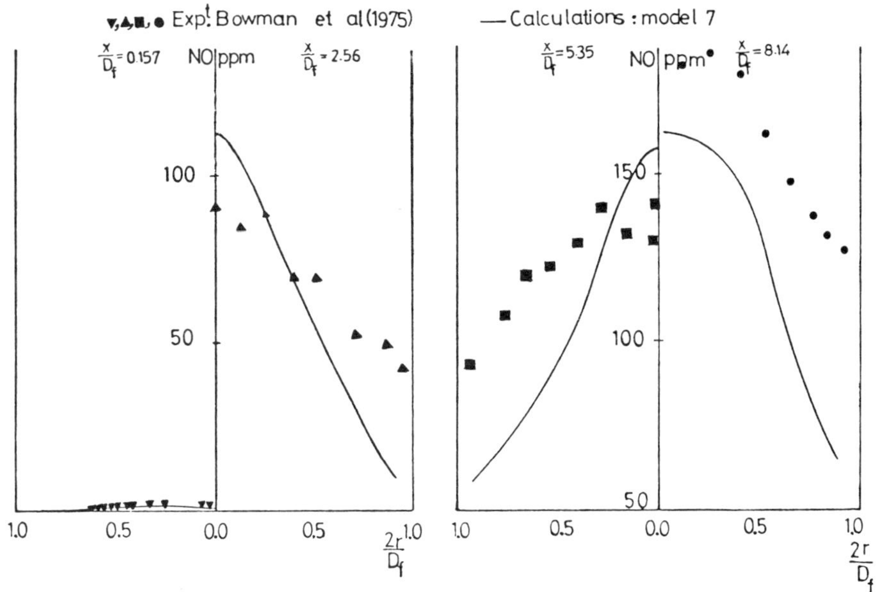

Figure 6.31 Measured and calculated profiles of NO_x concentrations in an axisymmetric furnace with zero swirl[35]

predicted. Figure 6.31 displays the comparisons between measured and calculated NO profiles in an axisymmetric furnace, which were reported by Khalil.[35]

It is now certainly clear that a number of further modifiications for the present model are required. They can be summarized as:

1. The inclusion of better physical modelling, which is achieved by taking into account the effect of turbulence–chemistry interaction.
2. Another step forward is to obtain the time-averaged rate of NO formation from the integral of the instantaneous rate and the local probability distribution. The shape of $P(M_{NO})$ can be obtained from the measured values or through the solution of transport equations.
3. It is also necessary to arrive at better modelling of prompt NO and to gain an accurate account of O-atom equilibrium.

6.3.3 CO Formation Model

The local concentration of the unburnt fuel can be determined from the solution of an appropriate conservation equation. The obstacle in the way of this approach is that the chemical reactions involved in the branch reactions of hydrocarbons to CO_2 and H_2O are numerous and complicated. That is why more complex approaches can be necessary. Previous calculations of unburnt hydrocarbons have been reported, for example, Borghi[72] and Singh,[73] and relate to diffusion and premixed flames in mixing layers. The following equations were considered:

$$CO + OH \rightleftharpoons CO_2 + H \tag{6.3.9}$$

$$CO + O \rightleftharpoons CO_2 \tag{6.3.10}$$

The Reaction 6.3.9 is very fast under normal conditions both in the flame region and in the post-flame region. Palmer *et al.*[74] indicated that Reaction 6.3.9 controlled the rate of disappearance of CO as

$$\bar{R}_{CO} = -8.14 \times 10^8 \, \bar{\rho}^2 \bar{M}_{CO} \bar{M}_{OH} \exp(-540/\bar{T}) \tag{6.3.11}$$

A complete model of CO kinetics must include the initial fuel kinetics which form CO as shown by Equation 6.2.41. A quasi-global scheme has been proposed by Edelman *et al.*,[75] where the initial hydrocarbon fuel was assumed to undergo an overall reaction of the type:

$$C_nH_m + \left(\frac{n}{2} + \frac{a}{2}\right) O_2 \rightarrow nCO + aH_2O + \left(\frac{m}{2} - a\right) H_2 \tag{6.3.12}$$

The rate parameters and the stoichiometric coefficients in this reaction were determined empirically.

The CO concentrations were obtained[35,41] from the solution of the set of Equations 4.2.41.

The approach of Gordon *et al.*[76] was based on the minimization of the change in Gibbs function for appropriate reactions. This leads to values of mass fractions of CO which are functions of the pressure, temperature, equivalence ratio and type of fuel. Since the pressure is known and the temperature and local-equivalence ratio are calculable, the concentration of CO can be determined with the aid of the computer program of reference 76.

The calculation of CO concentration was reported by Hutchinson *et al.*[41] for the confined town gas flame of reference 77; samples are shown in Figures 6.32 and 6.33. The formation mechanisms for CO were reviewed by Bowman[78] and Caretto.[79] Here again, as in the nitric oxide formation, the inclusion of the effects of turbulence–chemistry interactions is an urgent demand and particularly in high intensity combustors.

6.4 CLOSURE

This chapter has described the various available models that characterize the radiative heat transfer contribution to the total energy balance. For a radiation model to be most effective it must be easily coupled to the solution procedure of the flow and reaction equations. The differential equations of the radiation fluxes in the coordinate directions were solved, thus yielding the local values of these fluxes. The net source of heat due to thermal radiation was then obtained in terms of these fluxes; the source term was in turn substituted in the total enthalpy equation. The validity of the various models and the associated boundary conditions was assessed by way of comparisons with furnace flow measurements. The coupled flux model and the discrete ordinate model possessed two

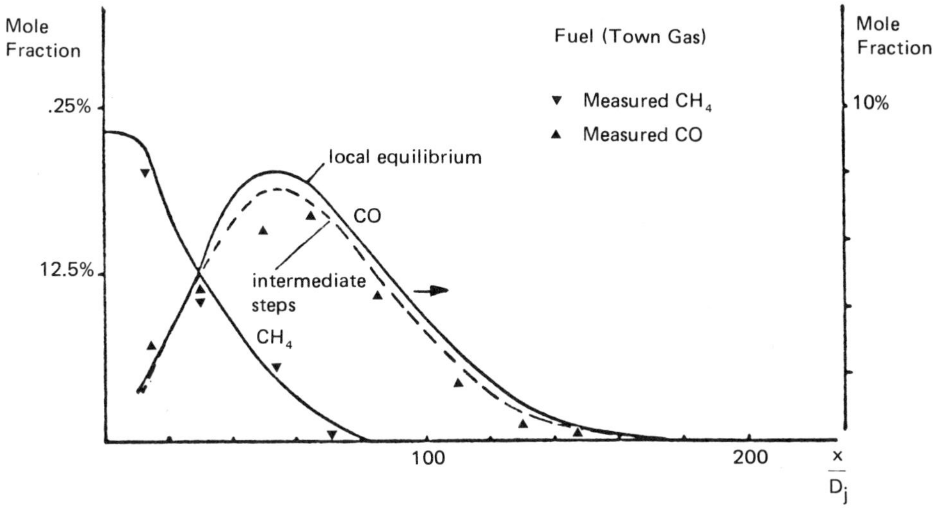

Figure 6.32 Measured and calculated CO concentrations in the furnace of reference 77, with town gas flame

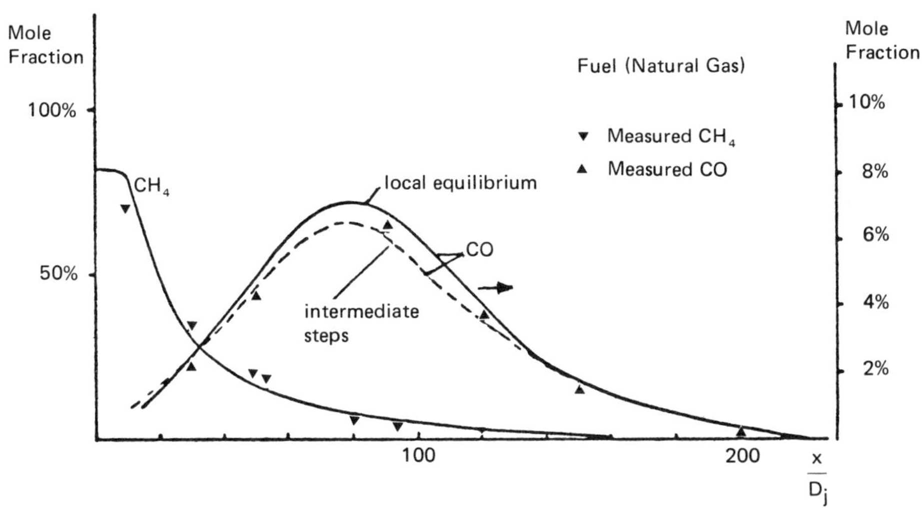

Figure 6.33 Measured and calculated CO concentrations in the furnace of reference 77, with natural gas flame

advantages over the other models, namely that they both represented adequately the radiative characteristics and were computationally economical.

Pollutant formation models were described and the various modelling assumptions were discussed in view of the available experimental evidence. A model that accounts for temperature and concentration fluctuations was described, and was found to predict adequately the local NO concentrations, when applied to furnace flows. A brief account of CO modelling assumptions was given.

References to Chapter 6

1. Pomraning, G. C. (1973) *Radiation Hydrodynamics* (Pergamon Press).
2. Truelove, J. S. (1974) Mathematical modelling of radiant heat transfer in furnaces. AERE Report, AERE-R7817.
3. Hottel, H. C. and Sarofim, A. F. (1967) Radiative Transfer (McGraw-Hill).
4. Hottel, H. C. and Sarofim, A. F. (1972) The status of calculations of radiation from non-luminous flames, *Proc. 4th Symp. on Flames and Industry: Predictive Methods for Industrial Flames.*
5. Johnson, T. R. and Beer, J. M. (1972) Radiative heat transfer in furnaces-further development of the zone method of analysis. *Proc. 14th Symp. (Int.) on Combustion*, p. 639.
6. Steward, F. R. and Cannon, P. (1971) The calculation of radiative heat flux in a cylindrical furnace using Monte Carlo Model, *Int. J. Heat and Mass Transfer*, **14**, p. 245.
7. Johnson, T. R. (1971) Application of the zone method of analysis to the calculation of heat transfer from luminous flames. PhD. Thesis, Sheffield University.
8. Whitacre, G. R. and McCann, R. A. (1975) Comparison of methods for the Prediction of Radiant Heat Flux Distribution and temperature. ASME Paper 75-HT-9.
9. Hottel, H. C. and Cohen, E. S. (1958) Radiant heat exchange in a gas filled enclosure: allowance for nonuniformity of gas temperature. *A.I.Ch.E. Journal*, **4**, p. 3.
10. Johnson, T. R. and Beer, J. M. (1972) The zone method analysis of radiant heat transfer; a model for luminous radiation. *Proc. 4th Symp. on Flames and Industry: Predictive Methods for Industrial Flames.*
11. Zuber, I. and Konecny, V. (1972) Mathematical model of combustion chambers for technical applications. *Proc. 4th Symp. on Flames and Industry: Predictive Methods for Industrial Flames.*
12. Lowes, T. M., Bartelds, H. and Heap, M. P. (1973) Prediction of radiant heat flux distribution. IFRF Doc. G 02/a/26.
13. Richter, W. and Quack, R. (1974) A mathematical model of a low-volatile pulverised fuel flame, in *Heat Transfer in Flames*, edited by Afgan (Scripta), p. 95.
14. Hyde, D. J. and Truelove, J. S. (1977) The discrete ordinates approximation for multi-dimensional radiant heat transfer in furnaces. AERE Report AERE-R 8502.
15. Bartelds, H., Lowes, T. M., Michelfelder, S. and Pai, B. R. (1973) The prediction of heat flux distribution in furnaces and its experimental testing. IFRF Doc. G 02/a/23.
16. Pai, B. R., Bartelds, H. and Michelfelder, S. (1976) Report on the M3 Trails – Part A: Cold heat sink. IFRF Doc. F 36/a/6.
17. Heap, M. P., Bartelds, H., Lowes, T. M. and Michelfelder, S. (1974) Prediction of radiant heat flux distribution, in *Heat Transfer in Flames*, edited by Afgan (Scripta).
18. Wu, H. L. and Fricker, N. (1971) An investigation of the behaviour of swirling jet flames in a narrow cylindrical furnace. *Proc. 2nd members Conference of the IFRF.* See also: *J. Inst. Fuel*, **49**, P. 144, (1976).
19. Selcuk, N., Siddall, R. G. and Beer, J. M. (1976) A comparison of mathematical models of the radiative behaviour of a large-scale experimental furnace. *Proc. 16th Symp. (Int.) on Combustion*, p. 53.
20. Khalil, E. E. and Truelove, J. S. (1977) Calculation of heat transfer in a large gas fired furnace, *Letters in J. Heat and Mass Transfer*, **4**, p. 353.
21. Khalil, E. E. (1979) Mathematical modelling of radiative heat transfer in axisymmetric furnaces, AIAA Paper 79-0099, AIAA 17th Aerospace Sciences Meeting.
22. Jeans, J. H. (1917) The equations of radiative transfer of energy, *Monthly Notes, Royal Astron. Soc.*, **78**, p. 28.
23. Kouganoff, V. (1963) *Basic Methods in Transfer Problems* (Dover).
24. Davison, B. (1957) *Neutron Transport Theory* (Clarendon Press).
25. Eddington, A. S. (1960) *The Internal Constitution of the Stars* (Dover).
26. Selcuk, N. and Siddall, R. G. (1976) Two flux spherical harmonic modelling of two dimensional radiative transfer in furnaces. *Int. J. Heat and Mass Transfer*, **19**, p. 313.
27. Chandrasekhar, S. (1960) *Radiative Transfer* (Dover).

28. Carlson, B. G. and Lathrop, K. D. (1968) Transport theory – the method of discrete ordinates, in *Computing Methods in Reactor Physics*, edited by Greenspan *et al.*, (Gordon and Breach).
29. Khalil, E. E. and Truelove, J. S. (1977) Calculation of heat transfer in a large gas fired furnace. AERE Report AERE-R 8747.
30. Hasenack, J. (1977) Private communication.
31. Khalil, E. E. (1980) Mathematical modelling of combustion processes in combustors. 1st Asian Fluid Mechanics Conference (India).
32. Schuster, A. (1905) Radiation Through a Foggy Atmosphere. *Astrophysics Journal*, **21**, p. 1.
33. Hamakar, H. C. (1947) Phillips Research Reports, No. 2, 55, 103, 112, 420.
34. Lockwood, F. C. and Spalding, D. B. (1971) Prediction of a turbulent duct flow with significant radiation. *Proc; Thermodynamics Colloquium*, (Société Française de Physique, Evian).
35. Khalil, E. E. (1977) Flow and combustion in axisymmetric furnaces Ph.D. Thesis, London University.
36. Truelove, J. S. (1976) Differential equation models of radiative transfer, *AERE report AERE-R 8364*.
37. Gosman, A. D. and Lockwood, F. C. (1973) Predictions of the influence of turbulent fluctuations on flow and heat transfer in furnaces. Imperial College, Dept. of Mechanical Engineering, Report HTS/73/52.
38. Khalil, E. E., Spalding, D. B. and Whitelaw, J. H. (1975) The calculation of local flow properties in two dimensional furnaces. *Int. J. Heat and Mass Transfer*, **18**, p. 775.
39. Hutchinson, P., Khalil, E. E., Whitelaw, J. H. and Wigley, G. (1976) The calculation of furnace flow properties and their experimental verification. *J. Heat Transfer*, **98**, p. 276.
40. Hutchinson, P., Khalil, E. E., Whitelaw, J. H. and Wigley, G. (1975) Influence of burner geometry on small scale furnaces performance. *Proc. 2nd European Symp. on combustion*, p. 675.
41. Hutchinson, P., Khalil, E. E. and Whitelaw, J. H. (1976) The calculation of wall heat transfer rate and pollution formation in axisymmetric furnaces. *Proc. 4th Members Conference of the IFRF.*
42. Patankar, S. V. and Spalding, D. B. (1972) Mathematical models of fluid flow and heat transfer in furnaces, a review. *Proc. 4th Symp. on Flames and Industry: Predictive Methods for Industrial Flames.*
43. Serag Eldin, M. A. (1977) The numerical prediction of the flow and combustion processes in a three-dimensional can combustor. Ph.D. Thesis, London University.
44. Serag Eldin, M. A. and Spalding, D. B. (1979) Computations of three dimensional gas turbine combustion chamber flows. *J. Engineering for Power*, **101**, p. 326.
45. Turan, A. (1978) A three dimensional model for gas turbine combustors. Ph.D. Thesis, Sheffield University.
46. Gibb, J. and Jenner, R. (1976). Unpublished Work, Central Electricity Generating Board.
47. DeMarco, A. G. (1974) A new flux method for the calculation of radiation heat transfer in three dimensioned combustion chambers. M.Sc. Thesis, London University.
48. DeMarco, A. G. and Lockwood, F. C. (1975) A new flux model for the calculation of three dimensional radiation heat transfer. *La Rivista du Combustibili*, **5**, p. 184.
49. Hutchinson, P., Khalil, E. E. and Whitelaw, J. H. (1977) Measurement and calculation of furnace flow properties. *AIAA Journal of Energy*, **1**, p. 212.
50. Steward, F. R., Osuwan, S. and Picot, J. J. (1972) Heat transfer measurements in cylindrical test furnaces. *Proc. 14th Symp. (Int.) on Combustion*, p. 651.
51. Khalil, E. E., Hutchinson, P. and Whitelaw, J. H. (1980) The calculation of the flow and heat transfer characteristics of gas fired furnaces. *Proc. 18th Symp. (Int.) on Combustion*, p. 1927.
52. Lockwood, F. C. and Shah, N. G. (1976) An improved flux model for the calculation of radiation heat transfer in combustion chambers. ASME. Paper 76-HT-55.

53. Lockwood, F. C. and Shah, N. G., (1980) A new radiation solution method for incorporation in general combustion prediction procedures. *Proc. 18th Symp. (Int.) on Combustion*, Paper 138.
54. Schefer, R. W. and Sawyer, R. F. (1976) Lean premixed recirculating flow combustion for control of oxides of nitrogen. *Proc. 16th Symp. (Int.) on Combustion*, p. 119.
55. Takagi, T., Ogaswara, M., Fujii, K. and Daizo, M. (1974) A study on nitric oxide formation in turbulent diffusion flames. *Proc. 15th Symp. (Int.) on Combustion*, p. 1051.
56. Sadakata, M. and Beer, J. M. (1976) Spatial distribution of nitric oxide formation rates in a swirling turbulent methane–air flame. *Proc. 16th Symp. (Int.) on Combustion*, p. 93.
57. Thompson, D., Brown, T. D. and Beer, J. M. (1972) Formation of NO in a methane–air flame. *Proc. 14th Symp. (Int.) on Combustion*, p. 787.
58. Ghezzi, U., Ortolani, C. U. and Pasini, S. (1975) Nitric oxide formation in gas turbine combustion chambers. *Proc. Italian Flame Days* (San Remo).
59. Quan, V., Kliegel, J. R. and Peters, R. L. (1975) Predicted effects of fluid dynamic parameters on nitric oxide formation in turbulent jet diffusion flames. *Combustion and Flame*, **25**, p. 67.
60. Quan, V., Marble, F. E. and Kliegel, J. R. (1972) Nitric oxide formation in turbulent diffusion flames. *Proc. 14th Symp. (Int.) on Combustion*, p. 851.
61. Jones, W. P. (1975) The effect of temporal fluctuations in temperature on nitric oxide formation. *Combustion Science and Technology*, **10**, p. 93.
62. Zeldovich, Ya. B., Sadovnikov, P.Ya and Frank-Kamenetskii, D. A. (1974) Oxidation of nitrogen in combustion (Translation by M. Shelef). Academy of Sciences of USSR, (Institute of Chemical Physics), Moscow.
63. Iverach, D., Basden, K. S. and Kirov, N. Y. (1972) Formation of nitric oxide in fuel-lean and fuel-rich flames. *Proc. 14th Symp. (Int.) on Combustion*, p. 767.
64. Barrere, M. (1973) Modelisation des foyers de turboreacteur en vue de l'étude de la pollution. AGARD meeting, CP 125.
65. Fenimore, C. P. (1970) Formation of nitric oxide in premixed hydrocarbon flames. *Proc. 13th Symp. (Int.) on Combustion*, p. 373.
66. Desoete, G. (1974) Overall reaction rates of NO and N_2 formation from fuel nitrogen. *Proc. 15th Symp. (Int.) on Combustion*, p. 1093.
67. Desoete, G. (1974) Overall kinetics of nitric oxide formation in flames. Survey Paper, Flame Chemistry Panel Meeting of the IFRF.
68. Pompei, F. and Heywood, J. B. (1972) The role of mixing in burner-generated carbon monoxide and nitric oxide. *Combustion and Flame*, **19**, p. 407.
69. Mizutani, Y. and Katsuki, M. (1976) Emissions from gas turbine combustors. *Bulletin of J.S.M.E.*, **19**, p. 1360.
70. Khalil, E. E. (1981) Numerical computation of combustion generated pollutants. Proc. Italian Flame Days (*La Rivista du Combustibili*).
71. Richter, W. and Chae, J. O. (1976) Prediction of natural gas flames including a simple model of NO formation. *Proc. 4th Members Conference of the IFRF.*
72. Borghi, R. (1973) Etude théorique de l'évolution résiduelle des produits polluants dans les jets de turboreacteurs. AGARD meeting, CP 125.
73. Singh, V. P. (1975) A study of turbulent flames stabilised in a high-velocity, high temperature flow. *Combustion Science and Technology*, **11**, p. 181.
74. Palmer, H. B. and Beer, J. M. (1974) *Combustion Technology, some Modern Developments*, (Academic Press).
75. Edelman, R. B. and Fortune, O. F. (1969) A quasi-global chemical kinetic model for the finite rate of combustion of hydrocarbon fuels with application to turbulent burning and mixing in hypersonic engines and nozzles. AIAA Paper 69–86 (7th Aerospace Sciences Meeting).
76. Gordon, S. and McBride, B. J. (1971) Computer program for calculation of complex chemical equilibrium composition, rocket performance, incident and reflected shocks and Chapman Jouguent detonations. NASA-SP 273.

77. Lenze, B., Milano, M. E. and Gunther, R. (1975) THe mutual influence of multiple jet diffusion flames. *Combustion Science and Technology*, **11**, p. 1.
78. Bowman, C. T. (1975) Kinetics of pollutant formation and destruction in combustion, *Progress in Energy and Combustion Science*, **1**, p. 33.
79. Caretto, L. S. (1975) Mathematical modelling of pollutant formation. *Progress in Energy and Combustion Science*, **1**, p. 1.

CHAPTER 7

NUMERICAL TECHNIQUES AND SOLUTION PROCEDURES

7.1 NUMERICAL TECHNIQUES

7.1.1 Preview

By now it must be clear that to solve analytically the conservation equations governing the flow field is indeed very difficult, particularly when combustion takes place. We should note that the real drive which made the modelling of reacting flows possible was the growth of available computer power, with algorithms that could be used to solve the governing equations. These are, in turn, complemented by physical models of adequate generality. Naturally, it is essential to discuss the techniques for the solution of the equations, to evaluate whether the encountered problems are user-dependent or problem-dependent. This assists in arriving at the prognosis of the problem, which of course leads to its eventual resolution. It follows that a clear understanding of the nature of numerical techniques should be achieved; otherwise numerical errors would be mistakenly considered as a disadvantage inherent in the physical model.

Numerical errors can qualitatively alter the flow fields. Although these numerical errors and difficulties, in all probability, are eventually solved more rapidly than many physical modelling problems, failure to recognize these errors as a major source of discrepancy between prediction and experiments can lead to a whole series of abortive theoretical modelling investigations.

The first step, in the development of a numerical procedure for solving the governing differential equations, is to superimpose a grid distribution on the flow domain. This is done to discretize the differential equations on all the grid points of the flow field, and to obtain 'equivalent' algebraic expressions called the 'finite difference equations'. The accuracy of the set of the finite difference equations, which approximate the partial differential equations, is totally dependent on the formulation of these difference equations and on the number of the grid nodes which represent the flow field.

To solve the governing equations numerically, there are three options to choose from: the finite difference method, the finite element method, and the Taylor series expansion method. The first and second methods were examined critically by Antonopoulos.[1] He showed that in most cases the two methods lead to the same resultant equations. However, the third method is less general than either the finite difference or the finite elements methods. This is because, apart from offering less physical insight in the derivation of the finite difference equations, the reciprocity requirement for the fluxes, at locations midway between grid nodes, usually leads to the central difference formulation. This formulation is an inaccurate representation of the fluxes at high Peclet numbers; this method, therefore, was discarded. Of the two other alternatives, the finite element was discarded because of the difficulty in prescribing the boundary conditions. Hence, in this chapter, discussion is limited to the finite difference methods simply because, at present, they are the most advanced in terms of fluid mechanics applications.

Commonly, a numerical scheme is either designated 'explicit' or 'implicit'. This terminology is appropriate to parabolic and hyperbolic equations, which possess an irreversible time or time-like coordinate, and in which the solution can be obtained by forward matching.[2] The steady laminar boundary layer equations form a parabolic system, with the streamwise direction as the marching coordinate.[3]

Elliptic-type problems, such as those discussed here, can be solved as steady state solutions of transient or pseudo-transient systems. As such it is then possible to use the parabolic or hyperbolic algorithms to solve these transient systems. Consequently, the terminology of explicit and implicit is meaningful when applied to this iterative (relaxation) approach to solving elliptic-type problems.

In furnaces and combustion chambers, usually a recirculation zone is created to stabilize the flame, and obviously the mixing processes of the fuel and air cannot be ignored. The flow, therefore, should not be described by anything less than the full Navier–Stokes equations. These are derived from the conservation of mass, momentum and energy, including the chemical reaction source terms. The Navier–Stokes equations form a coupled non-linear system. Such coupling is clearly viewed in convection terms. In the governing equations, the convective terms appear as a first derivative multiplied by the mass flux in the derivative direction. As a result, the mass flux components couple the governing equations, i.e. the governing equations involve more than one dependent variable in each equation.

Gourlay[4] classified numerical procedures into three classes and these are reproduced in Table 7.1. Category A can be relatively slow on the computer in some situations. However, the difference between category A and category B is not clear cut.[5] Details of the comparison between the various categories can be found in reference 5. A summary of the classification of most of the numerical procedures which are appropriate to reacting recirculating flows is presented in Table 7.2 (McDonald[5]). Examples of category 'A'

TABLE 7.1

Gourlay's Classification of Numerical Procedures[5]

Model	Features
A	Fast (?) general purpose algorithms which produce answers with minimum expenditure of man and machine time, and which are of general applicability.
B	Classical finite difference methods which are more problem-dependent than category A, and which require more programming effort.
C	Methods which give guaranteed accuracy at the expense of considerable programming effort.

TABLE 7.2

Reacting Flow Codes According to Gourlay's Classification[5]

A	B	C
1. Imperial College PSI-OMEGA code Gauss-Scidel by equation S-2D-SPE-R	1. Anderson's MacCormack code Variant of Lax-Wendroff U-2D-E	1. 'Murphy's PSI-OMEGA code' Galerkin line block elimination S-2D-DBI-R-NR-INC
2. Imperial College TEACH code Line Gauss-Scidel by equation S-3D-SPI-R	2. Buggeln and McDonald's FREP code/ Relaxation block ADI S-2D-SBI-R	
3. Widhopf and Victoria Leap frog-Dufort-Frankel U-2D-SBI-E	3. Gibeling, McDonald and Briley's MINT code Time-dependent block ADI U-3D-SBI-I	
4. Los Alamos RICE/ALE Harlow-Amsden by equation U-2D-SPI-E/I	4. Dwyer-Sander adaptive code Yanenko-Gear block implicit U-ID-SBI-I	
	5. Chang's atmospheric dispersion code Block SOR-GEAR U-2D-SBI-I (Fluid motion specified)	
5. Pletcher's Dufort-Frankel scheme BL-2D-NR-E	6. Blottner's block implicit BL-2D-SBI-I	2. Kendall's PLIMP code* BL-2D-SBI-I

S = steady form of Navier–Stokes equations; U = unsteady form of Navier–Stokes equation; BL = steady form of boundary layer equations; 2D = two space dimensions; 3D = three space dimensions; SPI = source terms equation uncoupled but implicit; SPE = source terms equation uncoupled and explicit; SBI = source block implicit system *i.e.* coupled equations; E = fluid equations treated explicitly; I = fluid equations treated implicitly; R = relaxation.
*Non-reacting.

TABLE 7.3

Typical Computer Times per Time Step[5]

CDC computer; computer time per grid node and per iteration step

Code specification			Computer time, s
TEACH	3D	(incompressible, laminar non-reacting)	$0.2–0.6 \times 10^{-3}$
TEACH	2D	(compressible, reacting, turbulent)	1.0×10^{-3}
MINT	3D	(reacting, compressible, turbulent) (non-reacting, compressible, turbulent)	2.5×10^{-3} 0.7×10^{-3}
NASA	TEACH	(incompressible, laminar, non-reacting)	1×10^{-3}
NASA	CHORIN	(incompressible, laminar, non-reacting)	0.08×10^{-3}

are numerous, such as Patankar *et al.*,[6] Gosman *et al*,[7] Widhopf *et al.*,[8] Rivard *et al.*[9] and Pletcher,[10] while the stream-function/vorticity Navier–Stokes scheme of Murphy[11] was categorized as 'C'. McDonald concluded that there is some improvement in efficiency as one goes from 'A' to 'B' to 'C'. Weighed against this increased efficiency is either the increase in the complexity of the procedure, or the poorly understood properties of the scheme for the non-model problems which arise in real flows. He provided detailed discussion of the merits and limitations of the various categories in terms of accuracy and residual history.

Before going into the details of the present numerical procedure, an indication of the cost of running computer programs is given. A large portion of the labour per time step is involved in forming the coefficients of the discrete equations rather than in the integration process itself; hence it is roughly independent of the numerical scheme used to integrate the fluid equations. The cost of running a system with stiff chemical equations may exceed the cost of performing the integration of the equations. Table 7.3 indicates quoted run time for various numerical procedures, these run times have been converted into run time per iteration per node per equation. The usefulness of this table, as a guideline to run time, is complemented by the knowledge of the number of iterations required to achieve the specified convergence criteria. The convergence criteria, the density of the grid nodes and their arrangement, the use of artificial damping methods, presence of singularities, boundary and inlet conditions, and other related problems all affect the number of iterations necessary to reach convergence.[12,13] Typically, the number of iterations can vary between 50 and 500 depending on the complexity of the flow.

The findings of previous authors, for reacting and non-reacting flow modelling, are summarized in Tables 7.4, 7.5 and 7.6. These results are discussed from the point of view of the extent of agreement, number of grid nodes, turbulence models and combustion models. Table 7.4 includes five typical investigations of isothermal-two dimensional and axisymmetric flows; the reported agreement between the measurements and predictions varied from reasonable to good. These flows comprise the flow behind a disc of Pope *et al.*,[14] the channel wall obstruction of Durst *et al.*,[15] the flow of twin jet of Militzer *et al.*,[16] the flow downstream a backstep and fence of Castro,[17] and finally the

TABLE 7.4

Experimental–Theoretical Comparisons. Turbulent Non-reacting and/or Recirculating Flows[5]

Authors	Data	Code	Isothermal 2.D TU – model	Comments
Pope and Whitelaw	Wake flow LDA + HW	TEACH Cat. A	Multi-eqn models	Good mean flow predictions without recirculation. Predictions moderate with recirculation. Sensitivity to inlet conditions noted. Grid points approximately 400. No singularities present.
Durst and Rastogi	Channel wall obstruction LDA	TEACH Cat. A	2-eqn $k - \epsilon$	Good qualitative agreement moderate quantitative agreement on mean flow. Sensitivity to corner point singularities noted. Grid points approximately 250.
Militzer, Nicoll and Alpay	Twin jet	TEACH Raithby Cat. A	2-eqn	Good quantitative agreement after numerical model. Basic prediction drowned by truncation error-hence. Raithby mod. Grid points approximately 1000. No singularities present.
Castro	Backstep and fence	TEACH Cat. A	2-eqn	Great care with the numerics required to get reasonable predictions. Corner singularities particularly sensitive. Approximately 1600 grid points. Convergence sensitive to initial guess.
Gosman *et al.*[13]	Various cases	TEACH Cat. A	2-eqn $k - \epsilon$	Good qualitative agreement – approximately 400 nodes.

TABLE 7.5

Experimental–Theoretical Comparisons. Turbulent Recirculating Flow[5]. Variable Density/Reacting 2-D

Author	Data	Code	Turbulence	Comments
Khalil, Spalding and Whitelaw	Four furnaces	TEACH Cat. A	$k - \epsilon$	General agreement with data. Temperatures agreement 'moderate' 400 grid points used. Radiation included. No numerical problems cited. One step chemistry with turbulence.
Schulz	Hydrogen air	PSI-OMEGA Cat. A	Albegraic	Great care and considerable effort required to obtain reasonable calculations. Difficulties with non-uniform meshes, and corner singularities. Recirculating flow aspects poorly predicted. Otherwise mean flow prediction adequate. Convergence problems with density gradients.
McDonald and Buggeln	Methane air combustor (LDA)	FREP Cat. B	$k - \epsilon$	Good qualitative agreement on mean velocity. Poor agreement with simple one-step combustion model on temp. Sensitivity to numerical detail noted. Corner point singularities present. Stiff turbulence equations noted. Boundary conditions found important. Approx. 1000 grid points used.
Elliman, Fussey and Hay	Methane air combustor (sampling)	GASEL Cat. A	$k - \epsilon$	Good qualitative agreement on mean velocity. Poor agreement with simple one step chemistry on temp. Better with turbulence effect on chemistry. No corner singularities but difficulties in converging a species equation due to boundary conditions.
Hutchinson, Khalil and Whitelaw	Furnace (LDA)	TEACH Cat. A	$k - \epsilon$	Good qualitative agreement throughout turbulence chemistry model used. No singularities present. 400 grid points used. Skill in numerics and details required to obtain solutions. Initial conditions found important. Slow convergence of NO equations.

TABLE 7.6

Turbulent Non-Reacting Recirculating Flows, 3-D

Author	Data	Code	TU model
Serag Eldin	Cold furnace flow	3-D CHAM Cat. A	$k - \epsilon$ good agreement
Gibeling *et al.*	Cold flow	MINT Cat. B	$k - \epsilon$ poor agreement
Abou Ellail *et al.*	Cold furnace flow	3-D TEACH Cat. A	$k - \epsilon$ good agreement

flow downstream an axisymmetric sudden enlargement and bluff bodies of Gosman *et al.*[13]

Calculations for two-dimensional recirculating reacting flows were obtained with the aid of the numerical procedures of references 18, 19, 20, 21, 22. Their conclusions are summarized in Table 7.5.

For non-reacting three-dimensional flows, only a few predictions were compared with the corresponding experiments. These are reviewed in Table 7.6. The investigations exhibited in this table are those of Serag Eldin,[23] Gibeling *et al.*[24] and Abou Ellail.[25]

In Table 7.7 some of the turbulent reacting flow predictions, in three-dimensional flows, are shown. It is clear from these cases that all the problems mentioned earlier can and do exist when attempts are made to make detailed predictions in complex furnace geometries. However, whether in two-dimensional or three-dimensional flows, reported in Tables 7.4 to 7.7, the agreement is considerable and the degree of predictive success is encouraging.

7.1.2 Layout of the Chapter

The main feature of the procedure used here is the use of the SIMPLE algorithm of the semi-implicit method for pressure-linked equations proposed by Caretto *et al.*[18] and Patankar *et al.*[19] In the remainder of the chapter the various grid arrangements, finite difference equations and the solution algorithm are described. Problems of numerical stability and convergence, grid sizes and false diffusion are considered, and specifications of the boundary and inlet conditions are discussed.

7.2 NUMERICAL PROCEDURES FOR TWO-DIMENSIONAL FLOWS

7.2.1 Grid Arrangement

The numerical grid system consists of a set of orthogonal intersecting lines in the coordinate directions, with no restriction on the distribution of the grid lines in the solution domain. The intersection of these lines forms the grid nodes at which all the flow properties, except the velocities U and V, are stored. The axial velocity U is located

TABLE 7.7

Turbulent Reacting Recirculating Flows, 3-D

Author	Data	Code	TU-model	Comments
Abou Ellail, Gosman, Lockwood and Megahed	Methane furnace	3-D TEACH Cat. A	$k - \epsilon$	Good qualitative agreement on mean flow. Turbulence-chemistry model with one-step reaction. 450 grid points used. No singularities present. No numerical problems highlighted.
Gibeling, McDonald and Briley	Premixed vap. keros. combustor	MINT Cat. B	$k - \epsilon$	Only sketchy experimental data available. Simple hydrocarbon chemistry one-step reaction. Poor agreement on temperature. 2600 grid points used, which were insufficient according to authors. No singularities present. High cell Reynolds numbers highlighted as truncation error source.
Pai, Michelfelder and Spalding	Natural gas furnace	3-D TEACH Variant Cat. A	$k - \epsilon$	Good qualitative agreement on mean flow. Gas fully burnt on inlet to adiabatic flame temperature; 1200 grid points used for half the symmetric furnace. Major discrepancies attributed to coarse mesh. No singularities present.

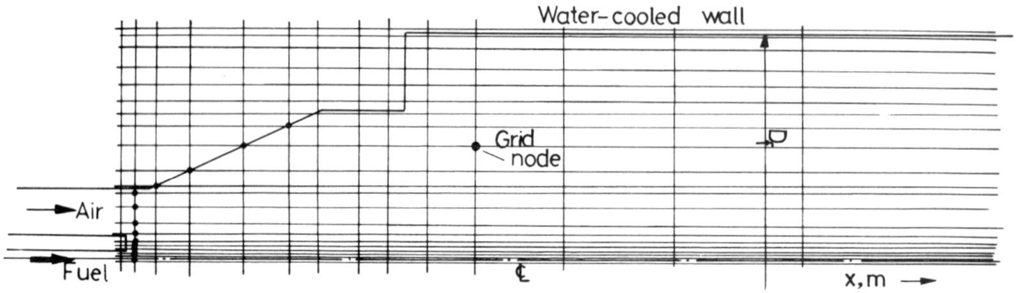

Figure 7.1 Numerical grid arrangement in an axisymmetric furnace

midway between the grid nodes in the axial direction, while the radial velocity V is located midway between grid nodes in the radial direction. Figure 7.1 shows the grid arrangement in the $x-y$ plane.

7.2.2 Finite Difference Equations

The partial differential equations, shown in previous chapters, can be represented by the general form:

$$\frac{\partial}{\partial x}\bar{\rho}\bar{U}\bar{\Phi}+\frac{1}{r}\frac{\partial}{\partial r}\bar{\rho}r\bar{V}\bar{\Phi}=\frac{\partial}{\partial x}\Gamma_{\Phi}\frac{\partial\bar{\Phi}}{\partial x}+\frac{1}{r}\frac{\partial}{\partial r}r\Gamma_{\Phi}\frac{\partial\bar{\Phi}}{\partial r}+\bar{S}_{\Phi} \qquad (7.2.1)$$

The finite difference counterpart of the general partial differential Equation 7.2.1 is derived by assuming that each variable is enclosed in its own control volume of 'cell' as shown in Figure 7.2. The grid node P is surrounded by two neighbouring nodes W and E in the west and east x-directions respectively, and two nodes N and S in the north and south y-directions respectively. The partial differential equation is integrated over the control volume, with the aid of assumptions that relate the nodal value of $\mathbf{\Phi}$ to the rates of generation/destruction of this entity within the cell, and to its transport by convection and diffusion across the cell boundaries. The former relation is represented, in a linearized form, as

$$S_{\Phi}=\int_{v}S_{\Phi}\,dv=S_{U}+S_{p}\Phi_{p} \qquad (7.2.2)$$

The latter relation is represented by an expression of the form

$$\rho U_{w}\frac{(\Phi_{p}+\Phi_{W})}{2}A-\Gamma_{\Phi_{w}}\frac{(\Phi_{p}-\Phi_{W})}{\delta x_{pW}}A \qquad (7.2.3)$$

This relation is valid when the quantity termed the Peclet number, $Pe_{W}=\rho U_{w}\delta_{pW}/\Gamma_{\Phi_{w}}$, is small. Otherwise, i.e. when Pe_{w} is large, the following relations stand:

$$\rho U_{w}\Phi_{W},\quad \text{when}\quad U_{w}>0 \qquad (7.2.4)$$

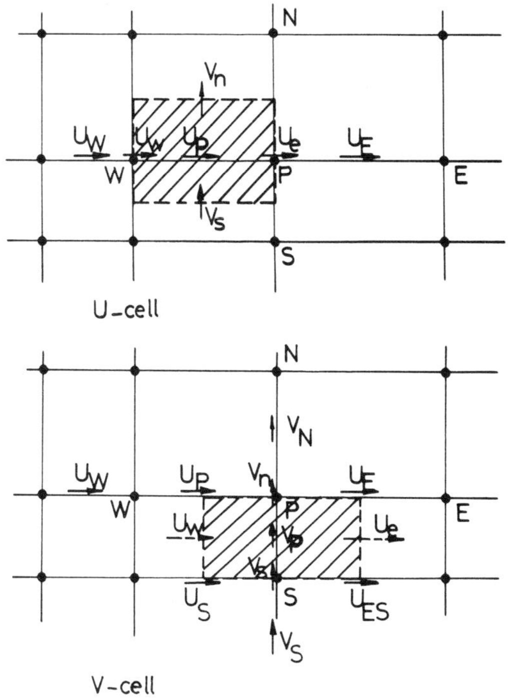

Figure 7.2 Finite difference grid for 2-D calculations

and

$$\rho U_w \Phi_p, \quad \text{when} \quad U_w < 0$$

Here the subscripts p and W refer to the central and west nodes respectively, and w denotes the intervening cell boundary. Assembling the above relations, and similar expressions for the remaining boundaries, yields the finite difference equation

$$A_p^\Phi \Phi_p = A_N^\Phi \Phi_N + A_S^\Phi \Phi_S + A_E^\Phi \Phi_E + A_W^\Phi \Phi_W + S_U^\Phi + S_p^\Phi \Phi_p \tag{7.2.5}$$

The coefficient A_p^Φ is expressed as:

$$A_p^\Phi = A_N^\Phi + A_S^\Phi + A_E^\Phi + A_W^\Phi \tag{7.2.6}$$

The coefficient A_n^Φ represents the net convective diffusion flux, $n = N, S, E$ and W; A_n^Φ is expressed as:

$$\begin{aligned} A_n^\Phi &= D_n^* - C_n \\ D_n^* &= \tfrac{1}{2}[D_n + | C_n | + | D_n - | C_n |] \\ D_n &= \Gamma_{\Phi,n} A/\delta x \\ C_n &= \rho \mathrm{U}_n A/2 \end{aligned} \tag{7.2.7}$$

and

$$S_\Phi^\Phi = S_U^\Phi + S_p^\Phi \Phi_p$$

where A is the area of the cell face normal to the x-direction.

Equations of this kind are formulated for each of the variables U, V, W, Φ at every cell. Appropriate modifications are made to the total flux Expressions 7.2.3 and 7.2.4 for the cells adjoining the boundaries of the solution domain, to account for the conditions imposed there. An equation for the remaining unknown, which is the pressure, is obtained by combining the continuity and momentum equations in the manner explained by Caretto *et al.*[18] and Patankar *et al.*[19] This entails connecting changes in pressure, denoted P', with changes in the velocities U and V by approximate formulae, which are derived from the momentum finite difference equations. This connection takes the form:

$$U_e = U_e^* + D_E^u(P_p' - P_E')$$

and (7.2.8)

$$U_w = U_w^* + D_W^u(P_w' - P_p')$$

Similarly for V_n and V_s. P' is the pressure correction and the Ds are the pressure-difference coefficients, which were stored during the setting up of the momentum equations. Substituting these formulae into the continuity equation yields a partial differential equation for P' similar to Equation 7.2.5, with S_U^Φ now representing the local continuity imbalance in the prevailing velocity field.

The effect of the pressure on the density should be included in the pressure correction equation; this does not necessarily mean that without this modification the pressure correction equation is only limited to uniform density flows. The effect of the temperature, concentration and pressure on the density has been accounted for by proper thermodynamic functions. However, the pressure correction equation is merely an intermediate device that leads to the final pressure distribution; the omission of some terms should have no effect on the correctness of the final solution.

When the effects of compressibility are predominant, for example in pressurized combustors or in supersonic flow problems, the above-mentioned treatment is likely to lead to numerical divergence.[20] The x-direction contribution to the net efflux of mass for the control volume of Figure 7.2, in this case, takes the form:

$$(\rho_e A U_e - \rho_w A U_w)$$

Now ρ_e and U_e are expressed as:

$$\rho_e = \rho_e^* + \left(\frac{1}{a^2}\right)(P_E' + P_p')/2 \qquad (7.2.9)$$

and

$$U_e = U_e^* + D_E^u(P_p' - P_E')$$

where a is the velocity of sound. The term $\rho_e U_e$ then is expressed as

$$\bar{\rho}_e \bar{U}_e = \rho_e^* U_e^* + [\rho_e^* D_E^u + (U_e^*/2a^2)]P_p' - [\rho_e^* D_E^u - (U_e^*/2a^2)]P_E' \quad (7.2.10)$$

The terms that involve P'^2 are neglected.

7.2.3 Solution Algorithm

The finite difference equations are solved iteratively employing the SIMPLE method (*S*emi-*I*mplicit *P*ressure *L*inked *E*quations) of Patankar and Spalding.[19] The procedure involves a series of guessing and correction operations for the coupling of the pressure with velocities. Two iteration sequences are employed: an inner sequence and an outer one.

The outer iteration sequence involves the cyclic application of the following steps:

1. A field of intermediate axial and radial velocities denoted by U^* and V^* is obtained by solving the associated momentum equations using the prevailing pressure field P^*.
2. Continuity is then enforced, by solving the equations for P' and thereby determining the required adjustments to the velocities and pressures through Equation 7.2.8.
3. The equations for the remaining variables are then solved in turn, and the whole process is repeated until a satisfactory solution is obtained, i.e. until the residuals (imbalance) in any of the finite difference equations is less than 10^{-4} of a specified inlet value of the entity in question.

The inner iteration sequence is employed to solve the equation sets for the individual variables. Solution is by a form of block iteration, in which a simple recurrence formula described by, for example, Gosman *et al.*[7] is used to solve simultaneously for the Φs along each grid line, in the line by line counterpart of the point Gauss–Seidal iteration. Complete convergence of the inner sequence is not necessary, and usually one to three applications of the block procedure suffice.

It should be noted that other solution procedures exist,[21,22,26] but the present SIMPLE procedure is the one which is more frequently applied to recirculating and reacting flows. The layout and the flow chart of the TEACH-T computer program are shown in Figure 7.3. The various subroutines are included; subroutine MAIN coordinates the various call operations. In the INIT subroutine all the geometrical information is specified and calculated and the initial flow field values are set out. The PROP subroutine calculates the densities and viscosities, temperatures and the probability density function. In the subroutines CALC Φ the coefficients of the difference equations and the source terms for each variable Φ are calculated. These subroutines also call modifications PROMOD, and solution LISOLVE. CALC-U, CALC-V and CALC-W refer to U, V and W the momentum equation solver. PRINT denotes printout routine, and PROMOD contains all modifications for wall and boundary conditions. The various miscellaneous details such as convergence and stability, grid size influence, numerical accuracy and false diffusion will be discussed later in the chapter.

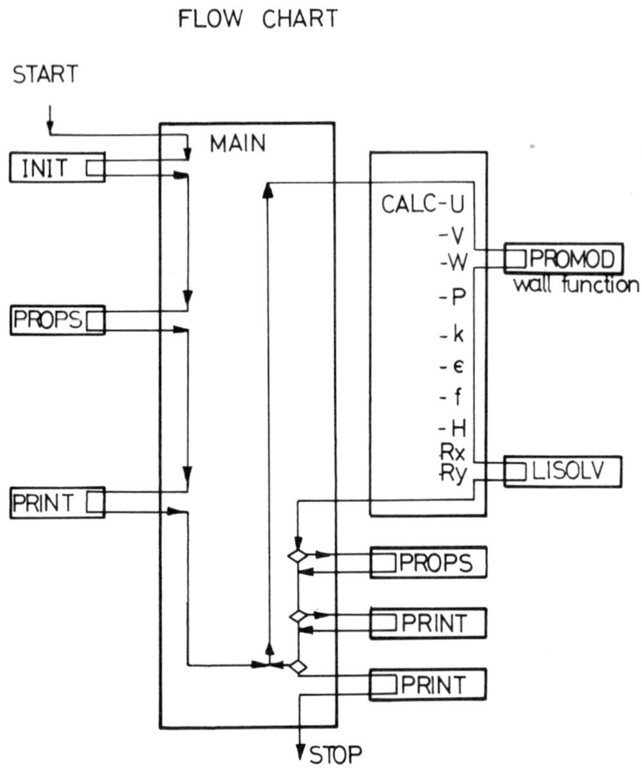

Figure 7.3 Flow chart of a 2-D computational scheme

7.3 NUMERICAL PROCEDURES FOR THREE-DIMENSIONAL FLOWS

7.3.1 Grid Arrangement

In 3-D flows, the arrangement of a staggered grid is employed;[19] this is illustrated in Figures 7.4 and 7.5. The figures show a central point P surrounded by its neighbouring points E, W, S and N in the X–Y plane and H and L in a plane normal to that one, i.e. the plane Y–Z. All dependent variables except the velocity components are stored at the grid line intersections (nodes), while the velocity components are stored between these nodes, as discussed earlier for 2-D flows.

7.3.2 Finite Difference Equations

The various coefficients of the general equation for Φ, i.e. convective and diffusive as well as source terms, are calculated in a similar manner to that of Section 7.2.2. The

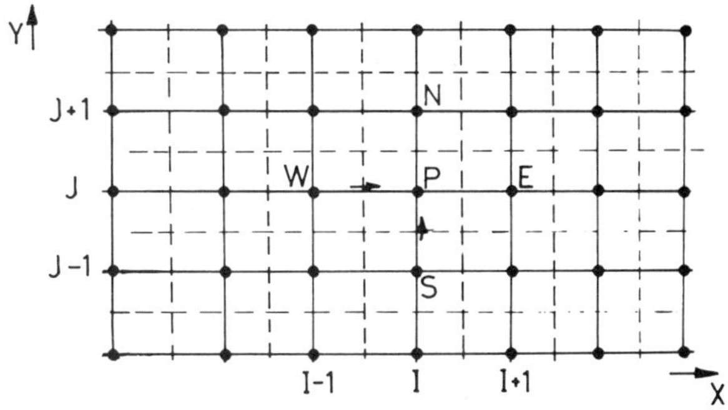

Figure 7.4 3-D finite difference grid (X–Y plane)

resultant finite difference equation is in the form:

$$A_p^\Phi \Phi_p = A_N^\Phi \Phi_N + A_S^\Phi \Phi_S + A_E^\Phi \Phi_E + A_W^\Phi \Phi_W + A_H^\Phi \Phi_H + A_L^\Phi \Phi_L + S_U^\Phi + S_p^\Phi \Phi_p \tag{7.3.1}$$

where the coefficient A_p^Φ is expressed as:

$$A_p^\Phi = A_N^\Phi + A_S^\Phi + A_E^\Phi + A_W^\Phi + A_H^\Phi + A_L^\Phi \tag{7.3.2}$$

Each of the coefficients A_n^Φ represents the net convection and diffusion flux, $n =$ N, S, E, W, H and L; A_n^Φ is expressed in the same way as Expression 7.2.7.

The presence of the pressure gradient as a source term in the momentum equations, requires the evaluation of the local pressures. The continuity equation coupled with the momentum equations yields the pressure correction equation. The incorporation of the P'-equation in the solution procedure proceeds through the following path:

1. Initially the momentum equations are solved with an assumed pressure field, thus yielding a first estimate of the velocity field. Continuity consideration at this stage is neglected.
2. The pressure field is then corrected using the pressure corrections. In the same way as for the 2-D flows, these pressure corrections are obtained from the solution of

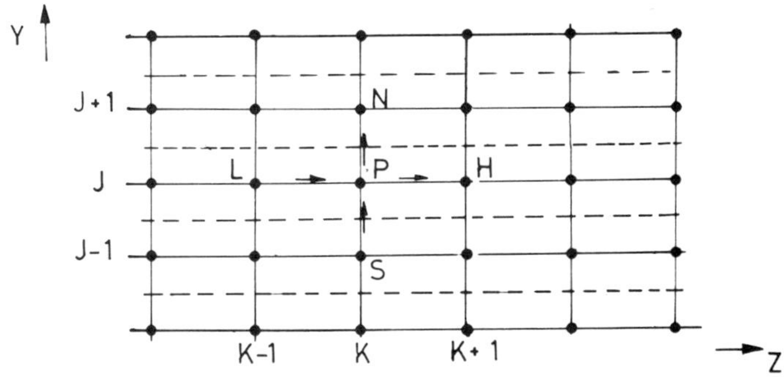

Figure 7.5 3-D finite difference grid (Y–Z plane)

the continuity-based pressure correction equation, in such a manner that when the momentum equation is solved again employing the updated pressure field, the subsequent velocities satisfy continuity. The correct velocities are calculated from modification to the initial guessed velocity field through the use of linearized form of momentum equations in the form:

$$\begin{aligned} U_e &= U_e^* + D_{\mathrm{E}}^{U}(p_p' - P_{\mathrm{E}}') \\ U_w &= U_w^* + D_{\mathrm{W}}^{U}(P_{\mathrm{W}}' - p_p') \\ \mathrm{W}_h &= \mathrm{W}_h^* + D_{\mathrm{H}}^{\mathrm{W}}(p_p' - P_{\mathrm{H}}') \\ \mathrm{W}_l &= \mathrm{W}_l^* + D_{\mathrm{L}}^{\mathrm{W}}(P_{\mathrm{L}}' - P_p') \end{aligned} \tag{7.3.3}$$

where p' is the pressure correction and the Ds are the pressure difference coefficients which are stored during setting up the momentum equations.

The partial differential equation for pressure correction is obtained in the form

$$A_p p' = A_{\mathrm{E}} p_{\mathrm{E}}' + A_{\mathrm{W}} p_{\mathrm{W}}' + A_{\mathrm{N}} p_{\mathrm{N}}' + A_{\mathrm{S}} p_{\mathrm{S}}' + A_{\mathrm{H}} p_{\mathrm{H}}' + A_{\mathrm{L}} p_{\mathrm{L}}' + \mathrm{S}_U \tag{7.3.4}$$

S_U is the mass imbalance resulting from the initial starred velocity field and is expressed as

$$\mathrm{S}_U = (\rho U_w^* A)_w - (\rho U_e^* A)_e + (\rho V_s^* A)_s - (\rho V_n^* A)_n + (\rho \mathrm{W}_l^* A)_n - (\rho \mathrm{W}_h^* A)_h \tag{7.3.5}$$

7.3.3 Solution Algorithm

In the preceding section, reference has been made to the momentum and pressure equation. The actual method of solving it is performed by the well known tridiagonal matrix algorithm.[19] It is not necessary to obtain an exact solution of the governing equations, because the coefficients of these equations are only tentative and require updating to account for the interlinkages and non-linearities. What is exactly needed is an iterative method of solving algebraic equations every time a few iterations are performed, and before the coefficients are recalculated. A double sweep of the TDMA, employed for 2-*D* flows, would be a triple sweep in 3-*D* computations; this would require relatively large computer storage and/or time.[27] An easy way out is to use a combination of a double sweep and a block correction operation.[23]

The procedure is as follows. First, attention is given to one *X*–*Y* plane, the coefficients are calculated, and the equations are solved by double sweep; then one proceeds to the next plane. The only defect of this procedure is that it takes very poor account of the interactions in the *Z* direction; and the convergence can be very slow. The central idea of the block procedure is to make uniform additions to the Φ values in an *X*–*Y* plane, such that the integral conservation equations are satisfied over the *X*–*Y* slabs of control volumes stacked in the *Z*-direction. Details of the block correction procedure can be found in reference 23. Very few computational methods for 3-*D* procedures have been described in the open literature; some of these are by Patankar and Spalding,[19] Serag Eldin *et al.*,[28] Serag Eldin,[23] Hjertager *et al.*[27] and Turan.[29]

7.4 MISCELLANEOUS DETAILS

The numerical solution is required to pass two acceptance tests; firstly, it must satisfy the finite difference equations (typically, the imbalance must be 0.01% or less), and secondly, it must be invariant with further increase in the number of grid nodes.

7.4.1 Convergence and Stability

The simultaneous and non-linear nature of the finite difference equations necessitates that special measures should be employed to procure numerical stability (convergence): one of these measures is the under-relaxation factors which relate the old and new values of the general dependent variable Φ as follows:

$$\Phi = \beta\Phi_{new} + (1 - \beta)\Phi_{old} \tag{7.4.1}$$

where β is the under-relaxation factor. The recommended[30] values of β are given in Table 7.8.

In swirling flow, the practice of gradual introduction of swirl was adopted, i.e. increasing the value of swirl number gradually with iteration to the specified value. Convergence problems associated with reacting flows were observed; the local mixture density was heavily under-relaxed to prevent divergence. The value of β in this case was 0.2. The residuals decreased monotonically after 50 iterations and satisfied the corresponding criteria of all residuals less than 10^{-4} at 200 to 500 iterations for 2-*D* flow configurations, and up to 1200 iterations for 3-*D* flows. The corresponding computational times were shown earlier in Table 7.3.

Examples of the variation of the residuals in *U*, *V*, *W* and continuity equations, with the number of iterations are shown in Figures 7.6, 7.7 and 7.8, for various recirculating flow situations. The figures demonstrate the effects of grid size and arrangement, swirl and recirculation and combustion on the convergence of the numerical solution algorithms.

7.4.2 Grid Independent Results

It is important to test the grid dependence of a solution, in order to economize on computer time by using the minimum number of nodes which can give a grid independent solution. A series of tests should be performed for each flow configuration to determine the influence of the number and distribution of grid nodes. Various grid distributions can be used, e.g. a linear progression where the ratio between any successive locations is

TABLE 7.8

Φ	U	V	W	P	k	ϵ	f	M_{fu}	H
$\beta_{initial}$	0.3	0.2	0.2	1	0.8	0.8	0.9	0.9	0.9
β_{max}	0.7	0.6	0.6	1	1.0	1.0	1.0	1.0	1.0

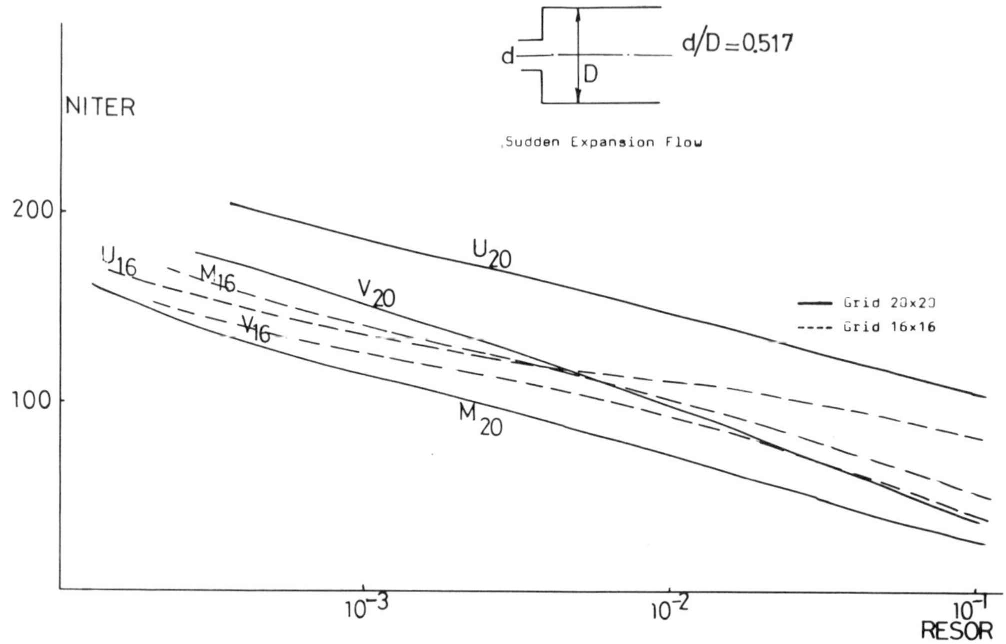

Figure 7.6 Variation of normalized residuals with number of iterations

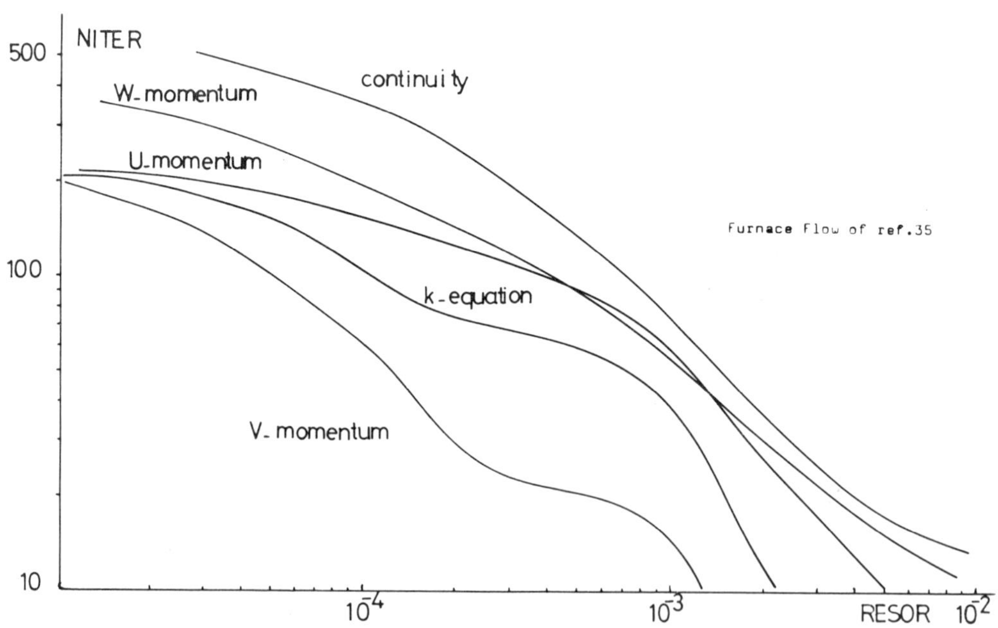

Figure 7.7 Variation of normalized residuals with number of iterations
Furnace flow of reference 35

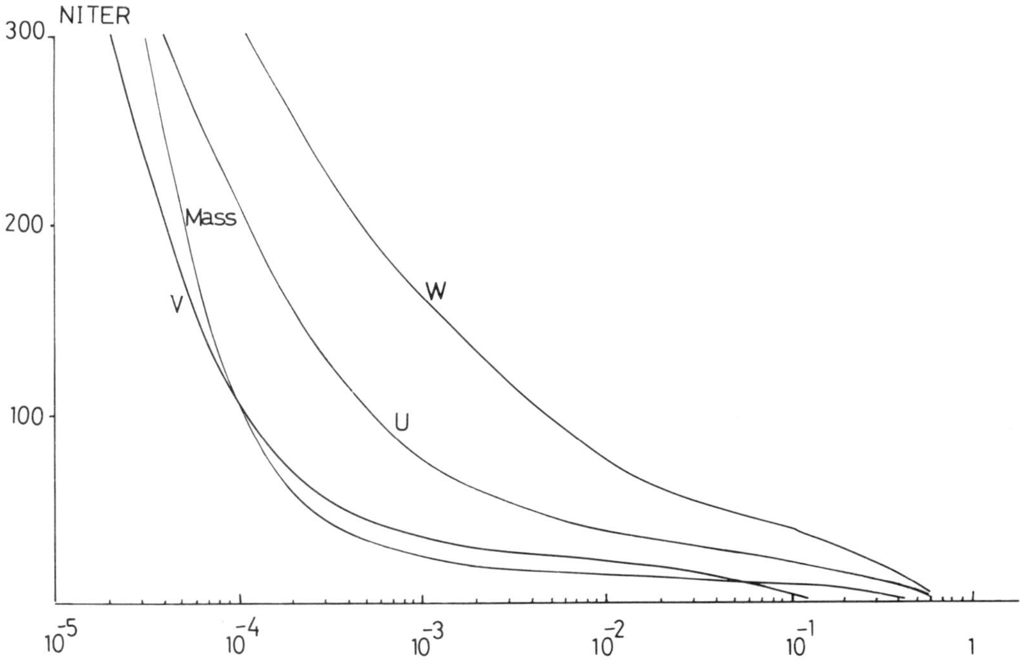

Figure 7.8 Variation of normalized residuals with iterations Turbulent swirling flame. Grid 20 × 20

less than 1.2; it is also possible to specify the grid nodes individually. An analysis of the errors resulting from the use of non-uniform grids, without the application of proper corrections, showed[12] that the error in the second derivative of Φ for example, is given as:

$$\text{Error} \propto \frac{1}{6}\frac{\partial^3\Phi}{\partial x^3}\left[\frac{\Delta x_+^3 - \Delta x_-^3}{\Delta x^2}\right] \tag{7.4.2}$$

where Δx_+, Δx_- are the distances between the nodes E and P, and P and W respectively, and Δx is the uniform spacing. The value of the velocity component at the edges of the control volume does not equal half the value of the two velocities at the two neighbouring nodes, as would be the case for uniform spacing, but is a weighted average of the two neighbouring velocities:

$$U_e = U_{\mathrm{E}}\left(1 - \frac{\Delta x_+}{\Delta x_+ + \Delta x_-}\right) + U_p\left(\frac{\Delta x_+}{\Delta x_+ + \Delta x_-}\right) \tag{7.4.3}$$

and for equal spacing this relation is reduced to $(U_{\mathrm{E}} + U_p)/2$.

The effect of grid size on the obtained results was investigated by, among others, Khalil[12,30] and Serag Eldin.[23] Samples of the investigated flows are shown in Figures 7.9 and 7.10. For sudden axisymmetric enlargement flow at $R_e = 6000$ and $d/D = 0.388$, grid sizes of 16 × 16, 20 × 20 and 25 × 25 are presented in Figure 7.9. The distributions obtained with 20 × 20 and 25 × 25 nodes were, for all intents and purposes the same, and the local flow properties were found to be grid independent. Figure 7.10

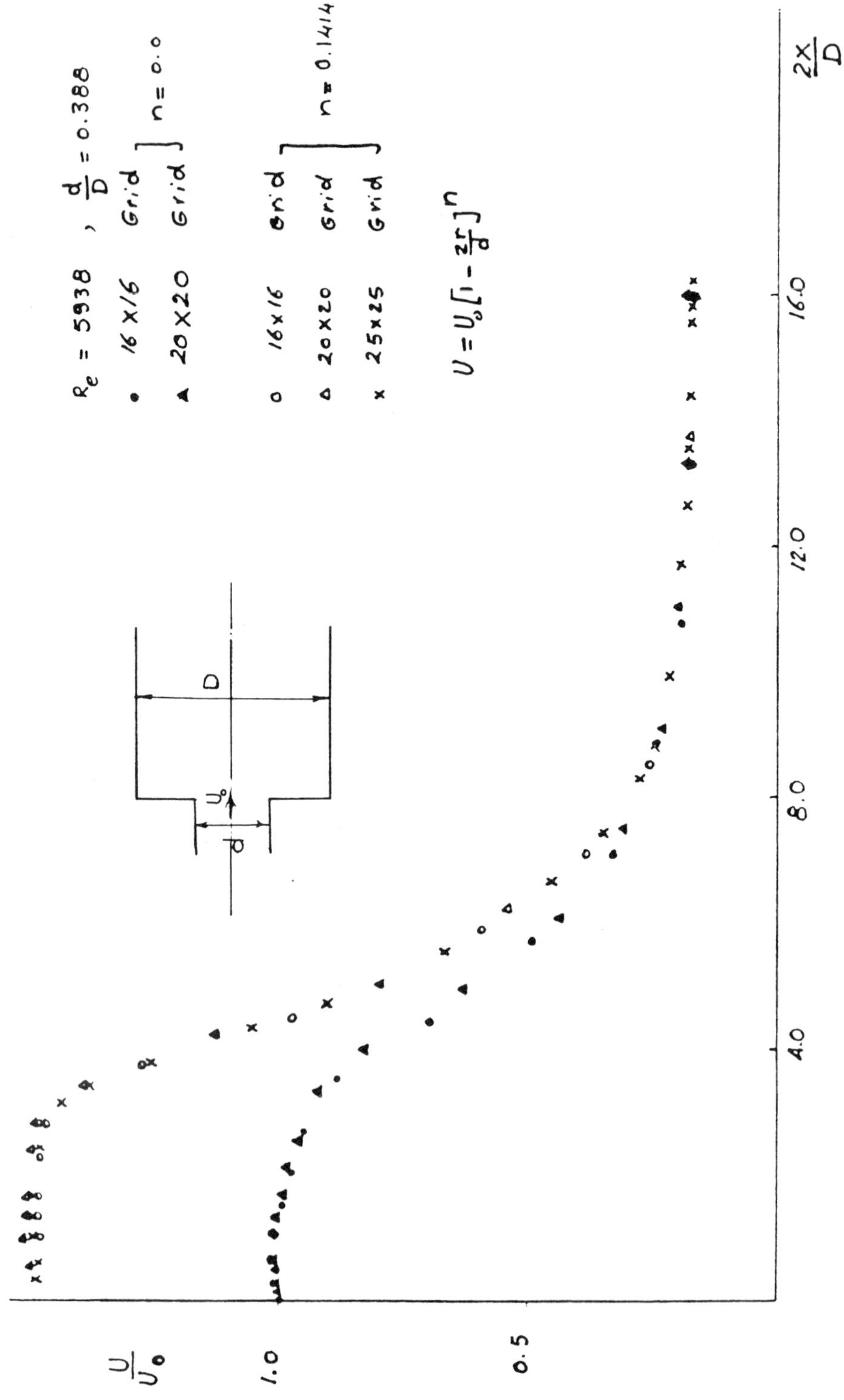

Figure 7.9 Effect of grid size on axial velocity distribution in a sudden expansion geometry[30]

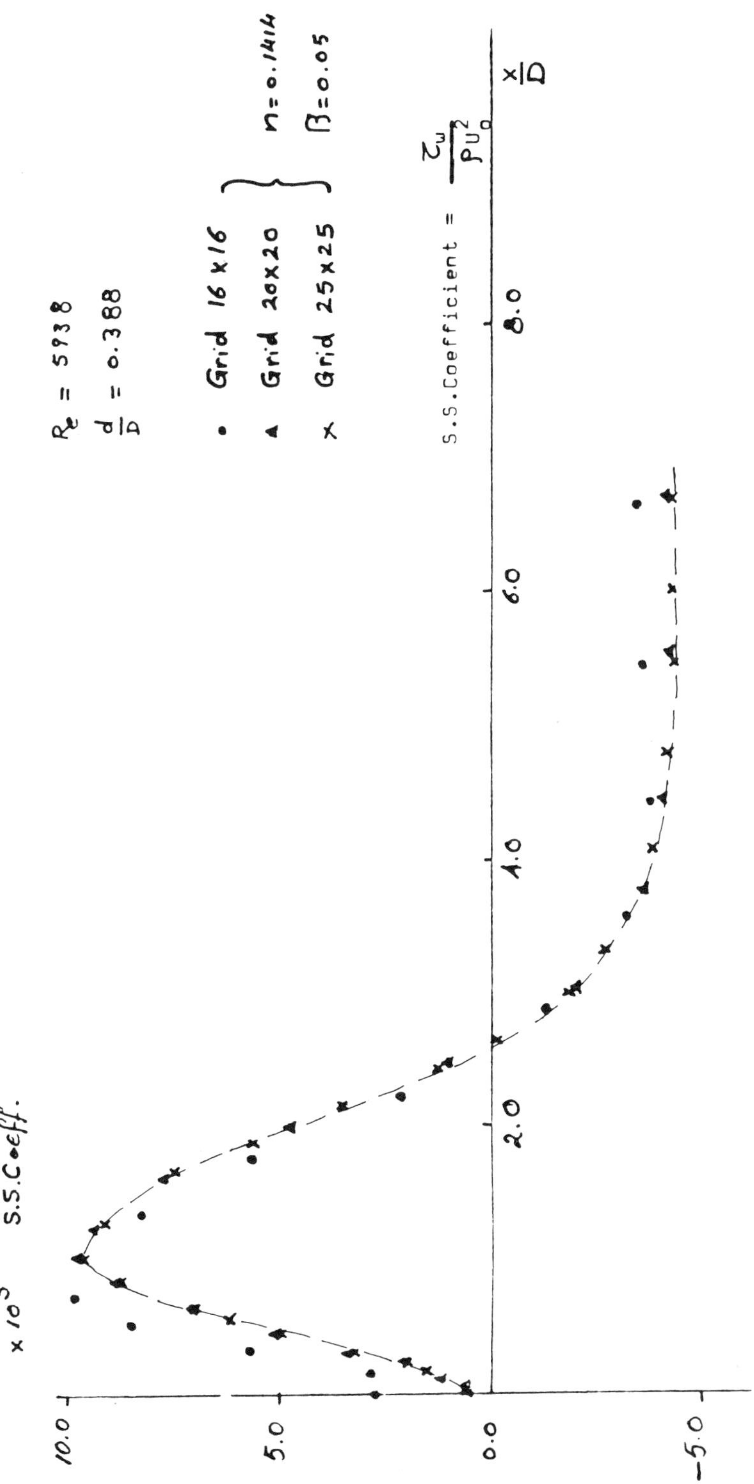

Figure 7.10 Effect of grid size on wall shear stress distribution in a sudden expansion geometry

illustrates the effect of grid size on the wall shear stress distribution. It is clear that the results obtained from 16 × 16 grid are grid dependent.

7.4.3 Numerical Accuracy

Three different schemes can be used to represent the convective and diffusive flux expressions in the finite difference equations.

These are the central, upwind and hybrid difference formulae. The central difference expressions give unrealistic results for $|Pe| > 2$. At large $|Pe|$, the upwind difference is preferable as reported by Gosman *et al.*[7] A combination of the two expressions, namely the hybrid formula, gives better representation of the combined fluxes for all the $|Pe|$; therefore it is used in the calculation of furnace flows.

One of the numerical errors associated with the finite difference representation of the conservation equations is false diffusion. It occurs because all the finite difference schemes evaluate Φ_p as a weighted mean of the surrounding Φs. The errors due to false diffusion diminish when the streamlines are parallel to the mesh and for small values of *Re*. An expression for the false diffusion was given by Gosman *et al.*[32] Its effect is important in laminar flows, but at high Reynolds number, i.e. in turbulent flows, the effective viscosity μ_{eff} is much larger than the laminar one and the effect of false diffusion diminishes.

7.4.4 Thermodynamic Properties

The density of mixtures of air, the combusting gases and the combustion products can be represented, with adequate precision for the present purposes, by the equation of state for a perfect gas:

$$\rho = P/(RT/\mathring{W}_{\text{mix}}) \tag{7.4.4}$$

where $\mathring{W}_{\text{mix}}$ and P are determined from the appropriate mass fractions and Dalton's Law of partial pressures. The molecular weight $\mathring{W}_{\text{mix}}$ is obtained from

$$\mathring{W}_{\text{mix}} = \frac{\mathring{W}_a}{\Sigma M_a} \tag{7.4.5}$$

and the heat of reaction of the fuel is represented by

$$H_{\text{fu}} = \Sigma M_a H_a$$

with the values of H_a, determined from JANAF chemical tables.[33]

The specific heat of the mixture was obtained from the equation

$$\bar{C}_p = \Sigma \bar{M}_a \bar{C}_{Pa} \tag{7.4.6}$$

where

$$\bar{C}_{Pa} = a_0 + a_1\bar{T} + a_2\bar{T}^2 + a_3\bar{T}^3 \tag{7.4.7}$$

with the values of the coefficients computed for each gas. The use of lower order polynomials has been found[12] to be significantly less satisfactory than the third order polynomials of Equation 7.4.7.

7.5 BOUNDARY AND INLET CONDITIONS

The selection of acceptable boundary conditions for aerodynamic flow problems is a major topic in itself.[34] Very few studies on combusting flows emphasize the importance of boundary conditions and associated problems; examples are those by Gosman *et al.*[13] and by Khalil.[30]

In order to economize on the grid nodes, often conditions adjacent to the solid wall are matched to some assumed boundary layer profiles of velocity, temperature and species. This practice is commonly known as employing wall functions, as discussed earlier in Chapter 3. In addition to the wall functions, boundary layer pressure approximation of zero pressure gradient normal to the wall is often invoked. As in most of the combusting flows, regions of interest are far from the wall, considerable lassitude in the wall boundary conditions is often tolerated.

In order to represent wall regions, the first requirement is to ensure that the near wall region is adequately resolved by the computational mesh. The following assumptions

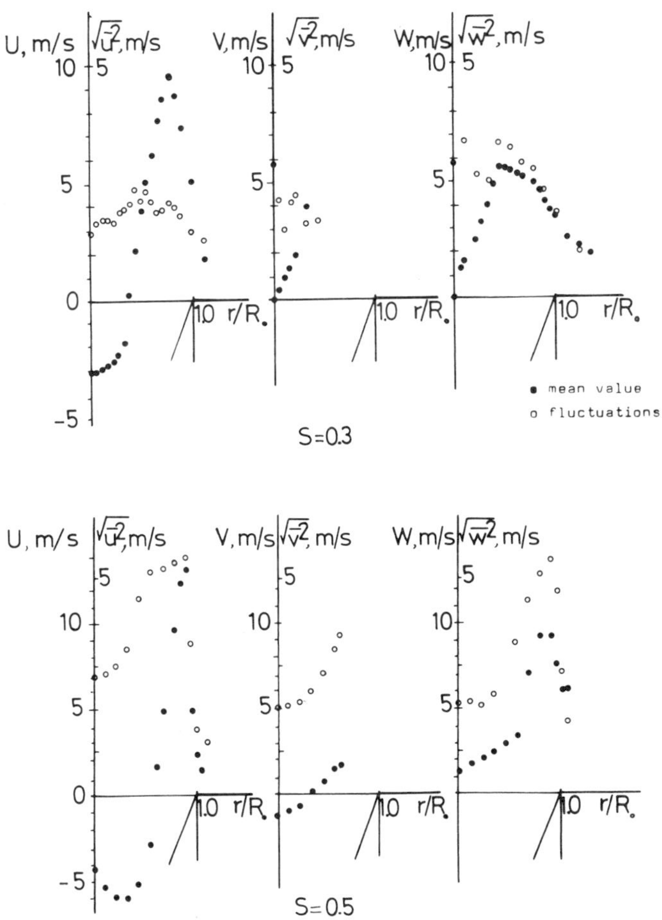

Figure 7.11 Inlet velocity profiles in non-reacting furnace flows[36]

could also be realistically made:

1. No slip at wall
2. Zero turbulent fluctuations
3. Specified wall temperature or an adiabatic wall
4. Zero normal gradient of species mass fractions
5. The normal pressure gradient is normally computed from the momentum equation applied to the wall.

The inlet boundary conditions are generally not well defined, and unfortunately the predicted solution can exhibit a considerable degree of sensitivity to poorly defined, but computationally required inlet information.[12,14,30] The sensitivity of the predictions to the streamline direction entering the computational domain has been noted in references 14 and 30; generally speaking this streamline direction is not known very precisely at the exit plane of the furnace or combustor. The problem of inlet conditions is twofold. First, what is the shape of the inlet profiles and secondly what is the magnitude of the variable at inlet? The influence of the various inlet profiles was investigated.[30] and it was found that a significant effect was observed on the mean flow pattern. Examples of various inlet velocity profiles are shown in Figures 7.11 and 7.12. It was concluded that

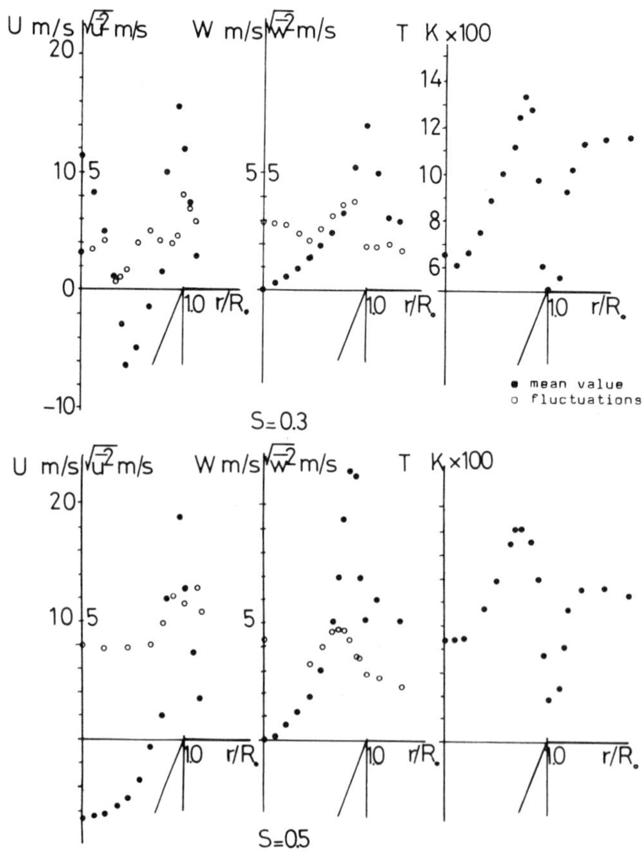

Figure 7.12 Inlet velocity and temperature profiles in reacting furnace flows[36]

TABLE 7.9

Inlet Flow Profiles

Variable	Inlet profile	
$\bar{U}$	$U_0\left(1-\frac{2r}{d}\right)^n, \quad 0.2 \geqslant n \geqslant 0$	pipe
	$U_0\left(1-\frac{R_a-r}{\left(R_a-\frac{D_n}{2}\right)}\right)^{0.1}$	annulus
$\bar{V}$	0	
$\bar{W}$	Solid body rotation; velocity is adjusted to reproduce measured swirl number	
k	$B\left(1+2\left(\frac{2r}{d}\right)^2\right)$	pipe
	$B\left(1+9\left(\frac{\lvert R_a-r\rvert}{\left(R_a-\frac{D_n}{2}\right)}\right)^{2.82}\right)$	annulus
ε	$C_\mu k^{3/2}/(0.03y_a)$	
$\bar{T}$	$T_0\left(1\pm\frac{2r}{d}\right)^n \quad 0.1414 \geqslant n \geqslant 0$ The sign depends on the wall/fluid temperature ratio.	

$R_a = 0.5y_a + D_n/2$
y_a = width of the annular gap of the burner
D_n = annulus inner diameter
U_0 = centreline velocity (of the annulus)
$B = 1.5(\sqrt{\bar{u}^2})^2_{\text{inlet}}$
T_0 = inlet temperature at centreline.

recommended inlet profiles can be given by the simple functions displayed in Table 7.9. However, it should be stressed that the inlet profiles should conform to measured values whenever available; otherwise these should be reasonably assumed. Measurements of inlet velocity and turbulence intensity profiles in combustor flows are scarce, but some detailed measurements were recently reported,[35–38] and were obtained with the aid of a laser Doppler Anemometer.

Exit boundary conditions have not been proved to be troublesome in practice. Here linear extrapolation from the internal nodes can be employed, i.e. specifying that the axial gradient is zero, that $\partial\Phi/\partial x$ equals zero, and that the overall conservation of mass and species is satisfied. The location of the exit plane was taken far downstream so as not to influence the upstream properties.

Along the axis of symmetry the gradients in the radial direction of all variables are equal to zero. Studies on the effect of corner and inclined burner quarl singularities as boundary conditions were reported by Buggeln *et al.*,[39] Schultz,[40] Hutchinson *et al.*[36] and Khalil.[30]

References to Chapter 7

1. Antonopoulos, K. (1975) A comparison of finite difference and finite element methods of solution of convective transport equation. M.Sc Thesis. London University.
2. Patankar, S. V. and Spalding, D. B. (1967) *Heat and Mass Transfer in Boundary Layer* (Academic Press).
3. Spalding, D. B. (1977) *Genmix – a General Computer Program for Two-Dimensional Parabolic Phenomena* (Pergamon Press).
4. Gourlay, A. R. (1971) Some recent methods for the numerical solution of time dependent partial differential equations. *Proc. Royal Society*, **323**, p. 219.
5. McDonald, H. (1979) Combustion modelling in two and three dimensions – some numerical considerations. *Progress in Energy and Combustion Science*, **5**, p. 97.
6. Patankar, S. V. and Spalding, D. B. (1967) A finite diference procedure for solving the equations of the two dimensional boundary layer. *Int. J. Heat and Mass Transfer*, **10**, p. 1389.
7. Gosman, A. D. and Pun, W. M. (1974) Calculation of recirculating flows. Imperial College, Dept. of Mechanical Engineering, Report HTS/74/2.
8. Widhopf, G. and Victoria, K. (1973) On the solution of the unsteady Navier–Stokes equations including multicomponent finite rate chemistry. *Comput. Fluids*, **1**, p. 159.
9. Rivard, W. C., Butler, T. D. and Farmer, O. A. (1974) The transient dynamics of chemically reactive gaseous mixtures with turbulence. *Proc. 4th Int. Conference on Numerical Methods in Fluid Dynamics.*
10. Pletcher, R. H. (1974) Prediction of transpired turbulent boundary layers. *J. Heat Transfer*, **96**, p. 89.
11. Murphy, J. D. (1977) An efficient solution procedure for the incompressible Navier–Stokes equations. *AIAA Journal*, **15**, p. 1307.
12. Khalil, E. E. (1977) Flow and combustion in axisymmetric furnaces. Ph.D. Thesis, London University.
13. Gosman, A. D., Khalil, E. E. and Whitelaw, J. H. (1979) The calculation of two-dimensional turbulent recirculating flows, in *Turbulent Shear Flows*, edited by Durst *et al.*, (Springer Verlag), p. 237.
14. Pope, S. B. and Whitelaw, J. H. (1976) The calculation of near wake flows. *J. Fluid Mechanics*, **73**, p. 9.
15. Durst, F. and Rastogi, A. K. (1977) Theoretical and experimental investigations of turbulent flows with separation. *Proc. Symp. on Turbulent Shear Flows.*
16. Militzer, J., Nicoll, W. B. and Alpay, S. A. (1977) Some observations on the numerical calculation of the recirculation region of twin parallel symmetric jet flow. *Proc. Symp. on Turbulent Shear Flows.*
17. Castro, I. P. (1977) Numerical difficulties in the calculation of complex flows. *Proc. Symp. on Turbulent Shear Flows.*
18. Caretto, L. S., Gosman, A. D., Patankar, S. V. and Spalding, D. B. (1972) Two calculation procedures for steady three dimensional flows with recirculation. *Proc. 3rd Int. Conference on Numerical Methods in Fluid Mechanics.*
19. Patankar, S. V. and Spalding, D. B. (1972) A calculation procedure for heat, mass and momentum transfer in three dimensional parabolic flows. *Int. J. Heat and Mass Transfer*, **15**, p. 1787.
20. Khalil, E. E. (1981) Combustion characteristics of high pressure gas fired combustors. AIAA Paper 81-0040 (19th Aerospace Sciences Meeting).
21. Patankar, S. V. and Spalding, D. B. (1972) A computer model for three-dimensional flow in furnaces. *Proc. 14th Symp. (Int.) on Combustion*, p. 604.
22. Spalding, D. B. (1974) A method for computing steady and unsteady flows possessing discontinuities of density. CHAM Ltd., London, Report CHAM/90/2.

23. Serag Eldin, M. A. (1977) The numerical prediction of the flow and combustion processes in a three dimensional can combustor. Ph.D. Thesis, London University.
24. Gibeling, H. J., McDonald, H. and Briley, W. R. Development of a three dimensional combustor flow analysis. AFAPL-TR-75-59.
25. Abou Ellail, M. M. M., Gosman, A. D., Lockwood, F. C. and Megahed, I. E. A. (1977) A three dimensional procedure for combustion chamber flows. AIAA Paper 77-138, AIAA 17th Aerospace Sciences meeting.
26. Spalding, D. B. (1972) A novel finite difference formulation for differential expressions involving both first and second derivatives. *Int. J. Numerical Methods in Engineering*, **4**, p. 551.
27. Magnussen, B. and Hjertager, B. (1975) A computation procedure for three dimensional flows. Inst. for Teknish varmelaere, Denmark.
28. Serag Eldin, M. A. and Spalding, D. B. (1978) Validation of three dimensional cold flow predictions inside a can combustor, in *Flow Mixing and Heat Transfer in Furnaces*, edited by Khalil *et al.* (Pergamon Press), p. 215.
29. Turan, A. (1978) A three dimensional model for gas turbine combustors. Ph.D. Thesis, University of Sheffield.
30. Khalil, E. E. (1981) Numerical computations of turbulent reacting combustor flows, in *Numerical Methods in Heat Transfer*, edited by Lewis *et al.* (John Wiley), p. 526.
31. Mathews, K. J. (1976) Gasel: a two-dimensional flame prediction model. CEGB. Report R/M/R232.
32. Gosman, A. D., Pun, W. M., Runchal, A. K., Spalding, D. B. and Wolfshtein, M. (1969) *Heat and Mass Transfer in Recirculating Flows* (Academic Press).
33. JANAF. (1971) Thermochemical Tables. National Bureau of Standards, NSRDS 37, Com-71, 50363.
34. Roache, P. J. (1972) *Computational Fluid Dynamics* (Hermosa).
35. Baker, R. J., Hutchinson, P., Khalil, E. E. and Whitelaw, J. H. (1974) Measurements of three velocity components in a model furnace with and without combustion. *Proc. 15th Symp. (Int.) on Combustion*, p. 553.
36. Hutchinson, P., Khalil, E. E. and Whitelaw, J. H. (1977) Measurement and calculation of furnace flow properties, *J. Energy*, **1**, p. 212.
37. Dovark, N. and Chigier, N. A. (1974. *Proc. 15th Symp. (Int.) on combustion*, p. 565.
38. Baker, R. J., Hutchinson, P. and Whitelaw, J. H. (1974) Preliminary measurements of instantaneous velocity in a 2 m square furnace using laser anemometry. *J. Heat Transfer*, **96**, p. 410.
39. Buggeln, R. C. and McDonald, H. (1977) Influence of aerodynamic phenomena on pollutant formation in combustion (theoretical studies). EPA 68-02-1873.
40. Schultz, J. (1977) An investigation of ducted two stream variable density turbulent jet mixing with recirculation. AFOSR-TR-76-1087.

APPENDICES

A1 CALCULATION OF THE SWIRL NUMBER

The most commonly used parameter to characterize swirling flows is the swirl number,[1] defined by the equation

$$S = \frac{2G_\phi}{G_x \cdot D_e} \tag{A1.1}$$

where

$$G_\phi = \text{the axial flux of the angular momentum}$$

$$= 2\pi \int_{R_1}^{R_2} W r \rho U r dr$$

$$G_x = \text{the axial flux of the linear momentum}$$

$$= 2\pi \int_{R_1}^{R_2} (U \rho U r dr + P r dr)$$

R_1 and R_2 are the inner and outer radii of the swirler and D_e is the exit throat diameter; P is the local static pressure. It should be noted that the swirl number is calculated from measured values of velocity and static pressure profiles at the burner exit. This is frequently not possible owing to the lack of detailed experimental results, and it has been shown[1] that the swirl number may be satisfactorily calculated from the geometry of most swirl generators.

Recent investigations[2] have calculated the swirl number with the aid of local velocity measurements obtained at the burner throat exit section. These measurements were obtained with the aid of a laser Doppler anemometer, and comprised mean axial and tangential velocities and the corresponding fluctuations.

References to Appendix 1

1. Beer, J. M. and Chigier, N. A. (1972) *Combustion Aerodynamics* (Applied Science Publishers).
2. Khalil, E. E. (1977) Flow and Combustion in Axisymmetric Furnaces. Ph.D. Thesis, London University.

A2 NUMERICAL TEST CASES

A. Isothermal Flows

Turbulent pipe flow
Re = 5×10^5
Computer time = 1 min

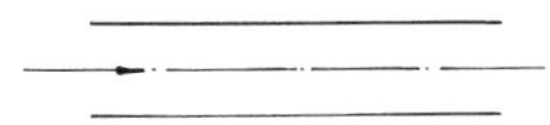

Turbulent swirling pipe flow
Re = 10^5
Swirl numbers of 0.55, 1.0, 3.0
Computer time = 1.78 min

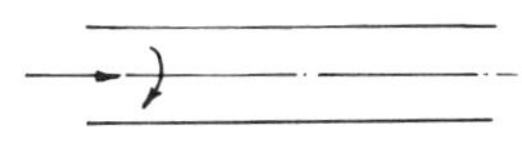

Sudden expansion turbulent flow
Re = 2×10^5
Diameter ratio up to 10.
Computer time = 2.87 min

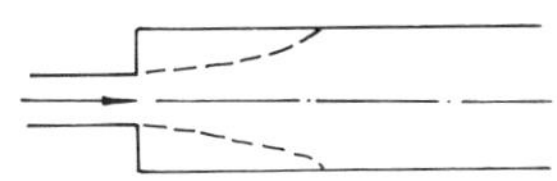

Sudden expansion turbulent swirling flow
Re = 2×10^5
Diameter ratio up to 10.
Computer time = 4.14 min

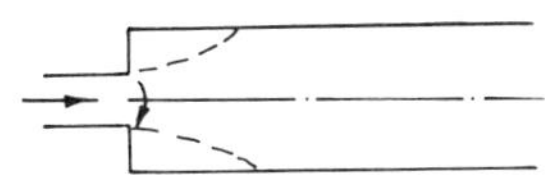

Flow behind a disc
Re = 4×10^5
Swirl numbers 0 up to 1.75
Computer time = 4.06 min

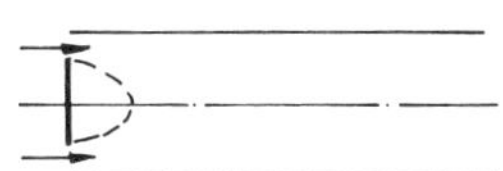

Flow behind a disc in a sudden expansion
Re = 0.4×10^5
Swirl number 0 up to 2.4
(HARWELL geometry, no swirl)
Computer time = 6.0 min

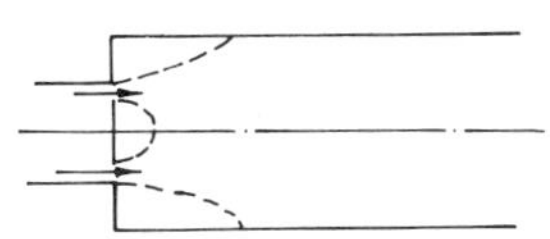

Co-flowing jets
Values of U_j/U_a = 0 up to 200.0
Swirl numbers up to 2.0
Any of the two jets can have swirl
Computer time = 6.5 min

Coaxial jets in a sudden expansion.
Re = 0.47×10^4
Swirl numbers of 0.0, 0.5, 1.0
(HARWELL burner number 1.)
Computer time = 8.5 min

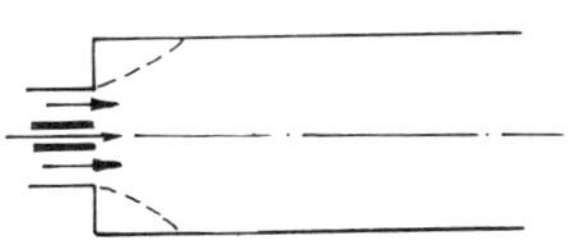

Coaxial jets in a sudden expansion
with a divergent air nozzle.
Re = 10^5
Swirl number of 0.52
(HARWELL geometry).
Computer time = 10.0 min

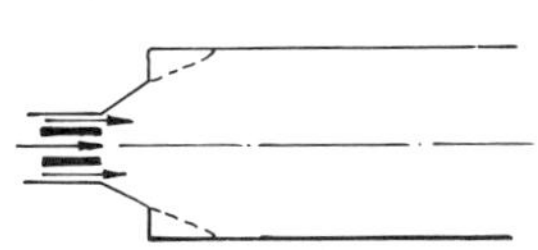

Sudden expansion flow with divergent fuel entry, divergent air nozzle.
Re = 10^5
Swirl number = 0.52
(HARWELL burner number 2.)
Computer time = 14.0

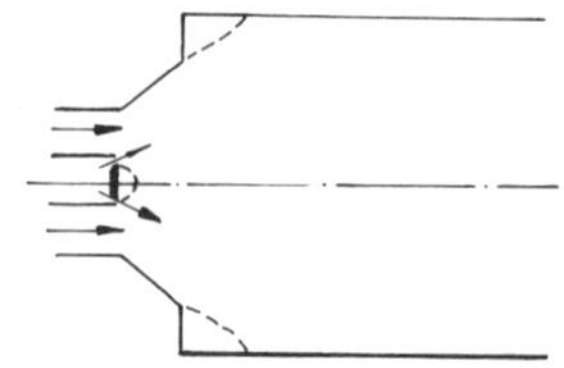

A3 COMPUTATIONS OF LARGE INDUSTRIAL FURNACES

A3.1 Introduction

In this appendix, the calculations of the flow characteristics of large furnaces and combustors are discussed; the agreement between these calculations and the available experimetns is evaluated. This contribution, albeit a brief one, extends the work described in the book to large-scale furnaces where the burner geometries are more complex than those normally used in small-scale laboratory furnaces. The problems associated with the modelling of large-scale combustors rather differ from those encountered in small-scale furnaces. For example, it is expected that the radiative heat flux contribution will be greater, and the amount of numerical procedure details involved will be larger. These features are appraised and quantified here. The comparisons shown here are intended to demonstrate the capability of the numerical methods, rather than to assess the validity of the various models which has been already done through the book.

Various large-scale furnace flow measurements are available in the open literature; it is worth noting that none of these sets of measurements are complete in the sense that they do not all provide the local velocity, temperature species concentration profiles as well as wall heat flux measurements and inlet profiles. A summary of the most complete sets of measurements is given in Table A3.1.

The measured properties in each furnace are summarized in Table A3.2 and comprise mean axial velocity, gas temperature, species concentrations and wall heat flux distributions. The corresponding inlet profiles were not included in these measurements except for the flow of Hasajnak,[6] where inlet velocity profiles at non-reacting conditions were reported.

Previous attempts to calculate the flow pattern and heat transfer in large furnaces and combustors were reported in references 7 to 18. The obtained agreement between the measured and calculated results varied from fair to good for the mean gas temperatures with a maximum discrepancy of the order of 10% of the local temperature, except of course in the vicinity of reaction zones where the discrepancy increased to 20%. The mean axial velocity and flow pattern predictions were in qualitative agreement with the measurements. In most cases, the existence of reversed flow regions was correctly predicted, but the size of the central recirculation zone was underpredicted. This was attributed to the defects of the turbulence model ($k-\varepsilon$) in highly swirling flows, and the insufficient information on the inlet profiles specifications. This discrepancy can be attributed as well to the precision of the measurements which were obtained by water-cooled probes.

TABLE A3.1

Furnace	Burner	Fuel	Load, MW	Excess air	D_f	L_f	D_1	D_2	D_3	D_4	S
Michelfelder *et al.*[1]	no quarl	C_3H_8	2.96	4%	2	6.25	0.019	0.06	0.175	0.383	0
Wu *et al.*[2]	with quarl	Nat. gas	1.06	0%	0.9	5.00	0.048	0.06	0.131	0.286	0 0.84
Wu[3]	with quarl	Nat. gas	1.044	–	0.6	5.0	0.0225	0.06	0.185	0.390	0.0 0.56 1.27
Rhines[4]	with quarl (premixed)	CH_4	5.479	4%	1.07	5.48	–	0.33	0.46	1.07	0
Shoffstall[5]	with quarl	CH_4	0.742	10%	1.50	4.50	0.037	0.05	0.45	0.6	0
Hasajnak[6]	with quarl	C_3H_8	12.88	33%	3.66	10.97	0.154	0.154	0.45	0.84	1.7 5.0

All dimensions are in metres.
Refer to figure A3.1 for geometry.

TABLE A3.2

Furnace	Velocities			Swirl number	Temperature $\overline{T}$	Wall temp.	Species concentration					Wall heat flux	
	$\overline{U}$	$\overline{V}$	$\overline{W}$				M_{fu}	M_{ox}	M_{CO_2}	M_{CO}	M_{NO_x}	q_w	$q_{convec.}$
Michelfelder *et al.*[1]	x	—	—	0	x	x	x	x	x	—	x	x	x
Wu *et al.*[2]	x	—	—	0, 0.84	x	x	x	x	x	x	—	x	x
Wu[3]	—	—	—	0, 0.56, 1.27	x	x	—	—	x	—	—	x	—
Rhines[4]	x	—	—	0	x	x	—	—	—	—	—	—	—
Shoffstall[5]	—	—	—	0	x	x	x	x	x	—	x	x	x
Hasajnak[6]	x	x	x	1.7, 5.0	x	x	—	—	—	—	—	x	x

TABLE A3.3

Furnace	Previous comput. reference no.	Turbulence model	Combustion model	Radiation model equations
Michelfelder *et al.*[1]	7	$m \cdot L$	4.2.1.1	–
	8	$k-\epsilon$	4.2.1.2	6.2.31–36
	9	$k-\epsilon$	4.2.1.2	6.2.31–36
	10	$k-\epsilon$	4.2.1.2	6.2.31–36
	11	$k-\epsilon$	4.2.1.3	6.2.31–36
	12	$k-\epsilon$	4.2.1.3	6.2.41–44
Wu *et al.*[2]	9	$k-\epsilon$	4.2.1.2	6.2.31–36
	13	$k-\epsilon$	4.2.1.3	6.2.41–44
	14	$k-\epsilon$	4.2.1.3	6.2.41–44
	15	$k-\epsilon$	4.2.3.2	6.2.41–44 and 6.2.23
	12	$k-\epsilon$	4.2.1.3	6.2.41–44
Wu[3]	16	$k-\epsilon$	4.2.3.2	6.2.41–44
	17	$k-\epsilon$	4.2.1.3	6.2.41–44
	15	$k-\epsilon$	4.2.3.2	6.2.41–44 and 6.2.23
Rhines[4]	12	$k-\epsilon$	4.2.2.1	6.2.41–44
Shoffstall[5]	12	$k-\epsilon$	4.2.1.3	6.2.41–44
Hasajnak[6]	12	$k-\epsilon$	4.2.3.2	6.2.23
	18		4.2.3.2	6.2.41–44

Predictions of species concentration profiles were in reasonable agreement with the measured values except near the burner, particularly so in furnaces with complicated burner geometry. The maximum discrepancy was reported to be less than 10%. The trends of the NO_x concentration profiles were correctly predicted, however the discrepancies in some locations were up to 100%. This is attributed to the significant role of temperature fluctuations in NO_x generation, and hence emphasizes the importance of adequate representation of the turbulence–chemistry interaction.

The wall heat flux calculations were reported to be in reasonable agreement with the corresponding measurements. The radiative heat transfer rates were predicted with a maximum discrepancy of less than 15%. Near the burner and at the furnace end walls, the reported discrepancies were of the order of 20%. Table A3.3 summarizes the previous numerical investigations that reported large furnace flow properties. The tabulated furnace flows have thermal loads varying from 0.74 to 12.9 MW. Figure A3.1 illustrates the furnace geometrical parameters and the associated boundary conditions.

A3.2 The Furnaces Considered

The present comparisons are represented in a sequence that allows the reader to assess the capability of the numerical procedure and the associated models and to understand

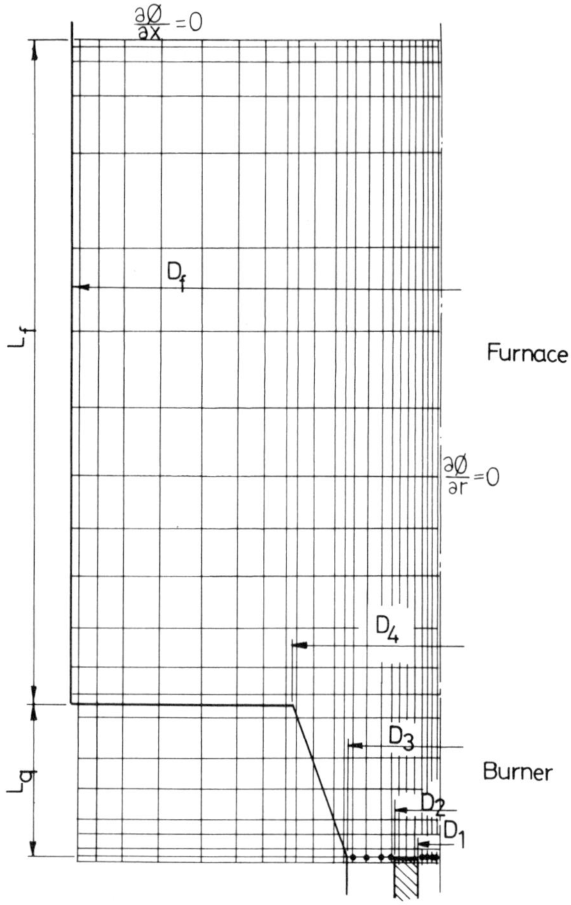

Figure A3.1 Furnace grid arrangement

the physical processes involved in furnace flows. Table A3.4 indicates the various furnaces considered here, together with the relevant numerical grid arrangement and the computer storage and time requirements.

In obtaining these calculations, the two-equation turbulence model was incorporated in the numerical procedure to represent the turbulent nature of the flow. The adequacy of such a turbulence model to furnace flows was assessed in Chapters 3 and 4. The combustion models, incorporated to represent the turbulent combustion characteristics and the associated turbulence–chemistry interactions, are indicated in Table A3.4. All the calculations take into account the effect of turbulent fluctuations and make use of the available probability distribution assumptions.

The source term in the stagnation enthalpy accounts for radiation heat transfer which is the major mode of heat transfer to the walls of large furnaces. The calculations reported in Table A3.4 were mainly obtained with the aid of the coupled four flux model. Some of these calculations however were also obtained with the aid of the discrete ordinate model.

TABLE A3.4

	Grid size	Nodes in quarl	Combustion model	Convergence	Computer specification	
					core	time, s
Michelfelder *et al.*[1]	400	–	4.2.1.3	0.008%	33*k*	780
Wu *et al.*[2]	484	85	4.2.1.3	0.01%	38*k*	1000
Wu[3]	460	85	4.2.3.2	0.01%	35*k*	950
Rhines[4]	484	90	4.2.2.1	0.01%	35*k*	800
Shoffstal[5]	484	82	4.2.1.3	0.009%	35*k*	810
Hasajnak[6]	484	96	4.2.3.2	0.01%	44*k*	1400

A3.2.1 *Flow of Michelfelder* et al.

The above investigations were obtained in the IFRF, square-sectioned horizontal furnace.[1] The velocity profiles were measured with water-cooled Pitot probe; while the reported mean gas temperatures were obtained with the aid of a suction pyrometer and were corrected for radiation and conduction losses. Species concentrations were measured by gas sampling techniques. The incident radiation flux was measured with the aid of ellipsoidal radiometer, and wall heat flux was obtained with heat flux meters. The calculations of mean axial velocity profiles, shown in Figure A3.2 are those of

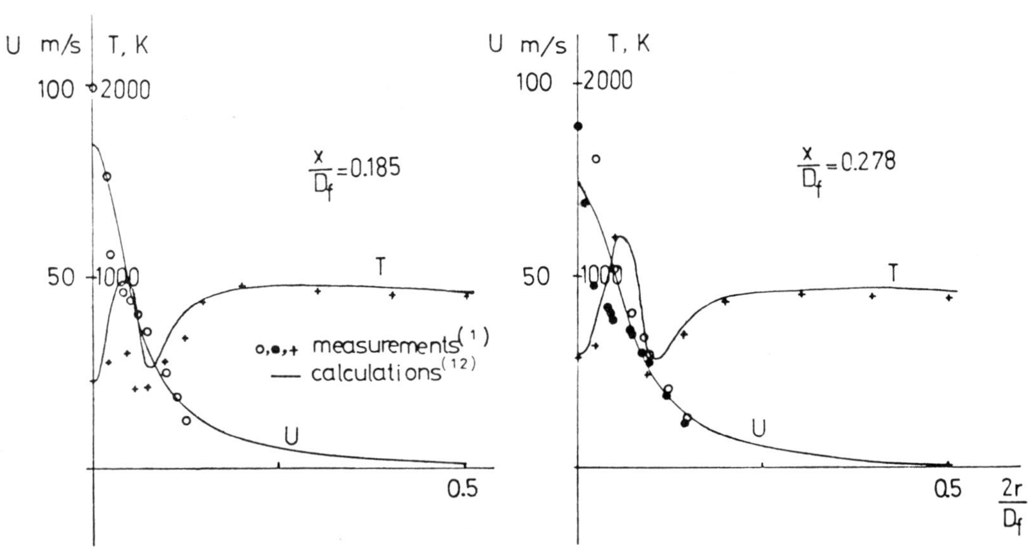

Figure A3.2 Measured and calculated profiles of mean axial velocity and temperature in the furnace of reference 1

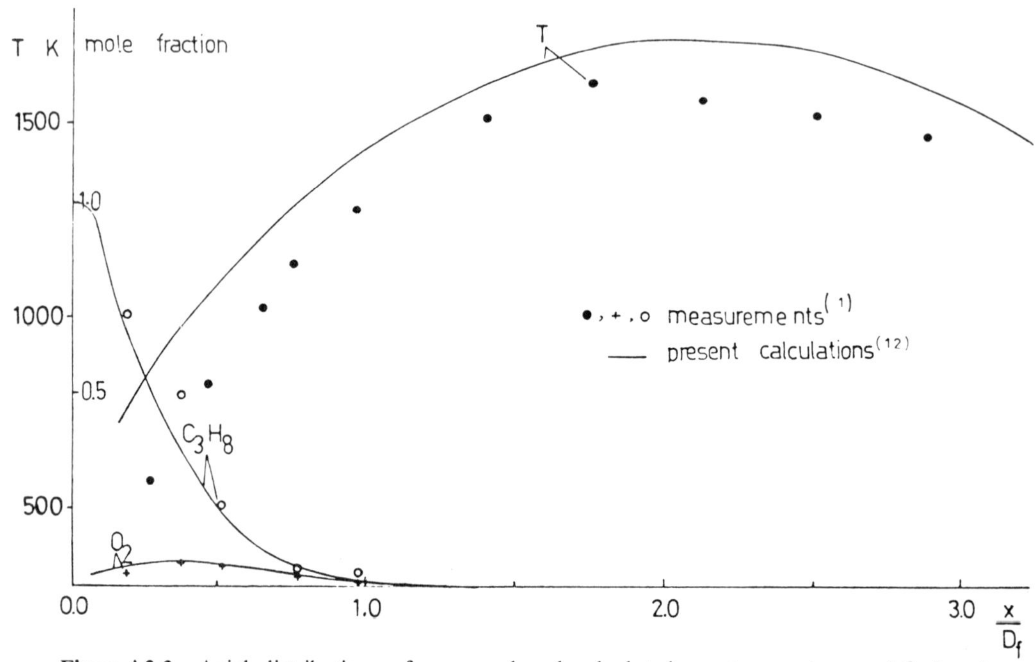

Figure A3.3 Axial distributions of measured and calculated gas temperature and fuel and oxygen concentrations in the furnace of reference 1

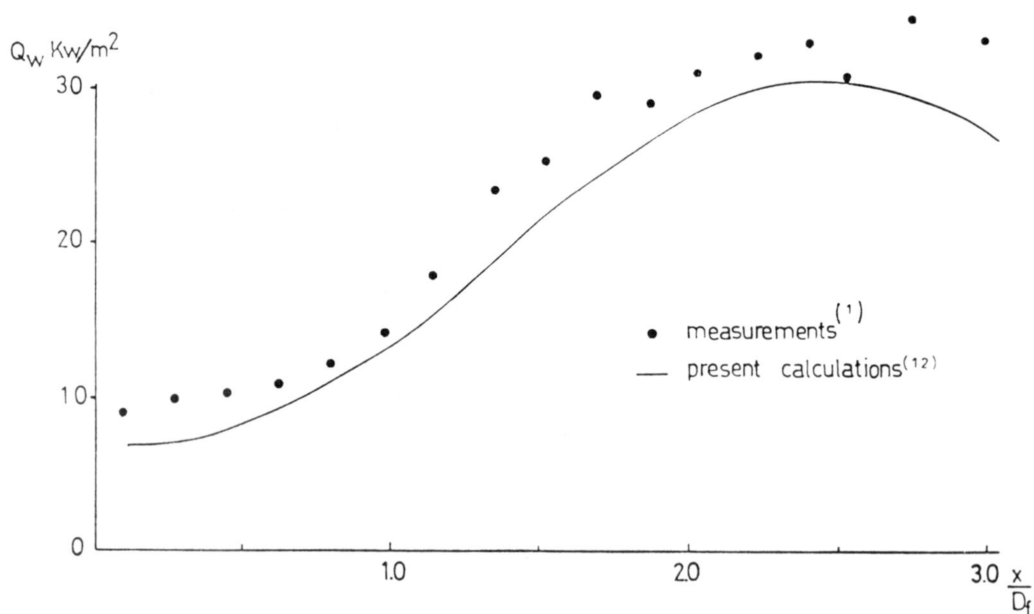

Figure A3.4 Measured and calculated wall heat flux distribution in the furnace of reference 1

Khalil *et al.*[12] and are compared with the corresponding measurements. The measurements were taken across the square-sectioned furnace. The solid and hollow circles represent the cross-stream profile. The solid lines represent the axisymmetrical predictions of the flow based on an equivalent furnace diameter that satisfied both the mass and energy balances. The flow pattern is adequately predicted, and indicated central forward jet flow. Due to the narrow flame configuration, most of the property changes are near the centreline. The gas temperature profiles are shown in the same figure and at two axial locations, these profiles show an off-centre peak in the reaction zone, followed by a trough in the wake of the annular air jet. The measured and calculated distributions are in reasonable agreement, with the maximum discrepancy of less than 15%. The observed narrow flame situation with large wall recirculation zones supports the use of cylindrical coordinates as the flame edges are away from the wall. Figure A3.3 represents the calculated centreline distributions of gas temperature and species concentrations, which are in reasonable agreement with the corresponding measurements.

The total wall heat flux distribution is shown in Figure A3.4; the measured and calculated distributions agree within 7% of the maximum value; the discrepancy could be due to soot formation with consequent radiation. On the whole, the above agreement is satisfactory and the calculations represent correctly the observed trends.

A3.2.2 *Flow of Wu* et al.

The present axisymmetric vertical furnace was segmented into ten water-cooled sections, each 0.5 m long. Various previous investigations were reported (see Table A3.3), that calculate the local flow and flame properties in this furnace, which is commonly used as a basic test case for furnace prediction procedure. The importance of this flow configuration stems from the fact that it actually resembles a fire-tube boiler furnace with L_f/D_f of 5.5. The expansion ratio D_f/D_3 is nearly 6.87 and results in considerable wall recirculation zone.

The flow pattern and local velocities were measured with the aid of a water-cooled five-hold hemispherical impact tube. The local gas temperatures were obtained with a suction pyrometer and gas sampling techniques were used to obtain the local gas concentration. The total heat flux to the furnace walls was measured by means of measurement of the cooling water for each segment of the furnace, and near the burner, using a Shell total heat flux meter. The radiative heat flux to the furnace walls was measured by means of an ellipsoidal radiometer.

The calculations shown here are those of Khalil *et al.*[12] The comparisons with experiments comprise mean axial velocity profiles, mean gas temperature profiles and wall heat flux distributions. Calculated and measured mean axial velocity profiles are shown in Figure A3.5 for the two swirl intensities of zero and 0.84. In both cases there exists a central recirculation zone which disappears at further downstream locations for $S = 0$, but persists for $S = 0.84$. Again, the size of the central recirculation zone was underpredicted by up to 20%. This can be attributed partially to the errors inherent in the measurements using water cooled Pitot probe.

The measured and calculated mean gas temperatures are shown in Figure A3.6, and the discrepancies observed were less than 150°K. The effect of swirl is twofold, one is to reduce the flame length, while the other is to increase its diameter.

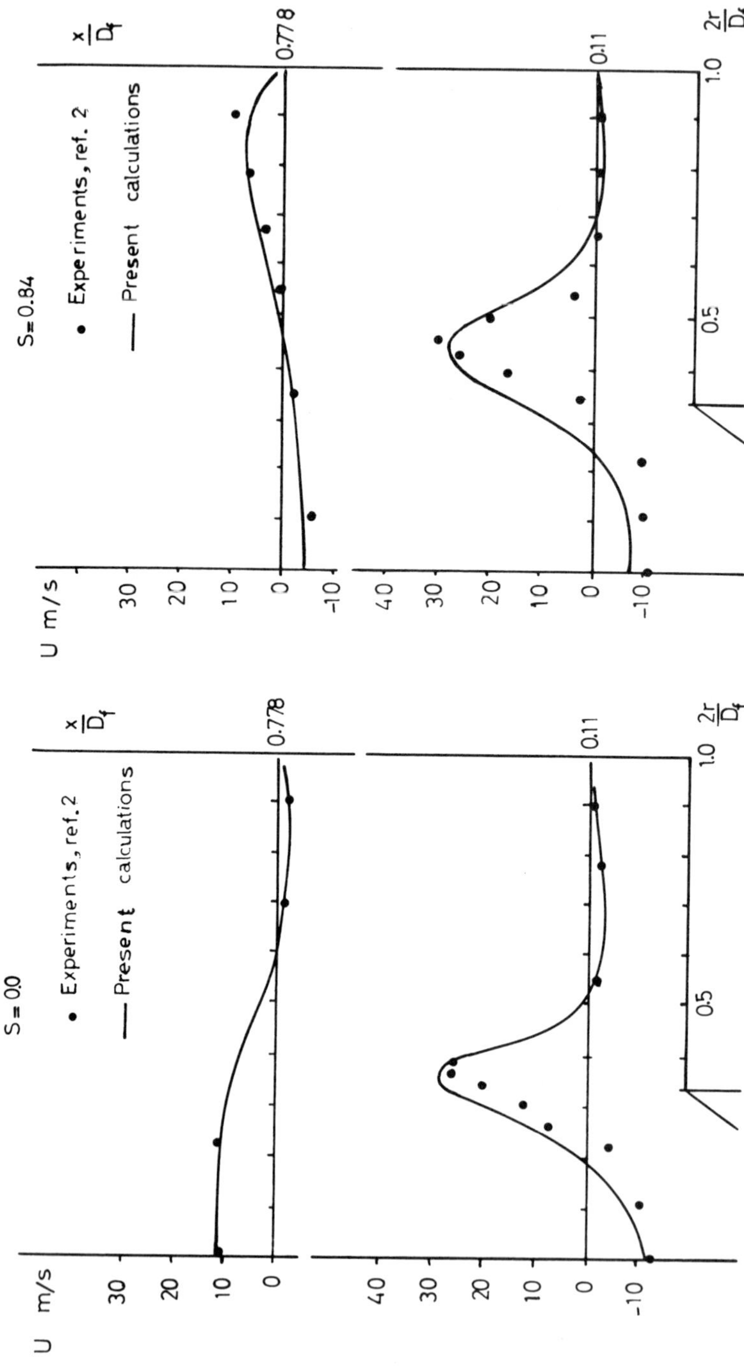

Figure A3.5 Measured and calculated profiles of mean axial velocity in the furnace of reference 2

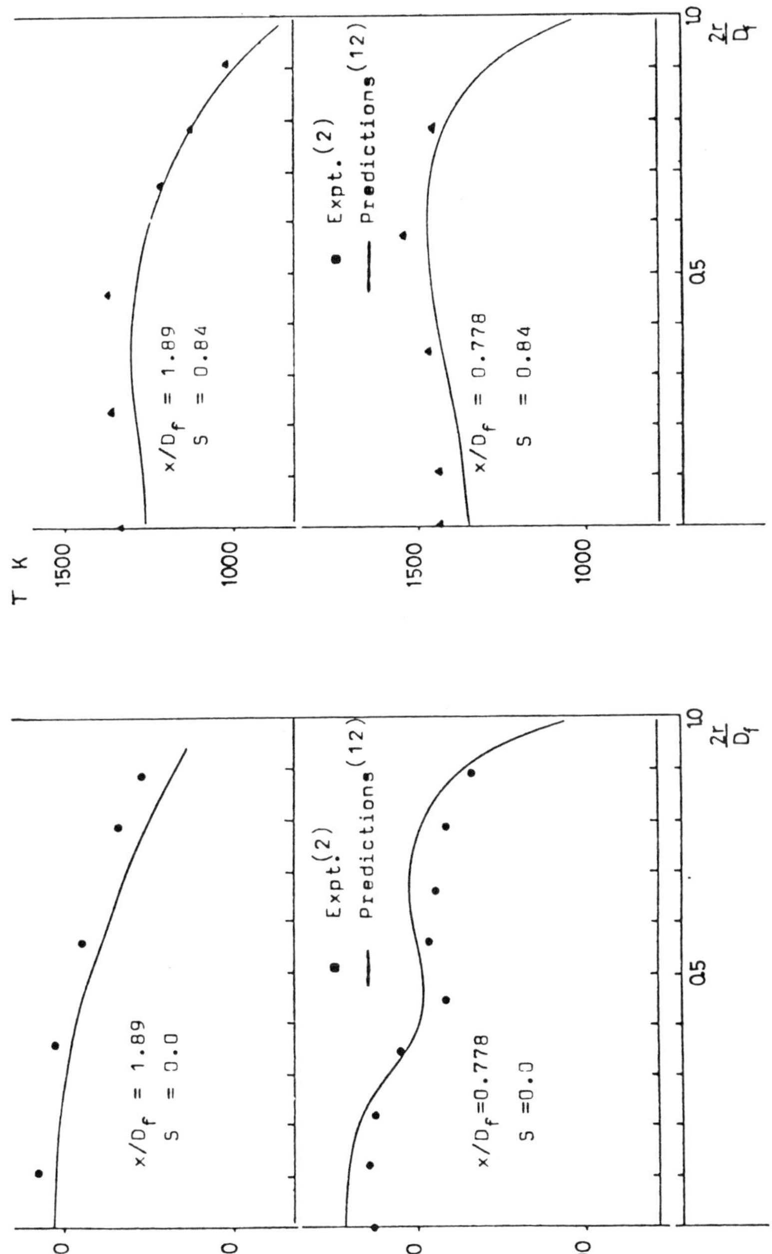

Figure A3.6 Measured and calculated mean gas temperature in the furnace of reference 2

Measurements of total wall heat flux distributions are shown in Figure A3.7 and are compared to the calculated distributions for the two swirl numbers. These distributions indicated clearly the effect of swirl on heat transfer as shown by the location and magnitude of the peak wall heat flux. At swirl number of zero, the peak heat flux is situated at x/D_f of 1.4, while at $S = 0.84$, the peak occurs at $0.4\ D_f$. These locations are consistent with reattachment point location. The convective wall heat flux contributed to 33% and 36% of the total wall heat flux for $S = 0$ and 0.84 respectively.

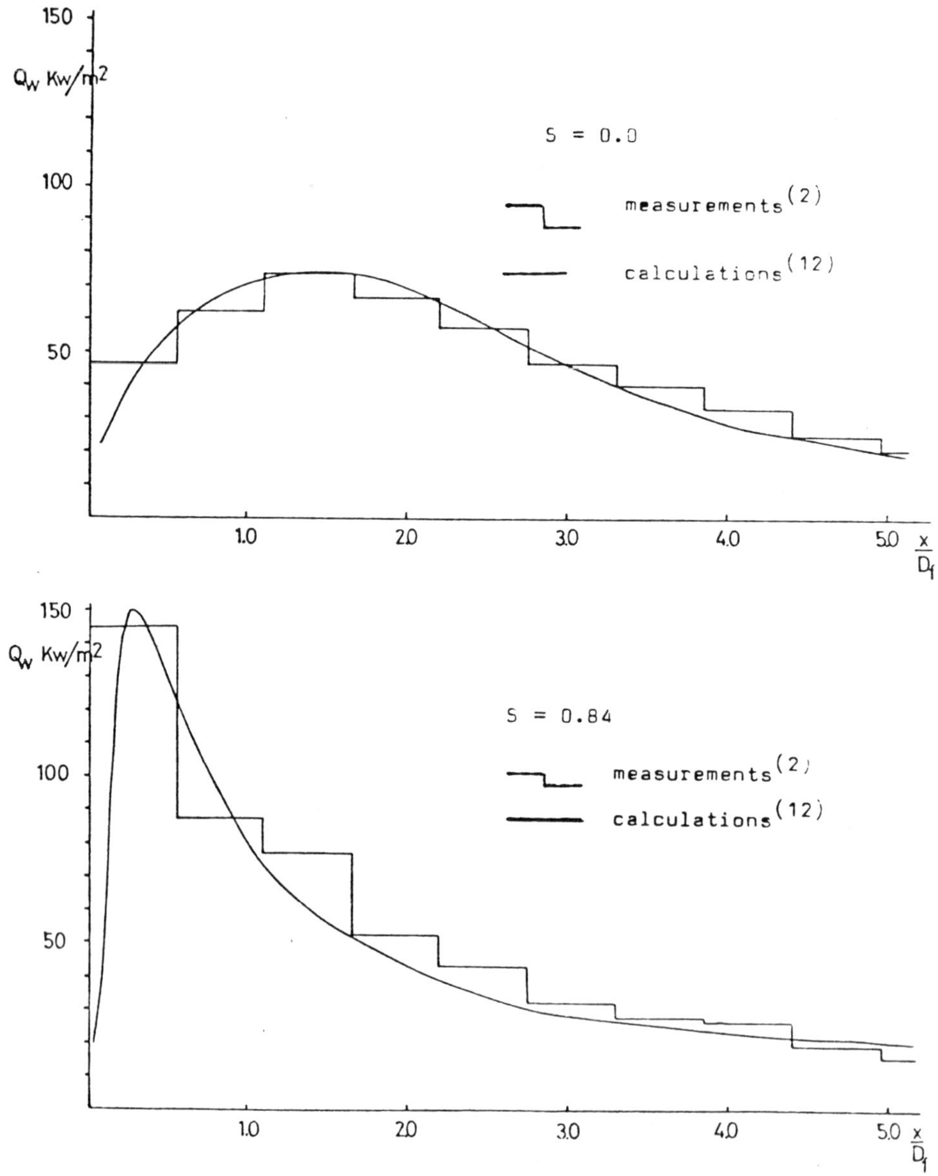

Figure A3.7 Measured and calculated wall heat flux distribution in the furnace of reference 2

A3.2.3 *Flow of Wu*

This investigation was carried out in an axisymmetric water-cooled narrow furnace which was segmented into ten water-cooled sections. Three air swirl numbers of 0, 0.56 and 1.27 were used. The measurements comprised centreline distributions of mean gas temperature, carbon dioxide concentration and wall heat flux.

Flame temperatures were measured by suction pyrometers having two concentric radiation shields surrounding the thermocouple. Gas analyses at various locations in the flame were made by an infra-red absorption analyser for CO_2 and CO. Both the cooled sampling probe and the suction pyrometer were mechanically traversible through the measuring ports. Total wall heat flux were obtained by measuring the cooling water load at each segment.

Figure A3.8 illustrates the distributions of mean gas temperature and carbon dioxide concentration along the furnace centreline for the three swirl numbers. Both measured and calculated distributions are in reasonable agreement. The measured and calculated total heat flux distributions along the furnace wall are also shown in Figure A3.8. The radiative heat flux to the wall contributed to nearly 65% of the total heat flux and the peak heat flux occurred at the locations coinciding with the reattachment point.

A3.2.4 *Flow of Rhines*

The measurements of Rhines[4] were obtained in a cylindrical, horizontal gas-fired water-cooled furnace whose dimensions are given in Table A3.1. Premixed air and fuel mixture were discharged through the burner into the furnace space. The measurements of axial velocity were obtained with the aid of water-cooled Pitot probes, and gas temperatures were measured by suction pyrometer.

Figure A3.9 demonstrates the comparison between measured and calculated axial velocity profiles at various locations; the calculations are those of Khalil *et al.*[12] The measured profiles exhibited asymmetry, which can be attributed to the furnace horizontal arrangement and the measuring precision. Neither the measurements nor the calculations exhibited corner recirculation zones. The obtained agreement was poor near the burner, but improved at further downstream locations.

The corresponding measured and calculated mean gas temperature profiles are shown in Figure A3.10. The measured temperatures near the burner were asymmetric and the obtained agreement was only qualitative. No measurements of local wall heat flux were reported. The average wall heat flux was measured as 120 kW/m^2, which is in reasonable agreement with the calculated value of 114.27 kW/m^2.

A3.2.5 *Flow of Shoffstall*

This furnace is a 1.5 x 1.5 m square-sectioned horizontal one. The burner has a complicated geometry with one third of the fuel (natural gas) issuing through a central axial jet, and the remainder issuing radially to impinge on the surrounding hot combustion air stream. The furnace walls were water-cooled by coils embedded in the wall. The mean gas temperature was measured by a suction pyrometer, and the gas and oxygen concentrations were obtained by gas sampling techniques. No measurements of wall heat flux were reported.

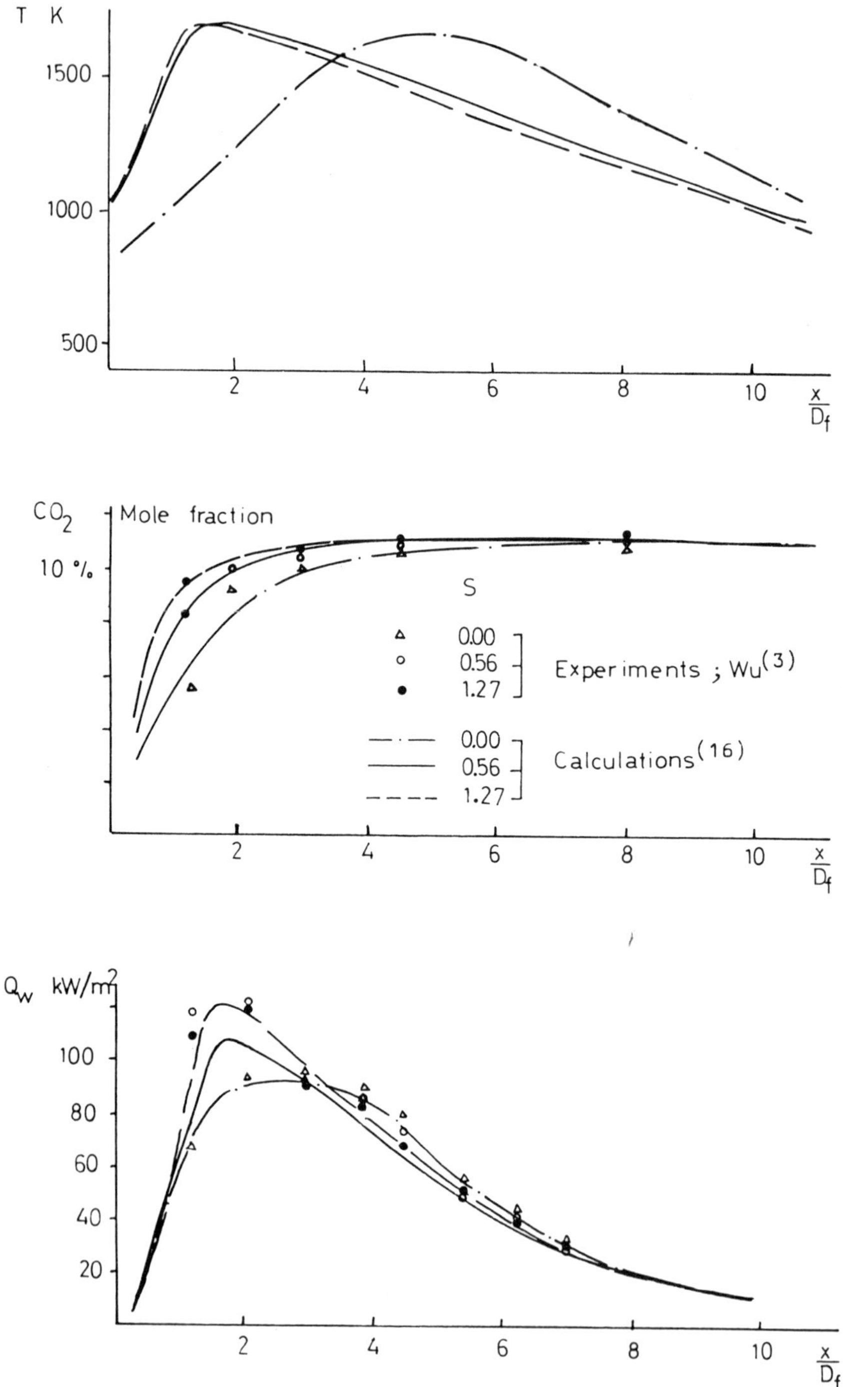

Figure A3.8 Measured and calculated mean gas temperature and carbon dioxide distributions along the centreline and the wall heat flux distribution for various degrees of swirl

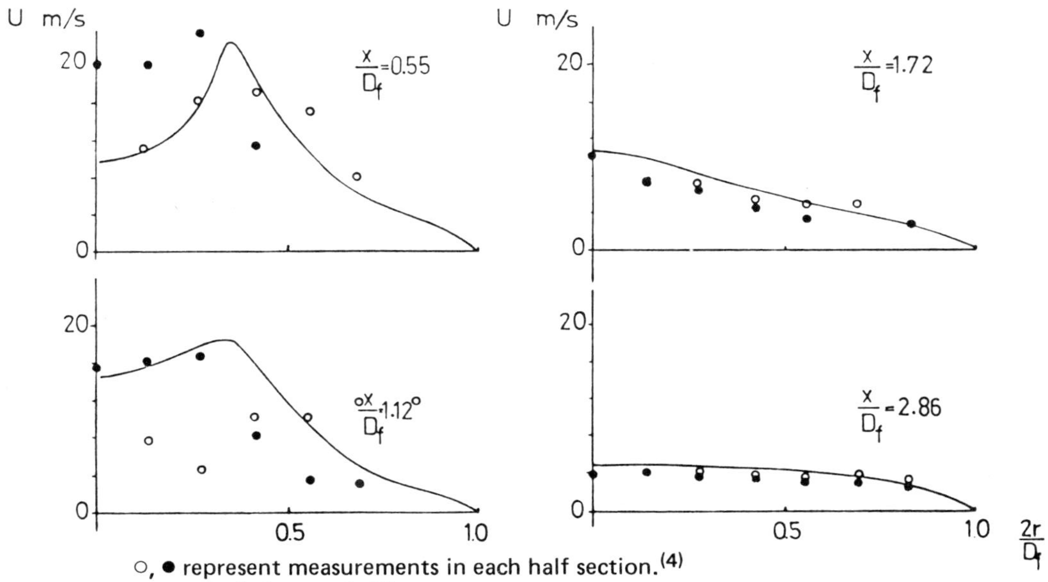

○, ● represent measurements in each half section.[4]
—— calculations, ref. 12.

Figure A3.9 Measured and calculated axial velocity profiles in the furnace of reference 4

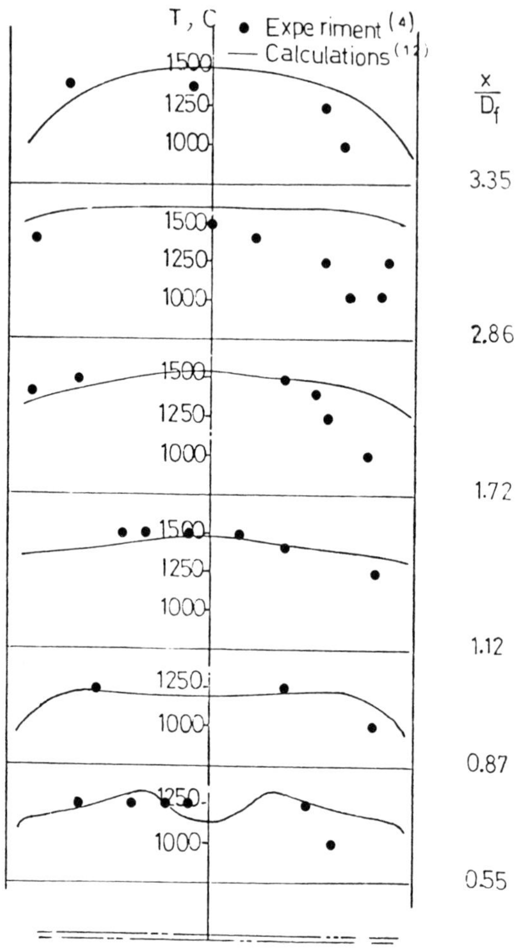

Figure A3.10 Measured and calculated gas temperature profiles in the furnace of reference 4

No velocity measurements were reported. However, the calculations of Khalil *et al.*[12] readily obtained them from the numerical procedure. The flow pattern is characterized by a corner recirculation zone and a narrow forward flow region, suggesting a narrow flame and supporting the use of an equivalent furnace diameter that conserves mass and energy. The calculated velocity profiles are shown in Figure A3.11. The measured and calculated gas temperature and species concentration profiles are shown in Figures A3.12 and A3.13 respectively. The maximum discrepancy is less than 200°K at 1500°K, the temperature troughs coinciding with the outer edge of the flame, i.e. in the wake of air stream jet. The gas temperature increases towards the wall due to the effect of recirculation. Profiles of fuel and oxygen concentrations are shown in Figure A3.13 and exhibited reasonable agreement; they confirm that the flame is narrow.

The centreline distributions of mean axial velocity, fuel and oxygen concentrations and gas temperature are all shown in Figure A3.14, where the comparison between measured and calculated properties is illustrated. The observed slow decay rate of the

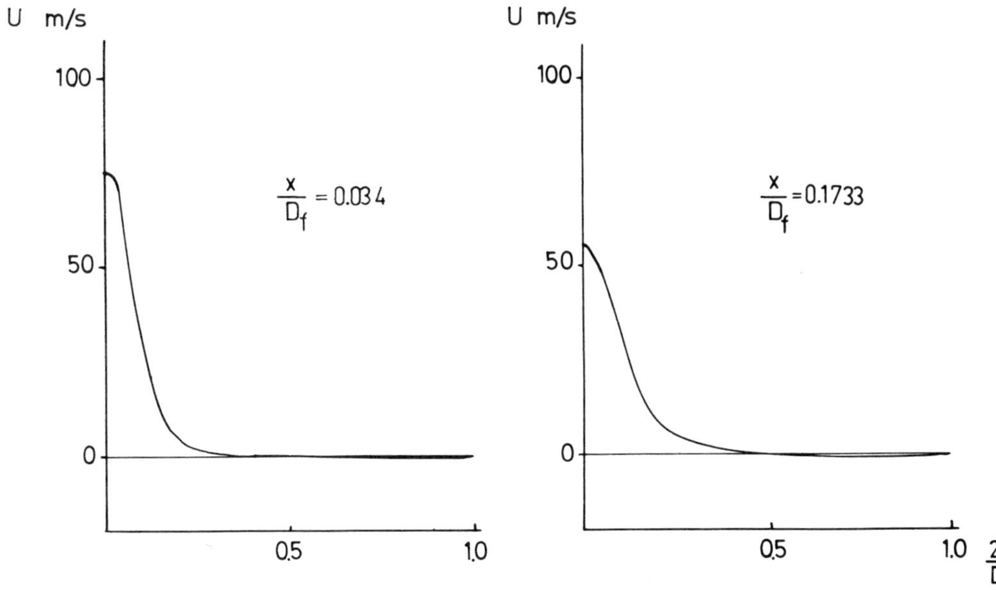

Figure A3.11 Calculated profiles of mean axial velocity in the furnace of reference 5

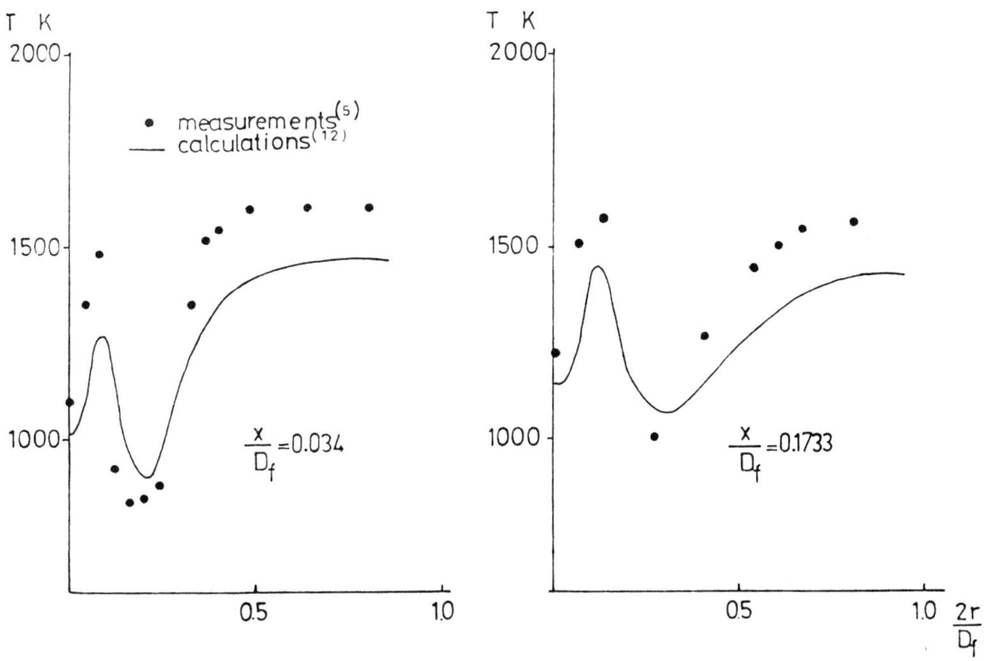

Figure A3.12 Measured and calculated profiles of mean gas temperature in the furnace of reference 5

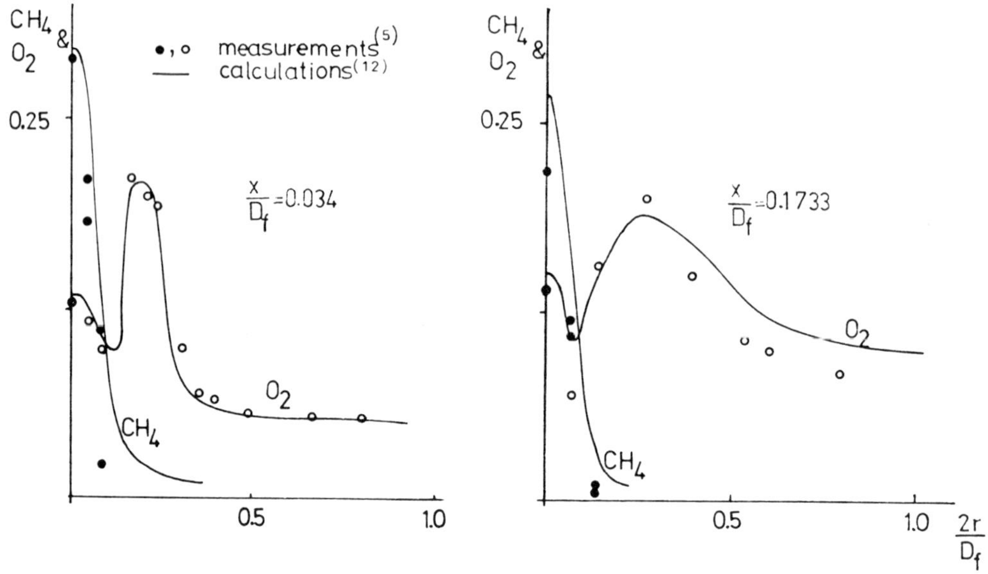

Figure A3.13 Measured and calculated profiles of oxygen and methane concentrations in the furnace of reference 5

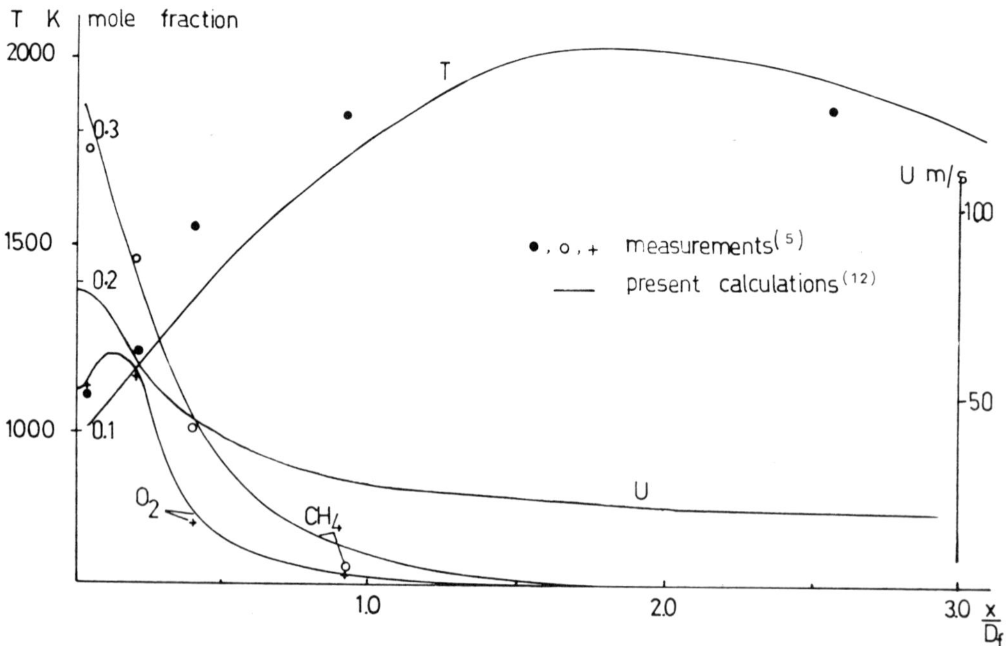

Figure A3.14 ·Measured and calculated axial velocity, temperature, species concentrations in the furnace of reference 5

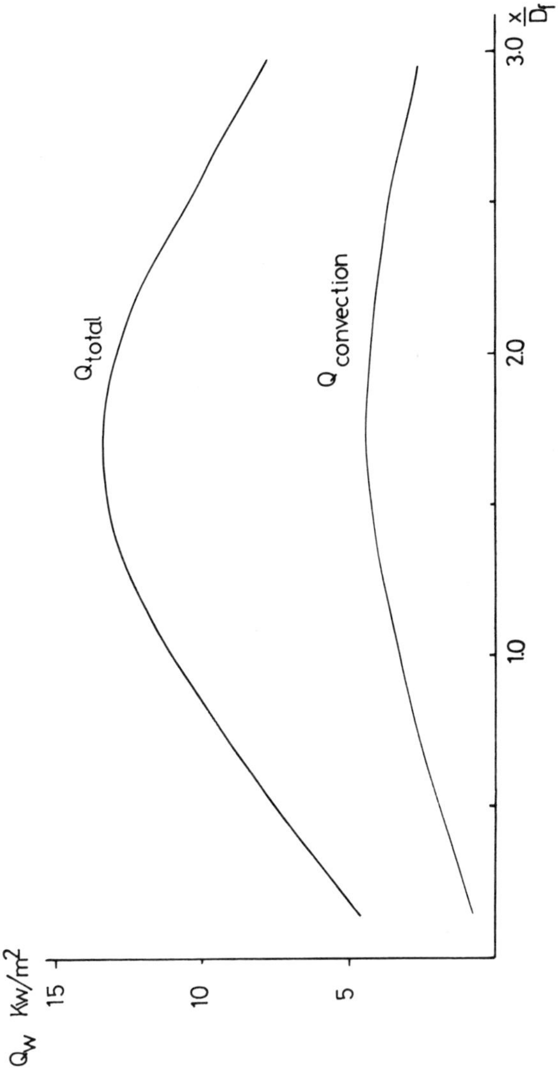

Figure A3.15 Wall heat flux distribution in the furnace of reference 5

velocity is due to the opposing effects of jet expansion and heat release. Figure A3.15 shows calculated wall heat flux and the convective heat transfer contribution is nearly 25% of the total flux. The average wall heat flux along the furnace was predicted as 11.4 kW/m^2.

A3.2.6 *Flow of Hasajnak*

This furnace is also square sectioned and horizontal. The flame is narrow, which supports the use of an equivalent furnace diameter that would conserve mass and energy. Measurements of mean velocity components near the burner were obtained with the aid of Pitot probes for non-reacting flow situations. The gas measurements were obtained with the aid of a suction pyrometer, and the wall heat flux was obtained from heat flux meters.

Figure A3.16 illustrates the measured and calculated profiles of U, V, W and P in the burner vicinity of $S = 1.7$. A large recirculation zone was formed just downstream from the burner exit due to the action of swirl and quarl angle. On the whole reasonable agreement was obtained. The measured and calculated gas temperature profiles were reported (12, 18) for two swirl numbers of 1.7 and 5.0 and are shown in Figures A3.17 and A3.18; the maximum discrepancy was less than 150°K.

The corresponding wall radiative heat flux distributions for the two swirl intensities are displayed in Figure A3.19. The calculations were obtained with the aid of two radiation models, the D–L and S_4 models described in Chapter 6. The radiative heat flux contributed to nearly 60% of the total wall heat flux. The average total wall heat flux was 45 kW/m^2, the calculated wall heat load was 5.88 MW and the corresponding measured value was 6.1 MW.

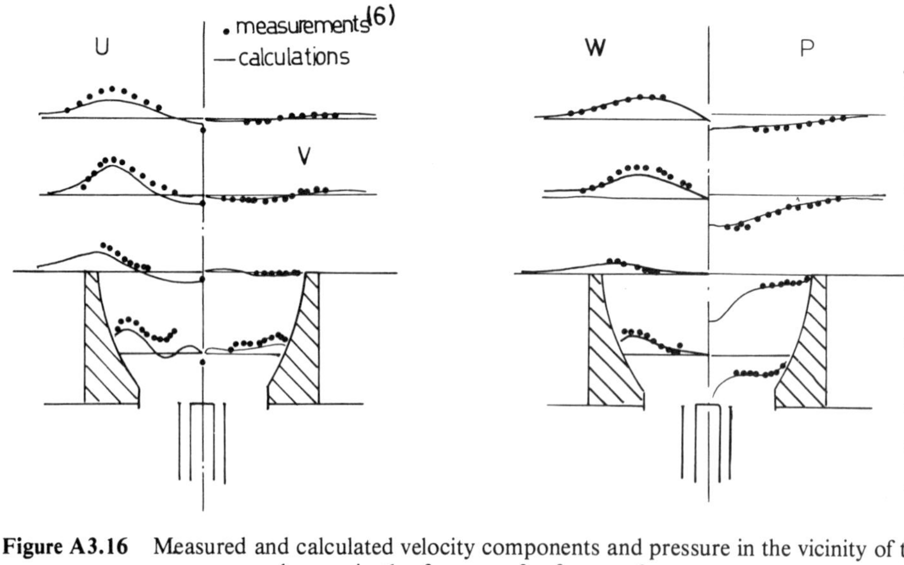

Figure A3.16 Measured and calculated velocity components and pressure in the vicinity of the burner in the furnace of reference 6

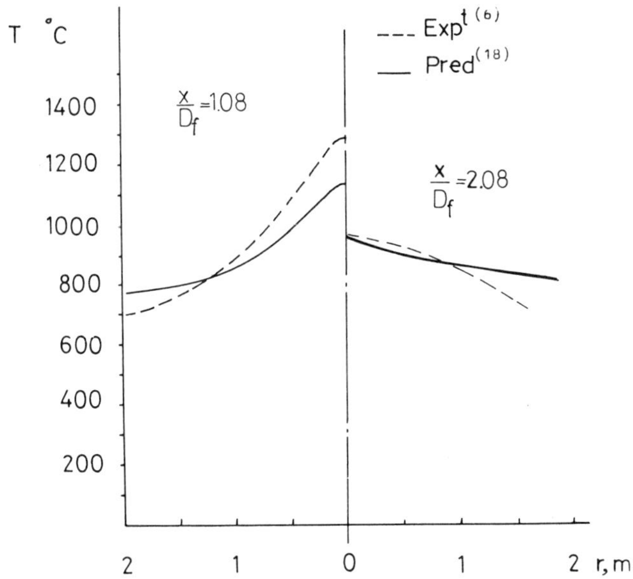

Figure A3.17 Measured and calculated gas temperature profiles in the furnace of reference 6; $S = 1.7$

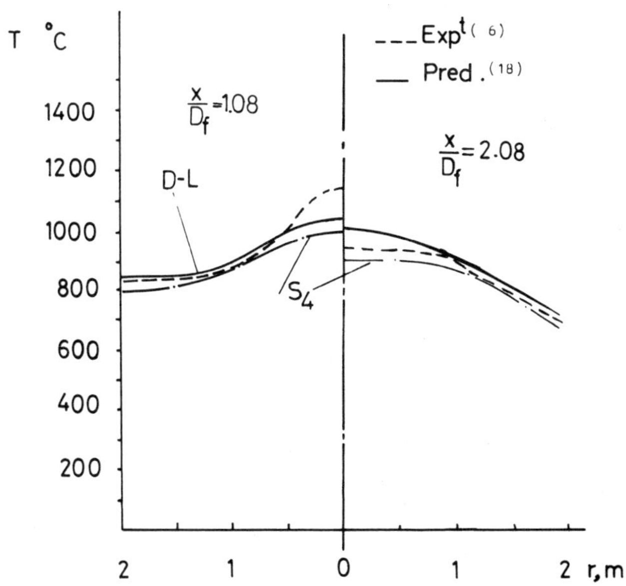

Figure A3.18 Measured and calculated gas temperature in the furnace of reference 6; $S = 5.0$

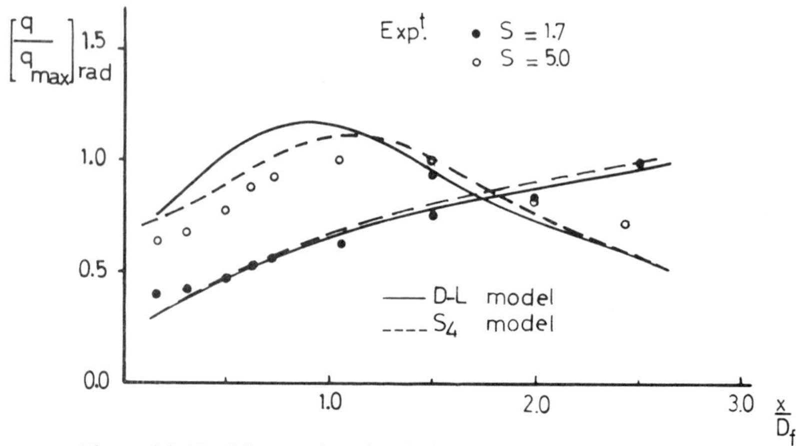

Figure A3.19 Measured and calculated radiative wall heat flux

A3.3 Closure

The achievements and shortcomings of the numerical solution procedure are discussed here in terms of applicability, convergence, accuracy and economy.

The procedure discussed in the book and applied here to large-furnace flows have a wide band of assertion applicability. The justification for this resides in the variety of phenomena which the procedure has already been shown to be able to handle. Application to large-scale furnaces of practical industrial utility crowns the efforts to justify the procedure as a design tool. It is worth noting that the solutions to all furnace problems discussed here and in Chapters 3–7 were obtained from the same straightforward procedure with few alterations to inlet and boundary conditions.

The convergence of the procedure is justifiably fast except in relatively few instances which have required special under-relaxation in very large property gradient variation locations.

The question of accuracy is quite often a major parameter to consider. Claims of accurate calculations were supported by comparison test cases, such as those discussed in the present book. The numerical accuracy depends on the grid size and arrangement and examination of such dependency is the subject of many investigations. The accuracy of the procedure depends, to a large extent, on the adequacy of the present knowledge of the basic physical processes. This is reflected in our attempts to formulate turbulence, combustion and radiation models that adequately represent the real processes.

The economy of the present procedure is viewed in terms of computer run times. These are not small in complicated furnace flows and increase with the complexity of the flows considered. Attempts to improve the speed of convergence are welcomed and new fast solution algorithms are required.

References

1. Michelfelder, S. and Lowes, T. M. (1974) Report on the M-2 trials. IFRF Report Doc. F 36/a/4.
2. Wu, H. L. and Fricker, N. (1971) An investigation of the behaviour of swirling jet flames in a narrow cylindrical furnace, 2nd Members Conference of the IFRF. See also: *J. Inst-Fuel*, **49**, p. 144 (1976).
3. Wu, H. L. (1969) Comparison of the performance of natural gas and oil flames in a cylindrical furnace. *J. Inst. Fuel*, **42**, p. 316.
4. Rhines, J. M. (1974) Aerodynamic and heat transfer characteristics of a proprietary burner fired with natural gas in a cylindrical furnace. *Proc. 3rd Members Conference of the IFRF.*
5. Shoffstal, D. R. (1976) Burner design criteria for control of NO_x from natural gas combustion. EPA-600/2-76-098.
6. Hasajnak (1976) Private communication.
7. Pai, B. R. and Lowes, T. M. (1972) The prediction of flow, mixing and heat transfer in the Ijmuiden furnace. IFRF Doc. G 02/2/22.
8. Gosmon, A. D. and Lockwood, F. C. (1973) Predictions of the influence of turbulent fluctuations on flow and heat transfer in furnaces. Imperial College, Dept. of Mechanical Engineering, Report HTS/73/52.
9. Khalil, E. E., Spalding, D. B. and Whitelaw, J. H. (1975) The calculation of local flow properties in two dimensional furnaces. *Int. J. Heat and Mass Transfer*, **18**, p. 775.
10. Richter, W. (1974) Prediction of heat and mass transfer in a pulverised fuel furnace. *Letters in Heat and Mass Transfer*, **1**, p. 83.
11. Hutchinson, P. Khalil, E. E. and Whitelaw, J. H. (1976) The calculation of wall heat transfer rate and pollution formation in axisymmetric furnaces. *Proc. 4th Members Conference of the IFRF.*
12. Khalil, E. E., Hutchinson, P. and Whitelaw, J. H. (1980) The calculation of flow and heat transfer characteristics of gas fired furnaces. *Proc. 18th Symp. (Int.) on Combustion*, p. 1927.
13. Khalil, E. E. (1977) Flow and combustion in axisymmetric furnaces. Ph.D. Thesis, London University.
14. Gosman, A. D., Lockwood, F. C. and Salooja, A. P. (1978) The prediction of cylindrical furnaces gaseous fuelled with premixed and diffusion burners. Imperial College, Dept. of Mechanical Engineering Report FS/78/36.
15. Khalil, E. E. (1979) Mathematical modelling of radiative heat transfer in axisymmetric furnaces, AIAA Paper 77-0099 (17th Aerospace Sciences Meeting).
16. Khalil, E. E. (1978) Aerodynamic and heat transfer characteristics of axisymmetric confined gaseous flames, *Proc. 6th Int. Heat Transfer Conference*, p. 25.
17. Khalil, E. E. (1978) Numerical computations of turbulent swirling flames in axisymmetric combustors, in *Flow, Mixing and Heat Transfer in Furnaces*, edited by Khalil *et al.*, (Pergamon Press), p. 231.
18. Khalil, E. E. and Truelove, J. S. (1977) Calculation of radiative heat transfer in a large gas fired furnace. *Letters in Heat and Mass Transfer*, **4**, p. 353.

A.4 PREDICTION METHODS – THE TOOLS TO COMBUSTOR DESIGN

A4.1 Introduction

This appendix is concerned with presenting some general predictions that investigate the parameters influencing the combustor performance. As the results shown are not compared to any experimental data, the purpose is to show what prediction methods can do, not how well they can do it. Among the parameters are the swirl intensity, expansion ratio, precombustion chamber and burner geometry, furnace length to diameter ratio, air to fuel ratio, fuel characteristics, spray cone angle and dilution. The general predictions are given here for two cases, namely:

1. Flow and flame characteristics in prechamber dump combustor.
2. Flow and flame characteristics in gas turbine combustors.

A.4.2 Flow and Flame Characteristics in a Prechamber Dump Combustor

In the combustor configuration under consideration, an annular swirling flow over a central hub enters a prechamber, followed by a sudden enlargement in the main chamber. Novick *et al.*[1,2] calculated the flow characteristics in such a combustor with the aim of investigating numerically the effects of:

1. Swirl intensity
2. Exit geometry
3. The use of double concentric reverse swirlers

The combustor geometry is shown in Figure A4.1; a non-uniform grid mesh was superimposed on a symmetrical half-section of the combustor.

The effect of swirl vane angle on the non-reacting flow field structure was investigated[1] by computationally varying the vane angle between zero and 75°. Figure A4.2 shows the flow pattern at various swirl angles; with no swirl, the corner recirculation zone filled the whole chamber cavity. As swirl was introduced, the size of the corner recirculation zone decreased with an increase in the size of the central recirculation zone formed in the wake of the centre-body (hub) at 45° swirl. Further increase in swirl vane angle resulted in a continued enlargement of the central zone and a decrease of the corner recirculation zone, as shown in Figure A4.2. This change of swirl intensity influenced greatly the aerodynamic pattern and particularly the central recirculation zone, a fact which is of importance in designing a stabilized flame.

The removal of the constricted combustor outlet has an effect on the flow field. Novick *et al.*,[1] reported parametric studies of this effect. Figure A4.3 shows the calculated streamlines for the non-reacting flow with swirl vane angle of 70°. This flow

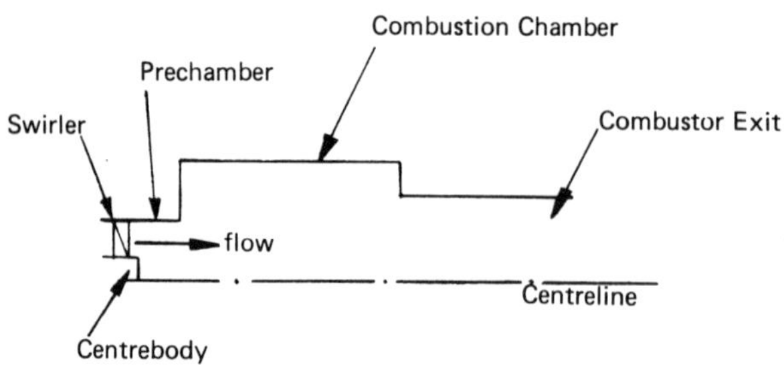

Figure A4.1 Combustor geometry

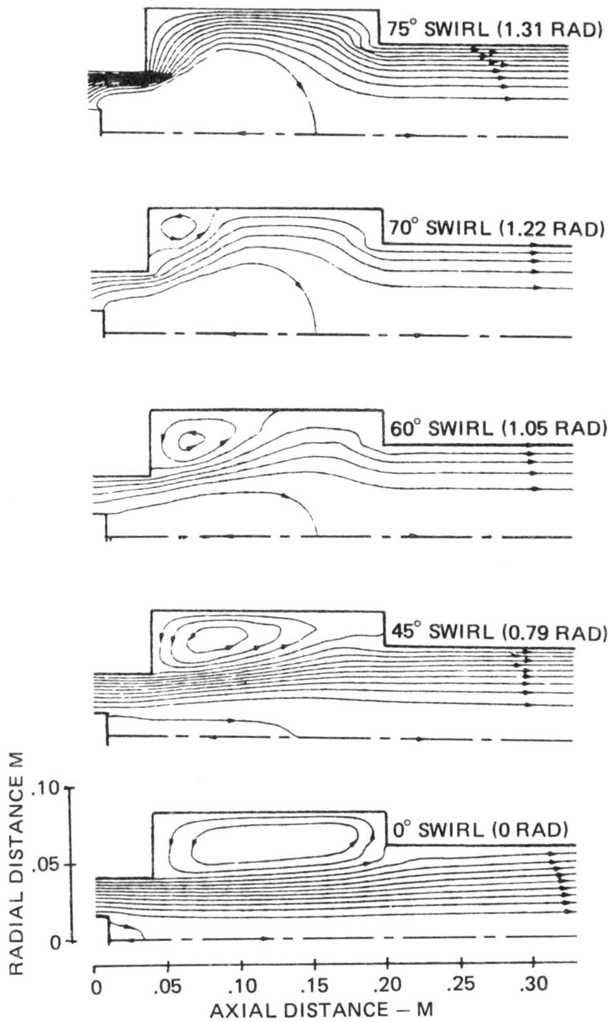

Figure A4.2 Calculated flow pattern at various swirl angles[1]

pattern when compared to the corresponding one in Figure A4.2 indicated a difference between the two patterns; the removal of the exit restriction diminished the effect of the chamber length to diameter ratio. The length of the central recirculation zone normalized by chamber length is unchanged as it is solely governed by the swirl intensity and chamber diameter.

The inlet cold flow of Figure A4.3 is replaced by a stoichiometric methane–air mixture, and the chemical reaction is allowed to proceed downstream from the inlet. Due to the presence of chemical reaction, the size of the wall recirculation zone is slightly decreased. The central recirculation zone is considerably smaller than for the

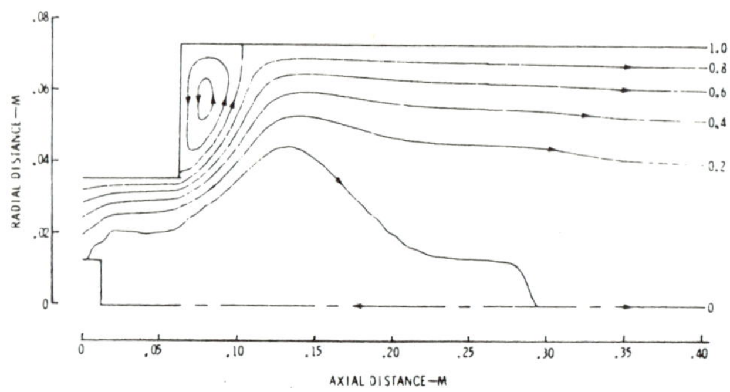

Figure A4.3 Calculated streamlines for non-reacting, 70° swirling flow[1]

non-reacting case. The reaction leads to a decrease in density and to an increase in the axial velocity which tends to shorten the recirculation zone.

The effect of swirl direction on the flow pattern was also investigated.[1,2] Equal-area double concentric reverse swirlers, each of 70° swirl angle were used; their effects on the flow pattern and flame characteristics are characterized[2] by:

1. A larger wall recirculation zone and a shorter central recirculation zone; this is attributed to the cancellation effect of the two counter-swirling inlet flows; the resulting effect resembles that of decreasing swirl number.
2. The two zones exhibit lower temperatures than those of the single swirler under reacting flow conditions; this is attributed to the reduced swirl intensity.

A4.3 Flow and Flame Characteristics in Gas Turbine Combustors

A sample of numerical computations of gas turbine combustion chamber is given here. The physical combustor configuration is shown in Figure A4.4. The air stream through the swirler entered the chamber with a finite degree of swirl. The primary, secondary and dilution air flows were each injected through six equally spaced circular orifices. Swithenbank *et al.*[3] reported calculations of the flow field and flame characteristics using the finite difference technique; the grid network comprised 27 transverse planes, 7 radial planes and 18 circumferential surfaces as shown in Figure A4.5. Due to the application of cyclic boundary conditions, only a 60° sector was considered. The air/fuel ratio was 68, the main swirl number was 0.8 and the total air flow rate was 0.0016575 kg/s.

Primary injection flow rate per orifice was 0.005421 kg/s, while the secondary injection flow rate was 0.006355 kg/s. Dilution air flow rate was 0.007820 kg/s per orifice and these injection flow rates are for a 60° sector. Three cases were investigated;[3] the first was that of a cold flow, while the second was the case of a hot flow with premixed

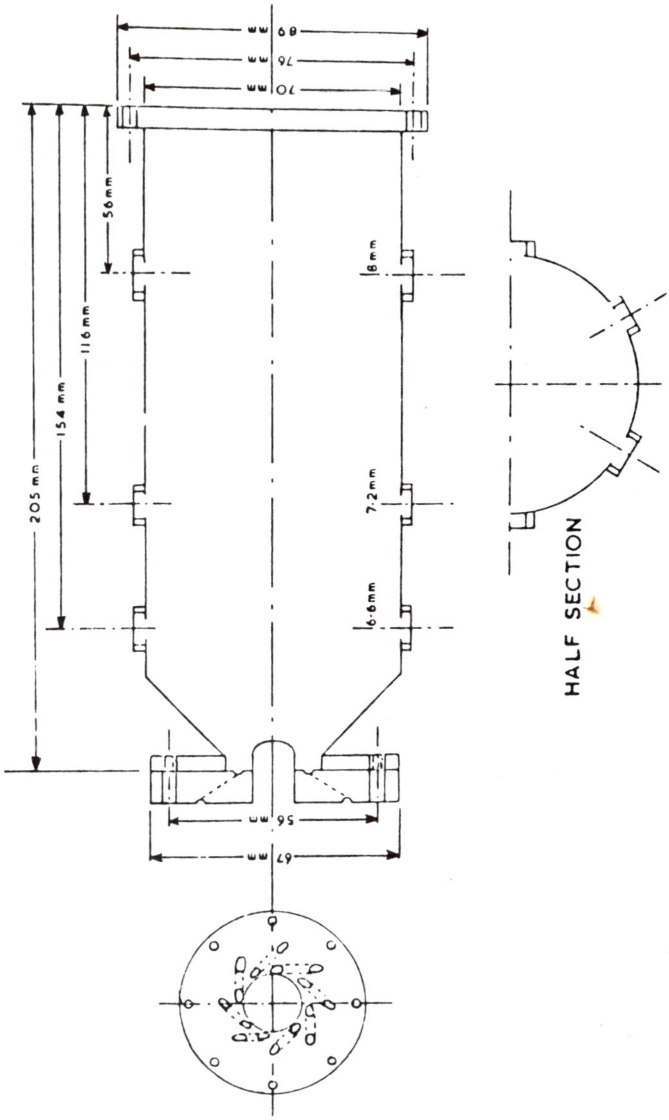

Figure A4.4 Combustor configuration[3]

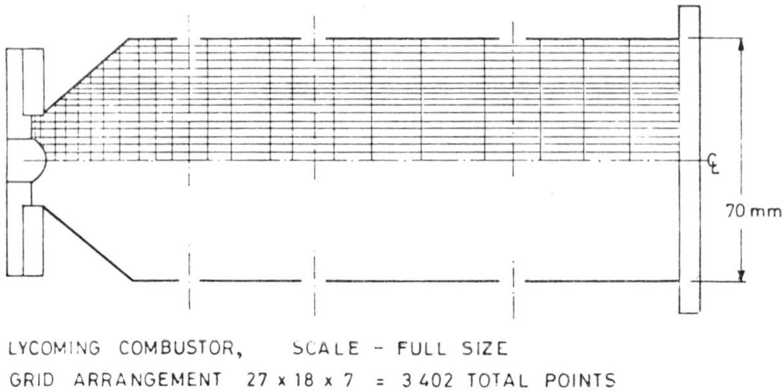

Figure A4.5 Finite difference grid arrangement for the combustor of Figure A4.4

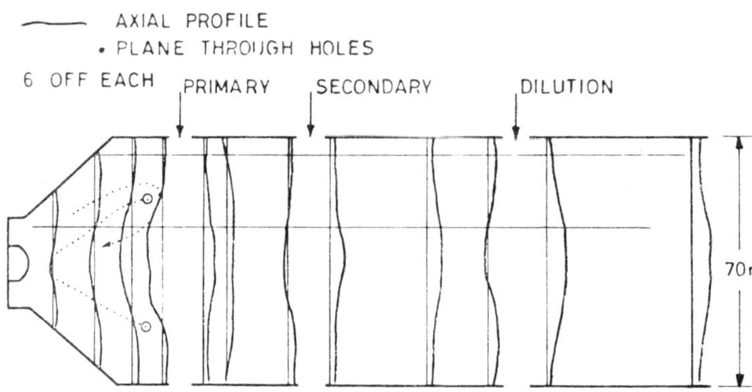

Figure A4.6 Calculated axial velocity profiles non-reacting flow situation[3]

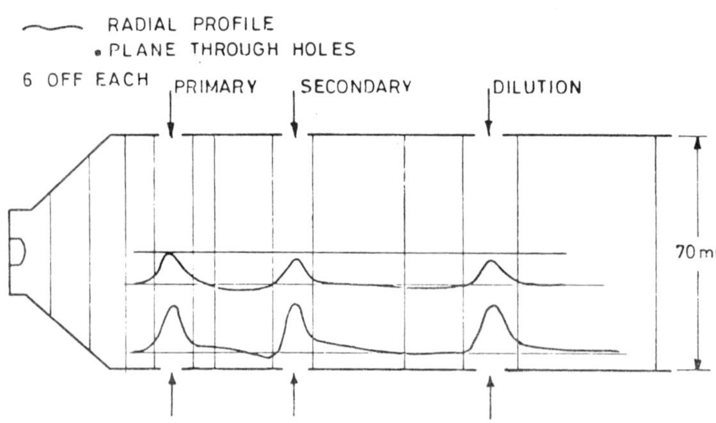

Figure A4.7 Calculated radial velocity profiles non-reacting flow situation[3]

prevapourized fuel introduced with the swirler air. The third case is for a liquid fuel spray introduced as a spray with an 80° included angle, from an axially located fuel nozzle, and an overall air/fuel ratio of 68.

Figure A4.6 shows the calculated axial velocity profiles obtained at various axial locations and in a longitudinal section through the centre of the injection holes. The velocity profiles indicated the existence of a central recirculation zone near the primary zone. Such a zone is very important for flame stabilization in reacting flows. Radial velocity component profiles are shown in Figure A4.7 and illustrate the decay of the radial jet velocity towards the combustor centreline.

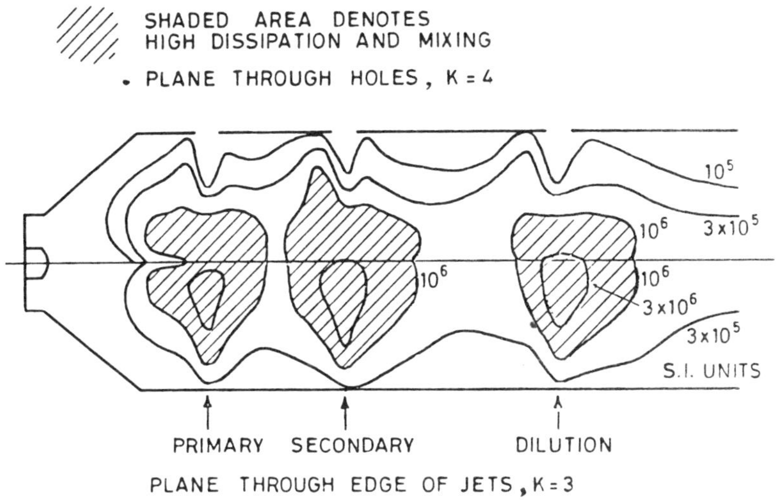

Figure A4.8 Calculated contours of turbulence dissipation rate in the combustor of Figure A4.4 under non-reacting flow conditions[3]

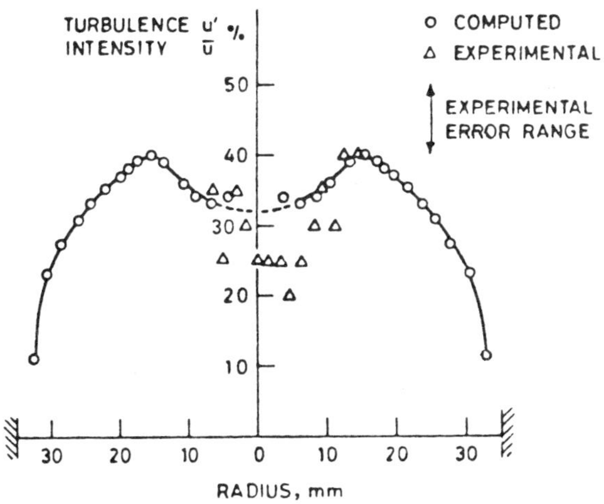

Figure A4.9 Measured and calculated exit turbulence intensity profiles[3]

Contours of turbulence dissipation rate were also predicted[3] and were plotted for an axial cross-section, as shown in Figure A4.8. Two sets of contours are shown and correspond to the sections through the jets, and at the edge of the jets. It can be noted that the dissipation (and mixing) were concentrated near the jets.

Figure A4.9 illustrates the level of agreement obtained for the hot premixed combustion. Profiles of local turbulence intensity are shown at the exit of the combustor. The agreement is remarkably good considering the complexity of the flow. The predicted profiles of the exit temperature for premixed pre-vapourized fuel are shown, in Figure A4.10, to be in qualitative agreement with the measurements.[3] The predicted velocity distributions along the hot combustor, taking into account the motion of the droplets are shown in Figure A4.11; the recirculation zone in the primary zone is clearly observed.

A4.4 Closure

The material presented in this appendix has illustrated the capability of numerical methods, solving the governing conservation equations, to adequately predict the details of the flow field and flame characteristics. No attempt is made to assess the validity of the procedures involved since this has been dealt with separately through the book and in other published work. Emphasis was placed on how much change in the flow pattern can be caused by a parametric change in swirl intensity, combustor dimensions or air to fuel ratio. Predictions of the turbulence characteristics and mixing patterns are very important for the studies involving reaction mechanism. Tracking of liquid fuel droplets assists the evaluation of the combustion efficiency and radiative heat transfer. Although only two examples are shown here, a number of other flows are reported in the open literature such as those of Boysan *et al.*,[4] Jones *et al.*,[5] Jones *et al.*[6] and Serag Eldin *et al.*[7]

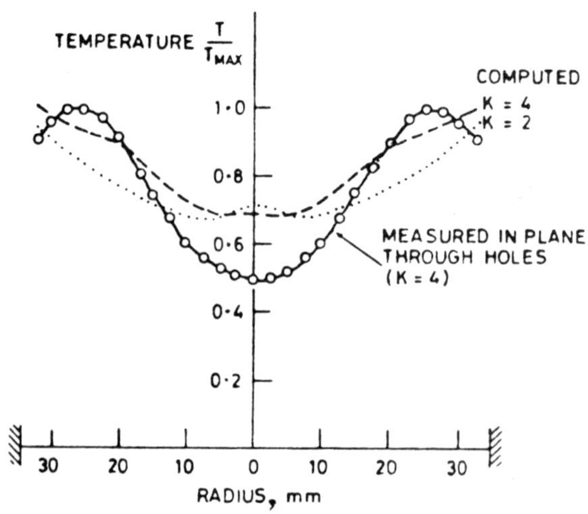

Figure A4.10 Measured and calculated exit temperature profiles in the combustor of Figure A4.4

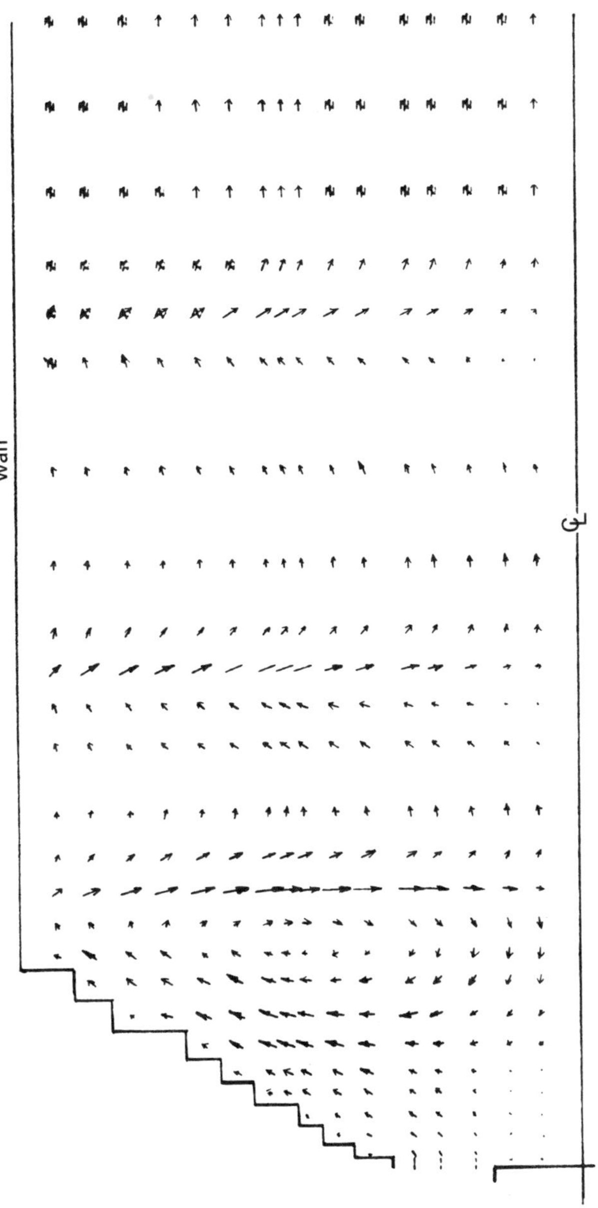

Figure A4.11 Predicted velocity vectors in the hot reacting flow in the combustor of Figure A4.4

References

1. Novick, A. S., Miles, G. A. and Lilley, D. G., (1979) Numerical Simulation of Combustor Flow Fields: A Primitive Variable Design Capability. *J. Energy*, **3**, No. 2, p. 95.
2. Novick, A. S., Miles, G. A. and Lilley, D. G., (1979) Modeling Parameter Influences in Gas Turbine Combustor Design. *J. Energy*, **3**, No. 5, p. 257.
3. Swithenbank, J., Turan, A. and Felton, P. G. (1978) 3-*D* 2-phase mathematical modelling of gas turbine combustors. University of Sheffield, Report HIC 303.
4. Boysan, F., Ayers, W. H., Swithenbank, J. and Pan, Z. (1980) Three dimensional model of spray combustion in gas turbine combustors. University of Sheffield, Report HIC 354.
5. Jones, W. P. and Priddin, G. H. (1978) Prediction of the flow field and local gas composition in gas turbine combustors. *Proc. 17th Symp. (Int.) on Combustion*, p. 399.
6. Jones, W. P. and McGuirk, J. J. (1980) Mathematical modelling of gas turbine combustion chambers. AGARD Proc. No. 275.
7. Serag Eldin, M. A. and Spalding, D. B. (1976) Prediction of flow and combustion processes in a three dimensional combustion chamber. Imperial College Report HTS/76/3.

A5 NON-FINITE DIFFERENCE SIMULATION METHODS

A5.1 Introduction

The real flows in practical combustors are turbulent, time dependent, three dimensional with mass transport and energy transfer with a possibility of the presence of two phase reactions. The application of the full elliptic partial differential equations of mass, momentum and energy can be, therefore, very complex and computer time consuming. At the first developing stages, simplified approaches for combustor design are still popular. The most common models include, for example, perfectly stirred reactors (PSRs), well stirred reactors (WSRs) and plug flow reactors (PFRs). Models differ in how these conceptions are interrelated to simulate the various aspects of mixing and reaction; a summary of the well-known models is listed[1] in Table A5.1.

A5.2 Analysis

The integral approach of Peters[2] assumes that the flow field in a sudden enlargement is dominated by the jet flow near the centreline. This approach is coupled to an assumed

TABLE A5.1

Some Simplified Approaches for Combustor Modelling

Approach	References	Date
Integral	Peters[2]	1969
Modular	Hamond and Mellor[3]	1970
	Fletcher and Heywood[4]	1971
	Swithenbank *et al.*[5]	1973
	Mosier and Roberts[6]	1974
Hybrid	Harsha and Edelman[7]	1978
	Swithenbank *et al.*[8]	1980

model to treat the recirculation zones. The jet flow calculations are based on an integral model for ducted coaxial turbulent mixing with boundary layer assumptions, and a cosine function for the cross-stream velocity profile. This method predicts the stagnation points quite well but overpredicts the peak reverse flow velocities.[1] The model does not account for the transport of species in recirculation zones, and finite rate chemical kinetics are not included within the integral analysis.

The modular approaches represent the combustor by a series of PSRs, WSRs and PFRs. The flow characteristics of such reactors are discussed in detail by, among others, Krambeck *et al.*,[9] Evangelista *et al.*[10] and Swithenbank *et al.*[5] The approach of Hamond *et al.*[3] represents the combustor flow by two PSRs. Fletcher *et al.*[4] employs a statistical mixing model for the primary zone utilizing a macromixed stirred reactor.

The approach of Swithenbank *et al.*[5] incorporates a series of WSRs and PFRs. Figure A5.1 shows a schematic modular arrangement of Swithenbank *et al.*;[5] in this instance the recirculating flow is fully accounted for. The approach of Mosier and Roberts divides the combustor into three PFRs; the initial portion consists of a central recirculating region surrounded by an outer streamtube of reacting flow. The secondary portion consists of a single streamtube dilution zone.

The hybrid method advances the level of complexity and expected realism. In this method, the combustor is divided into three major regions: a directed flow (treated by a

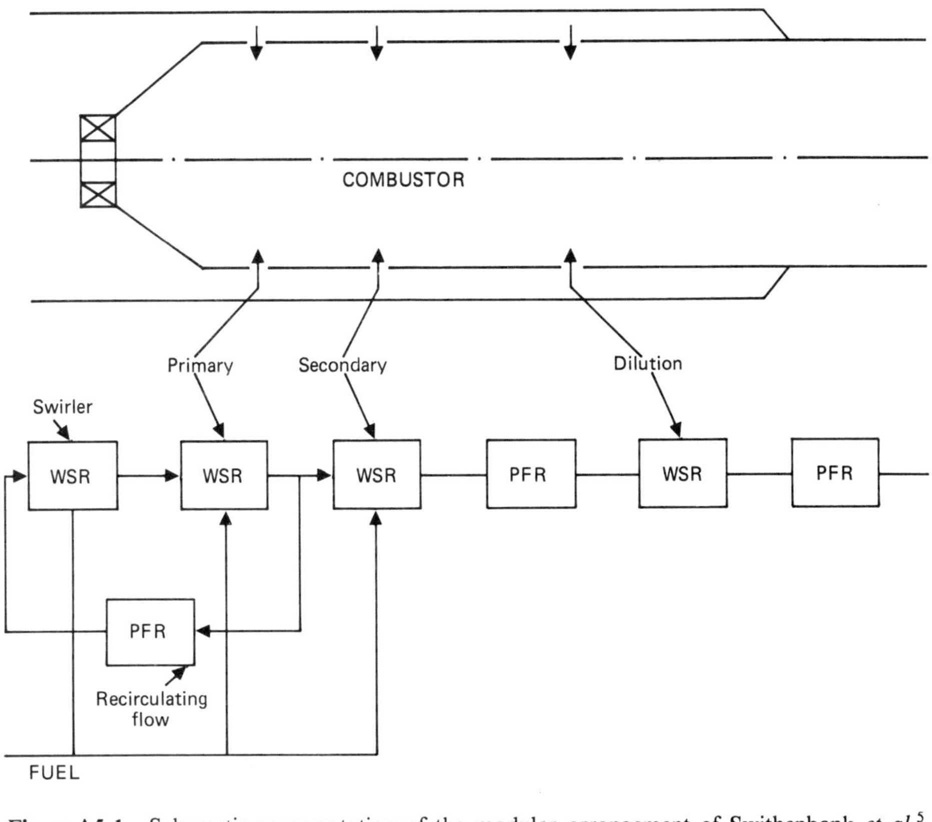

Figure A5.1 Schematic representation of the modular arrangement of Swithenbank *et al.*[5]

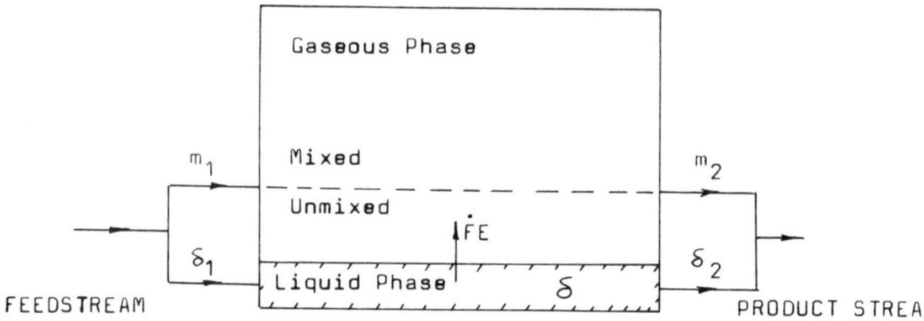

Figure A5.2 General steady state well-stirred reactor (WSR) composition[8]

parabolic marching procedure), a corner recirculation zone (represented by a WSR) and a turbulent shear layer region along the dividing streamline between the first two regions.[7] Fluxes of species and energy across the dividing streamline form the boundary conditions on the two computational regions. Full rate chemistry is included in both the directed flow and well-stirred reactor; henceforth the particle and droplet effects can be computed.

The introduction of two phase combustion interactions in reactor modelling was reported by Swithenbank *et al.*[8] Such introduction requires sub-models for a well stirred reactor which includes the internal processes of fuel evaporation, mixing and complex chemical kinetics. Although the processes of evaporation, mixing and reaction occur simultaneously, these were assumed to occur steadily in series. The fuel must first evaporate, mix and then finally react. The feedstream to, or the product stream from, any reactor is assumed[8] to be composed of any or all of the species considered by the reaction mechanism. Figure A5.2 shows the composition of the general two-phase steady-state reactor with the liquid phase shown coalesced for convenience. Transfer from the liquid phase to the gas phase is represented by the mean fuel evaporation rate $\dot{F}E$. The mean residence time of each phase in the reactor is the same. Such an assumption is incompatible with the existence of a relative slip velocity between the gas and droplets; nevertheless it simplifies the analysis and the calculation of fuel distribution around any particular reactor network. The analysis can be modified, however, to incorporate unequal phase residence times. Further details on stirred reactors can be found in reference 8.

References

1. Lilley, D. G., (1980) Prospects for computer modelling in ramjet combustors, AIAA/SAE/ASME 16th Joint Propulsion Conference, Paper AIAA-80-1189.
2. Peters, C. E., (1969) Turbulent mixing and burning of coaxial streams inside a duct of arbitrary shape. Arnold Engineering Development Center, TR-68-270.
3. Hamond, D. C., Jr. and Mellor, A. M., (1970) A preliminary investigation of gas turbine combustor modelling. *Combustion Science and Technology*, **2**, p. 67.
4. Fletcher, R. S., and Heywood, J. B. (1971) A model for Nitric Oxide Emission from gas turbine engines. AIAA Paper 71-123.

5. Swithenbank, J., Poll, I., Vincent, M. W. and Wright, D. D. (1973) Combustor design fundamentals. *Proc. 14th Symp. (Int.) on Combustion*, p. 627.
6. Mosier, S. A. and Roberts, R. (1974) Low-power turbo propulsion combustor exhaust emissions. *Analysis*, **III**, (AFAPL-TR-73-36).
7. Harsh, P. T. and Edelman, R. B. (1978) Application of modular modelling to ramjet performance predictions. AIAA Paper 78-944.
8. Swithenbank, J., Turan, A. and Felton, P. G. (1980) 3-dimensional 2-phase mathematical modelling of gas turbine combustors, in *Gas Turbine Combustor Design Problems*, edited by Levebvre, (Hemisphere McGraw-Hill), p. 249.
9. Krambeck, F. J., Shinnar, R. and Katz, S. (1967) Stochastic mixing models for chemical reactors, *I&EC Fundamentals,* **6**, p. 276.
10. Evangelista, J. J., Shinnar, R., and Katz, S. (1967) The effect of imperfect mixing on stirred combustion reactors. *Proc. 11th Symp. (Int.) on Combustion*, p. 901.

AUTHOR INDEX

SUBJECT INDEX